Freshwater Algae

Freshwater Algae

Identification, Enumeration and Use as Bioindicators

Second Edition

Edward G. Bellinger

Department of Environmental Sciences and Policy,
Central European University, Hungary

and

David C. Sigee

School of Earth, Atmospheric and Environmental Sciences,
University of Manchester, UK

WILEY Blackwell

Library of Congress Cataloging-in-Publication Data

Bellinger, Edward G., author.
   Freshwater algae : identification, enumeration and use as bioindicators / Edward G. Bellinger and David C. Sigee. – 2e.
     pages cm
   Includes index.
   ISBN 978-1-118-91716-9 (hardback)
  1. Freshwater algae.   2. Indicators (Biology)   3. Environmental monitoring.   I. Sigee, David C., author.   II. Title.
   QK570.25.B45 2015
   579.8′176–dc23

                                                                                                2014031375

A catalogue record for this book is available from the British Library.

Cover image: David Sigee
Cover design by Soephian Zainal

Set in 10/12pt Times by Aptara Inc., New Delhi, India.

Printed in the UK

Contents

Preface to the First Edition

Almost any freshwater or brackish water site will contain one or many species of algae. Although they are mainly microscopic and therefore not as visually apparent as larger aquatic organisms, such as higher plants or fish, algae play an equally important role in the ecology of these water bodies. Their presence can sometimes be noticed when they occur as dense populations, colouring the water and in some cases forming massive surface scum.

Freshwater algae constitute a very diverse group of organisms. Their range of shapes and beauty, when viewed through a microscope, has delighted biologists for more than a hundred years. They have an enormous range of size from less than one micrometre to several centimetres (for the stoneworts) – equalling the size span (10^4) for higher plants seen in a tropical rainforest. Algal morphology is diverse, ranging from single cells to complex colonies and filaments. Some species are capable of active movement. The term 'algae' embraces a number of phyla (e.g. Cyanophyta, Bacillariophyta and Chlorophyta) of chlorophyll-containing organisms with different growth forms and cytologies. Algae are important primary producers in both freshwater and marine systems. In many lakes and rivers, they generate biomass which is the foundation of diverse food chains. Although algae have beneficial impacts on aquatic ecosystems, they can also have adverse effects. When present in very large numbers they can produce 'blooms' that, on decomposition, deoxygenate the water – causing fish death and other ecological problems. Some algae produce toxins that are lethal to both aquatic and terrestrial organisms. It is important to be aware of these impacts and to monitor waters for the presence of these potentially harmful organisms. Algae can be used to flag up and assess a range of human and natural impacts in aquatic systems because of their often rapid response to changes in the environment. Examples include nutrient enrichment (eutrophication), industrial pollution and changes to the hydrological regime of the water body. Some groups of algae preserve well as fossils in geological deposits such as lake sediments, analysis of which gives us information on past environmental changes.

This book comes at a time of increasing concern over the widespread effects of human activities on the general environment of this planet. Monitoring shifts in algal population gives us an insight into these changes. We need to be able to assess the 'health' of aquatic systems such as lakes and rivers, since water is vital to both human and general ecosystem survival. Knowledge of algal population dynamics can help us develop effective management strategies for those systems. Included in this book are sections on the general features of the main freshwater algal groups with notes on their ecology, methods of sample collection and enumeration, using algae as indictors of environmental conditions and, finally, a key to the identification of the more frequently occurring genera. The authors have tried to combine descriptive material with original colour photographs and line drawings, where possible, to help the reader. We would also like to gratefully acknowledge the help and encouragement of colleagues and students, and particularly appreciate the direct contributions of postdoctoral workers and research students mentioned under Acknowledgements. We would also like to thank our families for their understanding and patience during the preparation of this text.

We hope that all those using the book will find it useful, and will enjoy the numerous colour photographs of these very beautiful organisms.

Preface to the Second Edition

Revisions of the first edition have been carried out to give a general update and to broaden the global perspective. In Chapter 4, particularly, a substantial number of new photographs have been contributed from the United States and China (see Acknowl-edgements), and various plates have been redrawn to provide greater detail. The key has been extensively modified to give greater clarity and to provide additional information on several genera.

Acknowledgements

We are very grateful to Andrew Dean (Tables 2.3 and 2.4), Matt Capstick (Figs. 4.8, 4.10–4.12, 4.42, 4.62, 4.64, 4.66, 4.68–4.70a and 4.73a) and Huda Qari (Fig. 2.8) for allowing us to present previously unpublished data.

We also thank Academic Press, American Health Association, Cambridge University Press, Journal of Plankton Research, McGraw-Hill, Phycologia and Prentice Hall for giving us permission to use previously published data.

With the incorporation of a substantial amount of new material into the second edition of the book, we would particularly like to thank two colleagues from the United States and the Republic of China for their contributions:

Dr. Robin A. Matthews (Western Washington University, Bellingham, WA) – Figs. 4.2a, 4.2b, 4.4b, 4.18, 4.19, 4.24d, 4.29, 4.36, 4.45, 4.49, 4.51a, 4.51b, 4.52a, 4.53, 4.54 and 4.55.

Dr. Gaohua Ji (Shanghai Ocean University, Shanghai, China) – Figs. 4.8, 4.31, 4.42, 4.47, 4.48, 4.53 and 4.63.

1

Introduction to Freshwater Algae

1.1 General introduction

Algae are widely present in freshwater environments, such as lakes and rivers, where they are typically present as microorganisms – visible only with the aid of a light microscope. Although relatively inconspicuous, they have a major importance in the freshwater environment, both in terms of fundamental ecology and in relation to human use of natural resources.

This book considers the diversity of algae in freshwater environments and gives a general overview of the major groups of these organisms (Chapter 1), methods of collection and enumeration (Chapter 2) and keys to algal groups and major genera (Chapter 4). Algae are considered as indicators of environmental conditions (bioindicators) in terms of individual species (Chapter 1) and as communities (Chapter 3).

1.1.1 Algae – an overview

The word 'algae' originates from the Latin word for seaweed and is now applied to a broad assemblage of organisms that can be defined both in terms of morphology and general physiology. They are simple organisms, without differentiation into roots, stems and leaves – and their sexual organs are not enclosed within protective coverings. In terms of physiology, they are fundamentally autotrophic (obtaining all their materials from inorganic sources) and

photosynthetic – generating complex carbon compounds from carbon dioxide and light energy. Some algae have become secondarily heterotrophic, taking up complex organic molecules by organotrophy or heterotrophy (Tuchman, 1996), but still retaining fundamental genetic affinities with their photosynthetic relatives (Pfandl *et al.,* 2009).

The term 'algae' (singular alga) is not strictly a taxonomic term but is used as an inclusive label for a number of different phyla that fit the broad description noted earlier. These organisms include both prokaryotes (cells lacking a membrane-bound nucleus; see Section 1.3) and eukaryotes (cells with a nucleus plus typical membrane-bound organelles).

Humans have long made use of algal species, both living and dead. Fossil algal diatomite deposits, for example, in the form of light but strong rocks, have been used as building materials and filtration media in water purification and swimming pools. Some fossil algae, for example *Botryococcus*, can give rise to oil-rich deposits. Certain species of green algae are cultivated for the purpose of extracting key biochemicals for use in medicine and cosmetics. Even blue-green algae have beneficial uses. Particularly, *Spirulina*, which was harvested by the Aztecs of Mexico, is still used by the people around Lake Chad as a dietary supplement. *Spirulina* tablets may still be obtained in some health food shops. Blue-green algae are, however, better known in the freshwater environment as nuisance organisms, forming dense blooms. These can have adverse effects in relation to toxin build-up

Freshwater Algae: Identification, Enumeration and Use as Bioindicators, Second Edition. Edward G. Bellinger and David C. Sigee.
© 2015 John Wiley & Sons, Ltd. Published 2015 by John Wiley & Sons, Ltd.

and clogging filters/water courses – affecting the production of drinking water and recreational activities.

1.1.2　Algae as primary producers

As fixers of carbon and generators of biomass, algae are one of three major groups of photosynthetic organism within the freshwater environment. They are distinguished from higher plants (macrophytes) in terms of size and taxonomy and from photosynthetic bacteria in terms of their biochemistry. Unlike algae (eukaryotic algae and cyanophyta), photosynthetic bacteria are strict anaerobes and do not evolve oxygen as part of the photosynthetic process (Sigee, 2004).

The level of primary production by algae in freshwater bodies can be measured as fixed carbon per unit area with time (mg C m^{-3} h^{-1}) and varies greatly from one environment to another. This is seen, for example, in different lakes – where primary production varies with trophic status and with depth in the water column (Fig. 1.1). Eutrophic lakes, containing high levels of available nitrogen and phosphorus, have very high levels of productivity in surface waters, decreasing rapidly with depth due to light absorption by algal biomass. In contrast, mesotrophic and oligotrophic lakes have lower overall productivity – but this extends deep into the water column due to greater light penetration.

Although algae are fundamentally autotrophic (photosynthetic), some species have become secondarily heterotrophic – obtaining complex organic compounds by absorption over their outer surface or by active ingestion of particulate material. Although such organisms often superficially resemble protozoa in terms of their lack of chlorophyll, vigorous motility and active ingestion of organic material, they may still be regarded as algae due to their phylogenetic affinities.

1.1.3　Freshwater environments

Aquatic biology can be divided into two major disciplines – limnology (water bodies within continental boundaries) and oceanography (dealing with oceans and seas, occurring between continents). This book focuses on aquatic algae present within continental boundaries, where water is typically fresh (non-saline), and where water bodies are of two main types:

- Standing (lentic) waters – particularly lakes and wetlands.

- Running (lotic) waters - including streams and rivers.

The distinction between lentic and lotic systems is not absolute, since many 'standing waters' such as lakes have a small but continuous flow-through of water, and many large rivers have a relatively low rate of flow at certain times of year. Although the difference between standing and running waters is not absolute, it is an important distinction in relation to the algae present, since lentic systems are typically dominated by planktonic algae and lotic systems by benthic organisms.

Although this volume deals primarily with algae present within 'conventional freshwater systems' such as lakes and rivers, it also considers algae present within more extreme freshwater environments such as hot springs, algae present in semi-saline (brackish) and saline conditions (e.g. estuaries and saline lakes) and algae present within snow (where the water is in a frozen state for most of the year).

1.1.4　Planktonic and benthic algae

Within freshwater ecosystems, algae occur as either free-floating (planktonic) or substrate-associated (largely benthic) organisms. Planktonic algae drift freely within the main body of water (with some species able to regulate their position within the water column), while substrate-associated organisms are either fixed in position (attached) or have limited movement in relation to their substrate. These substrate-associated algae are in dynamic equilibrium with planktonic organisms (Fig. 2.1), with the balance depending on two main factors – the depth of water and the rate of water flow. Build-up of phytoplankton populations requires a low rate of flow

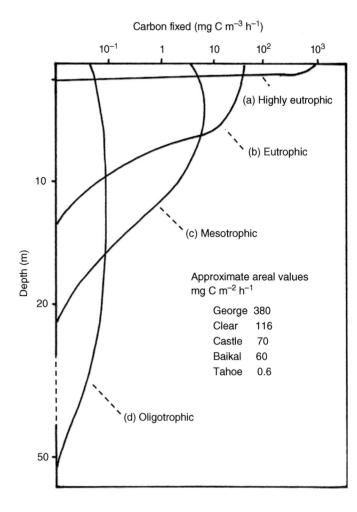

Carbon fixed (mg C m^{-3} h^{-1})

(a) Highly eutrophic

(b) Eutrophic

(c) Mesotrophic

Approximate areal values
mg C m^{-2} h^{-1}

George	380
Clear	116
Castle	70
Baikal	60
Tahoe	0.6

(d) Oligotrophic

Depth (m)

Figure 1.1 Examples of algal primary production in lakes of different trophic status, showing how rates of production typically change with depth. Examples of each lake type include (a) highly eutrophic; Lake George (Uganda). (b) eutrophic; Blelham Tarn (English Lake District), Clear Lake (USA), Erken (Sweden). (c) mesotrophic; Grasmere (English Lake District), Castle Lake (USA). (d) oligotrophic; Lake Tahoe (USA), Lake Baikal in part (Russia), Wastwater (English Lake District). Adapted from Horne and Goldman (1994)

(otherwise they flush out of the system) and adequate light levels, so they tend to predominate at the surface of lakes and slow-moving rivers. Benthic algae require adequate light (shallow waters) and can tolerate high rates of water flow, so predominate over phytoplankton in fast-flowing rivers and streams. Benthic algae also require adequate attachment sites – which include inorganic substrate, submerged water plants and emergent water plants at the edge of the water body. The distinction between planktonic and non-planktonic algae is ecologically important and is also relevant to algal sampling and enumeration procedures (see Chapter 2).

Planktonic algae

Planktonic algae dominate the main water body of standing waters, occurring as a defined seasonal succession of species in temperate lakes. The temporal sequence depends on lake trophic status (see Section 3.2.3; Table 3.3) with algae forming dense blooms in eutrophic lakes of diatoms (Fig. 1.16), colonial blue-green algae (Fig. 1.5) and late populations of dinoflagellates (Fig. 1.10). During the annual cycle, phytoplankton blooms correspond to peaks in algal biovolume and chlorophyll-*a* concentration and troughs in 'Secchi depth' – the inverse of turbidity (Fig. 2.8).

Benthic algae

Benthic algae occur at the bottom of the water column in lakes and rivers and are directly associated with sediments – including rocks, mud and organic debris. These algae (usually attached) may form major growths on inorganic surfaces or on organic debris, where they are frequently present in mixed biofilms (bacteria, fungi and invertebrates also present). Under high light conditions, the biofilm may become dominated by extensive growths of filamentous algae – forming a periphyton community (Fig. 2.23). Attached algae may also be fixed to living organisms as epiphytes – including higher plants (Fig. 2.29), larger attached algae (Fig. 2.28) and large planktonic colonial algae (Fig. 4.35). Some substrate-associated algae are not attached, but are able to move across substrate surfaces (e.g. pennate diatoms), are loosely retained with gelatinous biofilms or are held within the tangled filamentous threads of mature periphyton biofilms. (Fig. 2.29).

Many algal species have both planktonic and benthic stages in their life cycle. In some cases, they develop as actively photosynthetic benthic organisms, which subsequently detach and become planktonic. In other cases, the alga spends most of its actively photosynthetic growth phase in the planktonic environment, but overwinters as a dormant metabolically inactive phase. Light micrographs of the distinctive overwintering phases of two major bloom-forming algae (*Ceratium* and *Anabaena*) are shown in Fig. 2.7.

1.1.5 Size and shape

Size range

The microscopic nature of freshwater algae tends to give the impression that they all occur within a broadly similar size range. This is not the case with either free floating or attached algae.

In the planktonic environment (Table 1.1), algae range from small prokaryotic unicells (diameter < 1 μm) to large globular colonies of blue-green algae such as *Microcystis* (diameter reaching 2000 μm) – just visible to the naked eye. This enormous size range represents four orders of magnitude on a linear basis (×12 as volume) and is similar to that seen for higher plants in terrestrial ecosystems such as tropical rainforest.

Table 1.1   Size Range of Phytoplankton.

Category	Linear Size (Cell or Colony Diameter) (μm)	Biovolume[a] (μm³)	Unicellular Organisms	Colonial Organisms
Picoplankton	0.2–2	4.2×10^{-3}–4.2	Photosynthetic bacteria Blue-green algae *Synechococcus* *Synechocystis*	-
Nanoplankton	2–20	4.2–4.2×10^3	Blue-green algae Cryptophytes – *Cryptomonas* *Rhodomonas*	
Microplankton	20–200	4.2×10^3–4.2×10^6	Dinoflagellates *Ceratium* *Peridinium*	Diatoms *Asterionella*
Macroplankton	>200	$>4.2 \times 10^6$	-	Blue-green algae *Anabaena* *Microcystis*

[a]Biovolume values are based on a sphere (volume $4/3\Pi r^3$).

Planktonic algae are frequently characterised in relation to discrete size bands – picoplankton (<2 µm), nanoplankton (2–20 µm), microplankton (20–200 µm) and macroplankton (>200 µm). Each size band is characterised by particular groups of algae (Table 1.1).

In the benthic environment, the size range of attached algae is even greater – ranging from small unicells (which colonise freshly exposed surfaces) to extended filamentous algae of the mature periphyton community. Filaments of attached algae such as *Cladophora*, for example, can extend several centimetres into the surrounding aquatic medium. These macroscopic algae frequently have small colonial algae and unicells attached as epiphytes (Fig. 2.28), so there is a wide spectrum of sizes within the localised microenvironment.

Diversity of shape

The shape of algal cells ranges from simple single non-motile spheres to large multicellular structures (Fig. 1.2). The simplest structure is a unicellular non-motile sphere (Fig. 1.2b), which may become elaborated by the acquisition of flagella (Fig. 1.2c), by a change of body shape (Fig. 1.2a) or by the development of elongate spines and processes (Fig. 1.2d). Cells may come together in small groups or large aggregates but with no definite shape (Figs. 1.2d and 1.2e), or may form globular colonies that have a characteristic shape (Figs. 1.2f and 1.2g). Cells may also join together to form linear colonies (filaments), which may be unbranched or branched (Figs. 1.2h and 1.2i).

Although motility is normally associated with the possession of flagella, some algae (e.g. the diatom *Navicula* and the blue-green alga *Oscillatoria*) can move without the aid of flagella by the secretion of surface mucilage. In many algae, the presence of surface mucilage is also important in increasing overall cell/colony size and influencing shape.

Size and shape, along with other major phenotypic characteristics, are clearly important in the classification and identification of algal species. At a functional and ecological level, size and shape are also important in terms of solute and gas exchange, absorption of light, rates of growth and cell division, sedimentation

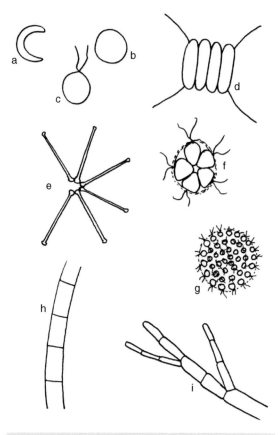

Figure 1.2 General shapes of algae. Non-motile unicells: (a) *Selenastrum*; (b) *Chlorella*. Motile unicells: (c) *Chlamydomonas*. Non-motile colony: (d) *Scenedesmus*; (e) *Asterionella*. Motile colony: (f) *Pandorina*; (g) *Volvox*. Unbranched filament: (h) *Spirogyra*. Branched filament: (i) *Cladophora*.

in the water column, cell/colony motility and grazing by zooplankton (Sigee, 2004).

1.2 Taxonomic variation – the major groups of algae

Freshwater algae can be grouped into 10 major divisions (phyla) in relation to microscopical appearance (Table 1.2) and biochemical/cytological characteristics (Table 1.3). Some indication of

Table 1.2   Major Divisions of Freshwater Algae: Microscopical Appearance.

Algal Division (Phylum)	Index of Biodiversity[a]	Typical Colour	Typical Morphology of Freshwater Species	Motility (Vegetative Cells/Colonies)	Typical Examples
1. Blue-green algae (**Cyanophyta**)	297	Blue-green	Microscopic or visible – usually colonial	Buoyancy regulation Some can glide	Synechocystis Microcystis
2. Green algae (**Chlorophyta**)	992	Grass-green	Microscopic or visible – unicellular or filamentous colonial	Some unicells and colonies with flagella	Chlamydomonas Cladophora
3. Euglenoids (**Euglenophyta**)	124	Various colours	Microscopic – unicellular	Mostly with flagella	Euglena Colacium
4. Yellow-green algae (**Xanthophyta**)	73	Yellow-green	Microscopic – unicellular or filamentous	Flagellate zoospores and gametes	Ophiocytium Vaucheria
5. Dinoflagellates (**Dinophyta**)	54	Red-brown	Microscopic – unicellular	All with flagella	Ceratium Peridinium
6. Cryptomonads (**Cryptophyta**)	15	Various colours	Microscopic – Unicellular	Mostly with flagella	Rhodomonas Cryptomonas
7. Chrysophytes (**Chrysophyta**)	115	Golden brown	Microscopic – unicellular or colonial	Some with flagella	Mallomonas Dinobryon
8. Diatoms (**Bacillariophyta**)	1652	Golden brown	Microscopic – unicellular or filamentous colonies	Gliding movement on substrate	Stephanodiscus Aulacoseira
9. Red algae (**Rhodophyta**)	22	Red	Microscopic or visible – unicellular or colonial	Non-motile	Batrachospermum Bangia
10. Brown algae (**Phaeophyta**)	2	Brown	Visible – multicellular cushions and crustose thalli	Non-motile	Pleurocladia Heribaudiella

[a]Biodiversity: number of species of freshwater and terrestrial algae within the British Isles.

Table 1.3 Major Divisions of Freshwater Algae: Biochemical and Cytological Characteristics.

Algal Division (Phylum)	Pigmentation[#]			Starch-Like Reserve	External Covering	Chloroplast Fine-Structure		Flagella (Vegetative Cells and Gametes)
	Chlorophylls	Carotenes	Diag.* carotenoids			Outer membranes	Thylakoid groups	
1. Blue-green algae (**Cyanophyta**)	**a**	β	**zea-**	Cyanophycean starch$^\alpha$	Peptidoglycan matrices or walls	0	0	0
2. Green algae (**Chlorophyta**)	**a, b**	α, β, γ	**viola-**	True starch$^\alpha$	Cellulose walls, scales	2	2–6	0–many. Similar (isokont)
3. Euglenoids (**Euglenophyta**)	**a, b**	β, γ		Paramylon$^\beta$	Protein pellicle	3	3	1–2 emergent
4. Yellow-green algae: (**Xanthophyta**)	**a, c_1, c_2**	α, β		Chrysolaminarin$^\beta$	Pectin or pectic acid wall	4	3	2 unequal (heterokont)
5. Dinoflagellates (**Dinophyta**)	**a, c_2**	β	**peri-**	True starch$^\alpha$	Cellulose theca (or naked)	3	3	2 unequal (heterokont)
6. Cryptomonads (**Cryptophyta**)	**a, c_2**	α, β	**allo-**	True starch$^\alpha$	Cellulose periplast	4	2	2 equal (isokont)
7. Chrysophytes (**Chrysophyta**)	**a, c_1, c_2, c_3**	α, β, ε		Chrysolaminarin$^\beta$	Pectin, plus minerals and silica	4	3	2 unequal (heterokont)
8. Diatoms (**Bacillariophyta**)	**a, c_1, c_2, c_3**	β, ε	**fuco-**	Chrysolaminarin$^\beta$	Opaline silica frustule	4	4	1 reproductive cell only
9. Red algae (**Rhodophyta**)	**a**	α, β		Floridean starch$^\alpha$	Walls with galactose polymer matrix	2	0	0
10. Brown algae (**Phaeophyta**)	**a, c_1, c_2, c_3**	β, ε		Laminarin$^\beta$	Walls with alginate matrix	4	3	2 unequal (heterokont) reproductive cells only

[#]Major pigments are shown in bold type.
*Diagnostic carotenoids, used for HPLC analysis (Fig. 2.11): zea- (Zeaxanthin: also present in chlorophytes, cryptophytes), viola- (violaxanthin), peri- (Peridinin), allo- (alloxanthin), fuco- (fucoxanthin, also present in chrysophytes).
Starch-like reserves $^\alpha$: α-1,4 glucan; $^\beta$:β-1,3 glucan.

the ecological and taxonomic diversity of these groups is given by the number of constituent species (Table 1.2) for freshwater and terrestrial algae in the British Isles (taken from John *et al.*, 2002), with green algae and diatoms far outnumbering other groups – reflecting their widespread occurrence and ability to live in diverse habitats. Diatoms in particular (over 1600 species) are ecologically successful, both as planktonic and benthic organisms. In addition to the above groups, John *et al.* (2002) also list other phyla – Raphidophyta (2 species), Haptophyta (5), Eustigmatophyta (3), Prasino-phyta (13) and Glaucophyta (2). Although these minor phyla have taxonomic and phylogenetic interest, they have less impact in the freshwater environment.

In terms of diversity, freshwater algae also have a major division into prokaryotes (blue-green algae) and eukaryotes (remaining groups) based on cell size, ultrastructure, antibiotic resistance and general physiology. Even within the eukaryote groups, fundamental differences in phenotype and molecular characteristics indicate evolutionary derivation from a range of ancestral types (polyphyletic origins).

1.2.1 Microscopical appearance

The colour of freshwater algae is an important aspect of their classification (Table 1.2), and ranges from blue-green (Cyanobacteria) to grass green (Chloro-phyta), golden brown (Chrysophyta, Bacillario-phyta), brown (Phaeophyta) and red (Rhodophyta). Variations in colour are shown in Fig. 1.3 and in the colour photographs of Chapter 4. The use of colour as a taxonomic marker can be deceptive, however, since the normal balance of pigments may vary. Green algae living on snow, for example, may have a pre-ponderance of carotenoid pigments – forming a 'red bloom' (Hoham and Duval, 2001).

The presence of chlorophylls and associated pigments is also variable (Sigee, 2004). Obligate heterotrophs, entirely dependent on uptake of organic molecules by organotrophy or phagotrophy, may have completely lost their chloroplasts (e.g. Fig. 1.11) and

appear colourless. Facultative heterotrophs (photo-organotrophs, mixotrophs) retain plastids and show green pigmentation. Even within a 'normal' ecological situation, the colour of a particular alga can show considerable variation (see, for example, *Anabaena*, Fig. 4.24).

Apart from colour, the other obvious characteristics viewed under the light microscope are overall size, whether the organism is unicellular or colonial and whether it is motile (actively moving) or non-motile. Within different groups, algae may be largely unicellular (euglenoids, dinoflagellates, cryptophytes), multicellular (brown algae) or a mixture of the two (other groups). Motility (single cells or entire colonies) is also an important feature, with some algal groups being entirely flagellate (dinoflagellates, cryptophytes) while others are a mixture of flagellate and non-flagellate organisms (green algae, xanthophytes). Other groups of algae are entirely without flagella, but are able to move by buoyancy regulation (blue-greens), gliding movements on substratum (blue-greens, diatoms) or are entirely non-motile (red and brown algae).

Resistant spores

While the above description of microscopical appearance relates particularly to actively growing vegetative cells, it should be remembered that during winter months the algae of temperate water bodies typically occur as thick-walled resistant spores. Colour is often obscured in these structures or entirely altered, making taxonomic identification difficult. Examples of resistant spores are illustrated for blue-green algae (e.g. *Anabaena* akinetes – Figs. 2.7 and 4.24c), green algae (e.g. *Haematococcus* cysts – Fig. 4.54) and dinoflagellates (e.g. *Ceratium* cysts – Fig. 2.7).

1.2.2 Biochemistry and cell structure

Major biochemical features of freshwater algae include pigmentation, food reserves and external covering (Table 1.3). Different groups have distinctive combinations of chlorophylls and carotenes, while only three groups (blue-greens, cryptomonads and red algae) have phycobilins. All pigmented algae have

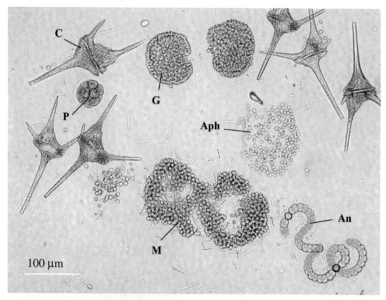

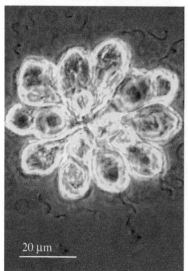

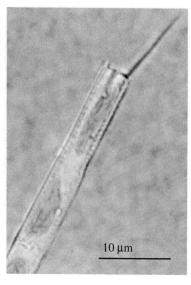

Figure 1.3 Colour characteristics of different algal groups. Top: Fresh lake phytoplankton sample showing colour differences between major algal phyla: Dinophyta (brown: C), Cyanobacteria (blue-green: An, Aph, M) and Chlorophyta (grass-green: P). Algal genera: An, *Anabaena*; Aph, *Aphanothece*; C, *Ceratium*; G, *Gomphosphaeria*; M, *Microcystis*; P, *Pandorina*. Bottom left: *Synura* (cultured alga, lightly fixed) showing golden brown colour of Chrysophyta. Bottom right: End of filament of *Aulacoseira granulata* var. *angustissima* (with terminal spine) from lake phytoplankton showing olive-green chloroplasts (Bacillariophyta).

chlorophyll-*a*, which can therefore be used for the estimation of total biomass (see Chapter 2). Diagnostic carotenoids have been particularly relevant in the application of high performance liquid chromatography (HPLC) for the identification and quantitation of major algal groups within mixed phytoplankton samples (see Section 2.3.3, Fig 2.11) – and have been used, for example, in the analysis of estuarine eutrophication (see Section 3.5.2).

Visualisation of key differences in cell structure normally requires the higher resolution of oil immersion (light microscopy), transmission electron microscopy (TEM) or scanning electron microscopy (SEM) and includes both internal (e.g. chloroplast fine structure) and external (e.g. location/number of flagella, cell surface ornamentation) features. Comparisons of light and electron microscopic images are shown in Figs. 1.4 (light/TEM) and 4.56 (light/SEM).

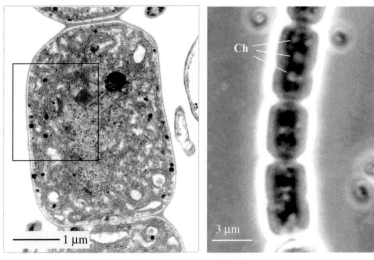

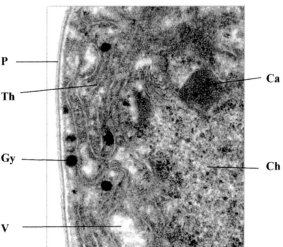

Figure 1.4 Prokaryote features of blue-green algae. Top right: Phase-contrast (light microscope) image of live filamentous colony of *Anabaena*, showing central patches of pale chromatin (Ch). Top left: Transmission electron micrograph of whole cell of *Anabaena*, showing central patch of granular chromatin. Bottom: Detail from top left, showing fine-structural features: Ch, central region of chromatin (no limiting membrane); Ca, carboxysome (polyhedral body); Gy, glycogen granule (cyanophycean starch); Th, peripheral thylakoid membranes; V, vacuole; P, thin peptidoglycan cell wall.

1.2.3 Molecular characterisation and identification

Although identification of algal taxa has traditionally been based on microscopical characteristics (morphology and colour), molecular techniques are being increasingly used to characterise algal communities. These are particularly useful where:

- No clear morphological characteristics are available. This has particularly been the case for unicellular blue-green algae.

- Algae are relatively inaccessible and difficult to visualise. This is the case for biofilms, where algae are enclosed in a gelatinous matrix, and in many cases are a relatively small component of a very heterogeneous community of organisms. This problem may also occur in highly radioactive environments, where handling and visual examination of samples can present a health hazard.

- Diversity is being studied within species, where strains are often distinguished in biochemical and genetic terms.

Ultimately, species definition and identification in both prokaryote and eukaryote algae may depend on molecular analysis, with determination of unique and defining DNA sequences followed by development of species-specific nucleotide probes from these (see next section).This approach would be particularly relevant in the case of blue-green algae, but there are a number of problems in relation to species-specificity in this group (Castenholz, 1992).

- Polyploidy (multiple genomic copies per cell) may occur, with variation between the multiple genomes. As many as 10 multiple genome copies have been observed in some blue-green algae.

- Horizontal gene transfer means that some DNA fragments are dispersed over a range of species.

DNA/RNA sequence analysis

Analysis of DNA sequences has been widely used for the identification of both blue-green (16S rRNA genes) and eukaryote algae (18S rRNA and chloroplast DNA). This technique has been used by Droppo *et al.* (2007), for example, to determine the taxonomic composition of biofilms – identifying bacteria, blue-green algae and some eukaryote unicellular algae. The technique involves:

- Collection of a sample of biomass from the entire microbial community.

- Obtaining a DNA sample. This may involve extraction from a mixed environmental sample such as biofilm or soil (Miller et al., 1999).

- Polymerase chain reaction (PCR) amplification of a specific nucleotide region – typically 16S or 18S rRNA genes, about 1500 bp in length.

- Separation of the amplified strands by denaturing gradient gel electrophoresis (DGGE), or purification of the PCR products using a rapid purification kit.

- DNA sequence analysis, using a standard sequencer or by applying the relatively new technique of pyrosequencing (Ronaghi, 2001). Comparison to known 16S and 18S rRNA gene sequences in a standard database such as NCBI leads to tentative species identification, normally requiring a match of at least 90%.

The *pattern* of bands in the DGGE gel gives an estimate of community complexity, while the *intensity* of individual bands (derived from particular species) provides a measure of species population size. In addition to providing taxonomic information where classical morphological characteristics do not apply, molecular identification also has the advantage that the whole of the microbial community (communal DNA sample) is being analysed in an objective way. Non-photosynthetic bacteria, Archaea and protozoa are also identified in addition to prokaryote and eukaryote algae (Droppo *et al.*, 2007; Galand *et al.*, 2008).

Potential limitations of molecular analysis are that identification may be tentative (often only to genus level) and taxonomic quantitation (relative numbers of different algae) is difficult. The technique has been particularly useful in relation to the biodiversity of marine blue-green unicellular algae and is now increasingly used with freshwater systems (Table 1.4). Molecular techniques can be used to probe particular enzymes (e.g. nitrogenase) as well as more general taxon-specific gene sequences.

Enzyme probes: Nitrogenase (*nif*) genes
Molecular probes have been used to monitor genes encoding both the large (*nif*K) and small (*nif*H) nitrogenase subunits.

Stancheva *et al.* (2013) monitored nitrogenase gene expression in benthic river algae by developing primers to the *nif*K gene, then using real-time reverse transcriptase PCR. The technique was used to validate the potential role of N_2-fixing blue-green algae as indicators of low ambient concentrations of inorganic nitrogen.

Foster *et al.* (2007) used DNA quantitative PCR (QPCR) technology to amplify and detect the presence of species-specific *nif*H genes in blue-green

Table 1.4   Molecular Identification of Algal Species in Aquatic Environmental Samples.

Environmental Samples	Techniques	References
Picoplankton in Lake Baikal	Direct sequencing	Semenova and Kuznedelov (1998)
Flagellate nanoplankton	SSU rRNA probes for *Paraphysomonas* (Chrysophyte)	Caron *et al.* (1999)
Mixed diatom populations and laboratory cultures	Large subunit rRNA probe for *Pseudo-nitzschia* (Diatom)	Scholin *et al.* (1997)
Estuarine river samples	Use of ITS-specific polymerase chain reaction (PCR) assays for *Pfiesteria* (Dinoflagellate)	Litaker *et al.* (2003)
Laboratory biofilm samples: blue-greens and unicellular eukaryotes	Amplification of 16s RNA genes, with denaturing gradient gel electrophoresis (DGGE)	Droppo *et al.* (2007)
Diazotrophic blue-green algae within the Amazon River freshwater plume	Quantitative PCR (QPCR) analysis of the *nif*H gene (encodes part of the nitrogenise enzyme)	Foster *et al.* (2007)
Arctic ecosystems: Freshwater stamukhi lake and inflow river	Sequencing of 18s rRNA genes to identify major cryptophyte river populations, plus minor lake populations of diatoms	Galand *et al.* (2008)
Lake and river flagellate samples	Population heterogeneity in *Spumella* (Chrysophyte) – SSU rRNA sequences	Pfandl *et al.* (2009)
Pond samples	Population heterogeneity in *Desmodesmus* (Chlorophyte) – ITS2 rRNA sequences	Vanormelingen *et al.* (2009)
Stream N_2-fixing cyanobacteria	Assessment of nitrogenase expression using real-time reverse transcriptase PCR	Stancheva *et al.* (2013)

SSU rRNA, small subunit ribosomal RNA; ITS, internal transcribed spacer.

algae. This gene encodes the iron-containing protein nitrogenase (the key enzyme involved in nitrogen fixation) and provides a marker for nitrogen-fixing (diazotrophic) algae in the freshwater environment. The technique was used to demonstrate that the blue-green algal symbiont *Richelia* (associated with the diatom *Hemiaulus hauckii*) was specifically linked to the Amazon freshwater outflow (river plume) in the Western Tropical North Atlantic (WTNA) ocean, and that the *H. hauckii–Richelia* complex could be used as a bioindicator for pockets of freshwater within the WTNA ocean.

Taxon-specific DNA sequences   The development of taxon-specific oligonucleotide probes from DNA sequence data, followed by *in situ* hybridisation, has considerable potential for the identification and counting of algae in environmental samples.

As with direct sequencing, this technique has particular advantages with small unicellular algae, where there are often relatively few morphological features available for identification. Caron *et al.* (1999) sequenced the small-subunit ribosomal genes of four species of the colourless chrysophyte genus *Paraphysomonas*, leading to the development of oligonucleotide probes for *Paraphysomonas imperforata* and *Paraphysomonas bandaiensis*.

Molecular probes have major potential for the detection of nuisance algae, particularly those that produce toxins. They have been used, for example, to distinguish toxic from non-toxic diatom species (Scholin *et al.*, 1997), where differentiation would otherwise require the time-consuming application of SEM and TEM. They have also been used for the rapid identification of *Pfiesteria piscicida*, a potentially toxic dinoflagellate that has been the cause of extensive fish mortalities in coastal rivers of the eastern USA. Litaker *et al.* (2003) used unique sequences in the internal transcribed spacer (ITS) regions ITS1 and ITS2 to develop PCR assays capable of detecting *Pfiesteria* in natural river assemblages. These have been successfully used to detect the potentially harmful organism in the St. Johns River system, Florida (USA).

1.3 Blue-green algae

Blue-green algae (Cyanobacteria) are widely occurring throughout freshwater environments, ranging in size from unicellular forms such as *Synechococcus* (barely visible under the light microscope; Figs. 2.16 and 4.31) to large colonial algae such as *Microcystis* (Fig. 4.34) and *Anabaena* (Fig. 4.24a). Large colonies of the latter can be readily seen with the naked eye and show a simple globular or filamentous form with copious mucilage. The balance of photosynthetic pigments present in blue-green algae (Table 1.3) varies with light spectrum and intensity, resulting in a range of colours from brown to blue-green (Fig. 4.24a). Phycobilin pigments are particularly prominent, with phycocyanin tending to predominate over phycoerythrin at low light levels, giving the cells the blue-green colour typical of these algae. It is thus an advantage to observe material from shaded as well as better illuminated situations wherever possible.

1.3.1 Cytology

Prokaryote status

The prokaryote nature of these algae is indicated by the small size of the cells (typically <10 μm in diameter) and by the presence of central regions of nucleoid DNA (not enclosed by a nuclear membrane). These nucleoid regions (Fig. 1.4) can be observed both by light microscopy (as pale central areas within living cells) and by transmission electron microscopy (as granular regions in chemically fixed cells) – where the absence of a limiting membrane can be clearly seen. Although these algae lack the cytological complexity of eukaryotes (no membrane-bound organelles such as mitochondria, plastids, microsomes and Golgi bodies), they do have a range of simple granular

inclusions such as carboxysomes, cyanophycin granules, polyphosphate bodies and glycogen particles (Fig. 1.4). Photosynthetic pigments are associated with thylakoid membranes, which are typically dispersed throughout the peripheral protoplasm.

As with other prokaryotes, cell walls are made up of a peptidoglycan matrix with an inner lipopolysaccharide membrane and are generally quite thin. Blue-green algae resemble Gram-negative bacteria in having an additional outer cell membrane. Some species also have a mucilage layer outside the cell wall which may be dense or watery, structured or unstructured. The outside layers of the cell wall can sometimes become stained straw coloured or brownish from iron and other compounds in the surrounding water as in *Scytonema* and *Gloeocapsa*.

The fundamental bacterial nature of these organisms distinguishes them from all other algae and determines a whole range of features – including molecular biology, physiology, cell size, cell structure and general morphology.

Gas vacuoles

In some species, gas vacuoles may be formed, appearing under the light microscope as highly refractive or quite dark structures. Gas vacuoles aid buoyancy in planktonic species, allowing the cells to control their position in the water column. Cells may then congregate at a depth of optimal illumination, nutrient concentration or other factor for that species. This is not necessarily at the surface - where the light intensity may be too great and cause photoinhibition or permanent cell damage. Movement up and down in the water column can enhance nutrient uptake as it allows the cells to migrate to depths where essential nutrients, for example phosphates, are more abundant and then up towards the surface for light energy absorption. When gas vacuole–containing species are present in a sample of water they often tend to float to the surface (this can cause problems when preparing the sample for cell counts; see Chapter 2).

In addition to gas vacuole–mediated movements of planktonic blue-greens within the water column, other types of motility also occur – including gliding movements of filamentous algae such as *Oscillatoria*) on solid substrata.

1.3.2　Morphological and taxonomic diversity

Blue-green algae are remarkable within the prokaryote kingdom for showing the range of size and form noted earlier, with some organisms forming quite large, complex, three-dimensional colonies. Limited differentiation (Figs. 4.24a-c) can occur within colonies, with the formation of heterocysts (nitrogen-fixing cells) and akinetes (thick-walled resistant cells). In filamentous forms the cells may be the same width along the filament or in some genera they may narrow or taper towards the end, even forming a distinct hair-like structure in the case of *Gloeotrichia* (Fig. 4.22). Clear branching occurs in the most complex forms, with a fundamental distinction between 'true branching' as in *Stigonema* (Fig. 4.20) and 'false branching' as in *Tolypothrix* (Fig. 4.21). True branching involves division of a single cell within the filament, giving rise to two or more daughter cells which themselves form branches. In contrast, false branching involves lateral extension of the filament without the single cell division and daughter cell development as above.

Freshwater blue-green algae can be divided into four main groups (Table 1.5) in relation to general morphology, presence/absence of specialised cells and the nature of branching in filamentous forms. These four groups form the basis for current taxonomy of this phylum, as adopted by John *et al.* (2002) and Komarek *et al.* (2003a, b).

- Chroococcales. The simplest blue-green algae, occurring essentially as solitary cells (no filamentous forms), typically enclosed by a thin layer of mucilage. The cells may remain as single cells or be aggregated into plate-like or globular colonies. Typically planktonic with some colonial forms (e.g. species of *Microcystis*) forming massive surface blooms containing individual colonies that are recognisable with the naked eye. They typically lack specialised cells, though one group (typified by *Chamaesiphon*) forms exospores.

- Oscillatoriales. Filamentous algae, lacking heterocysts and akinetes. These relatively simple algae occur as planktonic (some bloom forming) or benthic aggregations. In some cases, they form dense mats on mud or rocky substrata which secondarily detach as metaphyton into the main body of water (Fig. 2.1).

- Nostocales. A diverse group of filamentous algae, planktonic or benthic, with heterocysts and akinetes but not showing true branching. Filaments may be unbranched or show false branching. These algae are able to form large colonies by lateral association of filaments (bundles), 3D tangles or as radiating filaments from the centre of the sphere.

- Stigonematales. Filamentous algae, with heterocysts and akinetes and showing true branching. Structurally the most complex blue-green algae, with some thalli (e.g. *Fischerella*) differentiated into multiseriate/uniseriate basal filaments and uniseriate erect branches. Largely benthic algae, with genera such as *Stigonema* commonly attached to substrata in standing and flowing waters, detaching to form planktonic masses.

The range of size and form noted in Table 1.5 indicates a wide morphological diversity, which is useful in taxonomic identification. The relative importance of molecular versus morphological characteristics in relation to taxonomy reflects the debate as to whether these organisms should be treated as bacteria (cyanobacteria – Stanier *et al.*, 1978) or algae (blue-green algae – Lewin, 1976). Although they fundamentally resemble bacteria in their prokaryote features, they also differ from bacteria in carrying out photosynthesis coupled to O_2 evolution, and in the complexity of their morphology. Although current taxonomy is based primarily on phenotypic characters (see above), ultrastructural and molecular data provide useful supplementary information (Komarek *et al.*, 2003a, b). Current molecular analyses support the separation of non-heterocystous and heterocystous genera (Rudi *et al.*, 2000) and are particularly relevant in the case of unicellular blue-greens, where morphological features are not adequate. The

Table 1.5 Taxonomic and Morphological Diversity in Freshwater Blue-Green Algae.

Order: Major Morphotype	Colony Size and Form	Attached or Planktonic	Examples
Chroococcales			
(a) Unicellular to spheroid colonies, lacking specialised vegetative, resistant or reproductive cells	Single cells	Planktonic or attached to plant and substrate surfaces	*Synechococcus* (Fig. 4.31)
	Small colonies (4–32 cells)	Free-floating, tangled-up with filamentous algae or attached to substrates	*Chroococcus* (Fig. 4.33) *Gleocapsa* (Fig. 4.32)
	Flat plate of cells	Free-floating or sedentary	*Merismopedia*
	Large solid spherical colony	Planktonic	*Aphanocapsa* (Fig. 4.35) *Microcystis* (Fig. 4.34)
	Large hollow spherical colony	Planktonic	*Gomphosphaeria* (Fig. 4.30) *Coelosphaerium*
(b) Unicellular, forming exospores	Cells remain single or form multi-layered colonies	Attached to surfaces of aquatic plants, algae and inorganic substrate	*Chamaesiphon*
Oscillatoriales			
Filamentous algae, lacking heterocysts and akinetes	Elongate straight filaments	Planktonic or benthic Benthic mat	*Oscillatoria* (Fig. 4.27)[a] *Phormidium* (Fig. 4.29)
	Elongate spiral filaments	Planktonic or on mud surfaces	*Spirulina* (Fig. 4.26)
Nostocales			
Filamentous algae forming heterocysts and akinetes, but no true branching	Bundles of Elongate filaments	Planktonic or benthic	*Aphanizomenon* (Fig. 4.23) *Nostoc* (Fig. 4.25)
	3-D tangle of filaments	Planktonic	*Anabaena* (Fig. 4.24)
	Spherical colony of radiating filaments	Planktonic	*Gloeotrichia* (Fig. 4.22)
Stigonematales			
Filamentous algae forming heterocysts and akinetes, with true branching	Branched mass of filaments	Benthic or planktonic	*Stigonema* (Fig. 4.20)
	Differentiation into basal filaments and erect branches	Benthic	*Stauromatonema* *Fischerella*

[a]Planktonic species of *Oscillatoria* are placed in the genera *Planktothrix*, *Pseudanabaena* or *Limnothrix* by some authors.

classical (eukaryote) use of morphology to define species is also problematic in this group, since the species concept and definition of species are limited by a complete absence of sexual reproduction.

In the absence of sexual processes, reproduction in this group is by vegetative or specialised asexual means. Asexual spores (akinetes) consist of vegetative cells that are larger than normal. They generally have thickened walls and, in filamentous forms, are often produced next to heterocysts (Fig. 4.24 c). Baeocysts (small spherical cells formed by division of a mother cell) may be produced in some coccoid species and are released into the environment. In some filamentous forms, deliberate fragmentation of the filament can occur – releasing the fragments (hormogonia) to produce new filaments.

1.3.3 Ecology

Blue-green algae are thought to have arisen approximately 3.5 billion years ago (Schopf, 1993) during which time they have been the dominant form of life for about 1.5 billion years. As a result of this long evolutionary history, they have adapted to (and frequently dominate) all types of freshwater environment – including extreme conditions (thermal springs, desiccating conditions), brackish (semi-saline) conditions, high and low nutrient environments and planktonic/benthic habitats. The ability of blue-green algae to dominate freshwater environments is particularly important in standing waters, where algal blooms may result due to eutrophication (Section 3.2.3).

Algal blooms

In mid to late summer, eutrophic temperate lakes frequently develop massive populations of colonial blue-green algae. These may rise to the surface of the lake, forming a thick layer of algal biomass at the top of the water column, out-competing other algae and having major impacts on zooplankton and fish populations (Sigee, 2004). The ability of blue-greens to out-compete other freshwater algae has been attributed (Shapiro *et al.*, 1990) to a range of characteristics, including

- Optimum growth at high temperatures – *summer temperatures.*

- Tolerance to low light – *important within the dense algal bloom, and enabling algae to survive lower in the water column.*

- Tolerance of low N/P ratios – *allowing continued growth when N becomes limiting.*

- Depth regulation by buoyancy – *avoiding photoinhibition during the early phase of population increase, and allowing algae to obtain inorganic nutrients from the hypolimnion when the epilimnion becomes depleted in mid to late summer.*

- Resistance to zooplankton grazing – *limiting the impact of herbivory on algal growth.*

- Tolerance of high pH/low CO_2 concentrations – *allowing continued growth of blue-greens (but not other algae) at the lake surface during intense bloom formation.*

- Symbiotic association with aerobic bacteria – *bacterial symbionts at the heterocyst surface maintain the local reducing atmosphere required for nitrogen fixation and are also an important source of inorganic nutrients in surface-depleted waters (Sigee, 2004).*

Species preferences

Many blue-green algae can tolerate a wide range of environmental conditions, so strict ecological preferences are unusual in this group. *Microcystis aeruginosa* (Fig. 4.34), for example, can form massive growths in eutrophic lakes, but low numbers may also occur in oligotrophic waters (Reynolds, 1990). The distinction between planktonic and benthic organisms is also frequently not clear-cut. *Microcystis* is regarded as a typical planktonic alga, but may also occur as granular masses on lake bottoms, and (as with the majority of temperate lake algae) overwinters on lake sediments rather than in the water column. Many benthic algae become detached and either remain loose on the substratum or rise in the water column and become free-floating. Colonies of *Gloeotrichia*, for example, often detach from substrates and become planktonic (Fig. 4.22), where they grow and may reach bloom proportions.

In spite of these cautions, many blue-green algae can be characterised in relation to particular environments and whether they are typically attached or planktonic. Some examples are shown in Table 1.6, for both 'normal' and 'extreme' environments. As with other algae, ecological preferences typically relate to multiple (rather than just single) environmental factors. Colonial blue-green algae, which grow particularly well in high-nutrient lakes, are also suited to hard waters (high Mg, Ca) and to the alkaline conditions typical of these environments. Conversely, oligotrophic lakes, which support unicellular rather than colonial blue-greens, tend to be soft, slightly acid waters.

1.3.4 Blue–green algae as bioindicators

As with other algal groups, the presence or absence of particular species can be a useful indicator of

Table 1.6   Ecological Diversity of Freshwater Blue-Green Algae.

Major Ecosystem	Specific Conditions	Benthic Algae	Planktonic Algae
Standing waters – lakes and ponds	Meso- to eutrophic	*Gloeotrichia*: Attached to substrates. Detached colonies planktonic	Colonial blue-greens: *Microcystis, Anabaena*
	Oligotrophic		Unicellular blue-greens: *Synechococcus*
	Hard and soft waters		*Aphanothece*
	Soft, acid lakes	*Cylindrospermum*: often forming dark patches on submerged vegetation	*Tolypothrix*: Planktonic or in submerged vegetation
Wetlands	Soft waters – common in bogs	*Chroococcus* – Attached to substrates, mixed with tangles of filamentous algae or free-floating *Merismopedia* – typical of bog-water communities	
	Open water in bogs		*Aphanocapsa*
Running waters – streams and rivers	Low N concentrations, high N:P ratio	Benthic mats: *Nostoc, Calothrix* (Stancheva *et al.*, 2013)	*Oscillatoria rubescens*
Extreme environments			
Polar lakes, ponds, soils	Low temperature, low light (ice cover)	Benthic mats: *Calothrix, Phormidium. Leptolyngbya* (Martineau *et al.*, 2013)	
Thermal springs	45°C or greater	*Mastigocladus laminosus Oscillatoria* (pigmented) (Darley, 1982)	*Synechococcus lividus* (Stal., 1995)
Brackish waters – saline lakes	Range of salinity		*Aphanothece halophytica* (Yopp *et al.*, 1978).
Spent nuclear fuel storage ponds	High radioactivity, with range of radionuclides including ^{137}Cs and ^{60}Co.	Biofilm dominated by *Leptolyngbya* (Evans, 2013).	

ecological status (Table 1.6). The dominant presence of colonial blue-greens in lake phytoplankton (Fig 1.5.) provides a useful indicator of high nutrient status and these algae are a key component of various trophic indices (see Section 3.2.3). Conversely, populations of planktonic unicellular blue-green algae are indicative of oligotrophic to mesotrophic conditions. In streams, the relative abundance of benthic N_2-fixing (heterocystous) blue-greens can be used for rapid nutrient (N concentration) biomonitoring (see Section 3.4.7).

Changes in the population of colonial blue-greens in lakes and estuaries may also act as an indicator of climate change. An increase in the intensity of *Microcystis* blooms in San Francisco Bay (USA), for example, has been predicted with expected increases in water temperature and low stream flow during increasing incidence of droughts (Lehman *et al.*, 2013).

1.4   Green algae

In the freshwater environment, green algae (Chlorophyta) range in size from microscopic unicellular

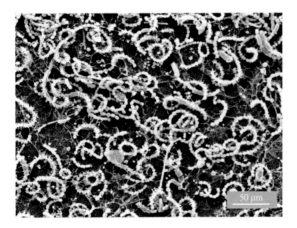

Figure 1.5 Mid-summer blue-green algal bloom in a eutrophic lake. SEM view of lake epilimnion phytoplankton sample (August), showing the dense population of filamentous *Anabaena flos-aquae* which totally dominated the algal bloom (chlorophyll-*a* concentration 140 µg ml^{-1}). Copious mucilage associated with this alga is seen as numerous fine strands, formed during the dehydration preparation process. See also Figs. 2.8 (seasonal cycle) and 4.24 (live *Anabaena*).

organisms to large globular colonies and extensive filamentous growths. They are characterised by a fresh green coloration due to the presence of chlorophylls-*a* and -*b*, which are not obscured by accessory pigments such as β-carotene and other carotenoids. In cases where algae have excessive light exposure, the carotenoid pigments may be photoprotective and occur at high levels, obscuring the chlorophylls and giving the alga a bright red colour. This is seen with the flagellate *Haematococcus* (Fig. 4.54), which frequently colours bird baths and other small pools bright red, and with the snow alga *Chlamydomonas nivalis* where areas of frozen landscape are similarly pigmented.

1.4.1   Cytology

In addition to their characteristic pigmentation and other biochemical features (Table 1.3), green algae also have a number of distinctive cytological aspects.

- Flagella, when present, occur in pairs. These are equal in length without tripartite tubular hairs and have a similar structure (isokont). Some genera have four or even eight flagella, but this is unusual.

- Chloroplasts vary in shape, size and number. In unicellular species, they tend to be cup-shaped (Fig. 4.55), but in filamentous forms may be annular (Fig. 4.19), reticulate (Fig. 4.17), discoid or ribbon-like (Fig. 4.13). They have a double outer membrane, with no enclosing periplastidal endoplasmic reticulum.

- Production and storage of the photosynthetic reserve (starch) occur inside the plastid, with granules frequently clustered around the pyrenoid. In all other eukaryote algae, the storage material occurs mainly in the general cytoplasm.

In motile species, an eyespot is frequently present, appearing red or orange (Fig. 4.38) in fresh specimens. The cell wall in green algae is made of cellulose.

1.4.2   Morphological diversity

Green algae are the most diverse group of algae, with about 17,000 known species (Graham and Wilcox, 2000). This diversity is reflected by the variation in morphology, with organisms being grouped into unicellular, colonial or filamentous growth forms (Table 1.7). The level of greatest morphological and reproductive complexity is represented by Charalean algae (e.g. *Chara, Nitella*), which can reach lengths of over a metre, have whorls of branches at nodes along the length of the thallus (Chapter 4, Plate 1) and have been frequently confused with aquatic higher plants such as *Ceratophyllum*.

In the past, this morphological diversity provided the taxonomic basis for green algal classification (Bold and Wyne, 1985) – with orders primarily being defined largely on a structural basis. These included the orders Volvocales (flagellate unicells and simple

Table 1.7 Range of Morphologies in Freshwater Green Algae.

Major Morphotype	Attached or Planktonic	Example
Flagellate Unicells	Planktonic	*Chlamydomonas* (Fig. 4.55)
Non-flagellate Unicells	Planktonic or benthic (often associated with periphyton)	Crescent shaped – *Selenastrum* (Fig. 4.78) Equatorial division – *Micrasterias* (Fig. 4.81) *Closterium* (Fig. 4.80)
Colonial	Planktonic	Net – *Hydrodictyon* (Fig. 4.1) Hollow sphere – *Volvox* (Fig. 4.40), *Coelastrum*, *Eudorina* (Fig. 4.39) Solid sphere – *Pandorina* (4.36) Plate of cells – *Gonium* (Fig. 4.38), *Pediastrum* (Fig. 4.47) Small linear colonies – *Scenedesmus* (Fig. 4.50) Branching colonies – *Dictyosphaerium* (Fig. 4.45)
Disk-shaped plate, one cell thick	Attached to higher plants (epiphyte) or to stones	*Coleochaete*, *Chaetosphaeridium*
Unbranched filaments	Typically Attached, but may detach to become planktonic	*Uronema*, *Microspora*, *Oedogonium* (Fig. 4.17), *Zygnema* (Fig. 4.15), *Spirogyra* (Fig. 4.13)
Branched filaments		*Chaetophora*, *Draparnaldia*, *Cladophora* (Fig. 4.6)
Large complex algae with whorls of branches	Attached	*Chara*, *Nitella*

colonial forms), Chlorococcales (unicells and non-motile coenobial colonies), Ulotrichales (unbranched filaments) and Chaetophorales (branched filaments). More recently, the application of cytological, comparative biochemical and molecular sequencing techniques has demonstrated the occurrence of extensive parallel evolution. On the basis that classification should reflect phylogeny, these original groupings are thus no longer valid, and a new classification is emerging where individual orders contain a mixture of unicellular, globular colonial and filamentous forms (Graham and Wilcox, 2000).

1.4.3 Ecology

Green algae are ecologically important as major producers of biomass in freshwater systems, either as planktonic (standing waters) or attached (running waters) organisms – where they respectively may form dense blooms and periphyton growths.

Algal Blooms

In mesotrophic and eutrophic lakes, green algae do not normally produce the dense blooms seen with diatoms and blue-green algae – but do become dominant or co-dominant in early summer, between the clear-water phase and mid-summer mixed algal bloom. In smaller ponds, filamentous *Spirogyra* and colonial *Hydrodictyon* frequently form surface blooms or scums, and in small garden ponds and birdbaths *Haematococcus* may form dense populations. When nutrient levels become very high, a switch may occur from colonial blue-green to green algae as major bloom formers. This is can be seen in some managed fishponds, where organic and inorganic nutrients are applied to enhance fish production by increasing carbon flow through the whole food chain. In the Třeboň Basin Biosphere Reserve (Czech Republic), application of lime (supply of carbonate and bicarbonate ions), organic fertilisers and manure leads to high pH and hypertrophic nutrient levels. In these conditions, a short diatom bloom

Table 1.8   Ecological Diversity of Freshwater Green Algae.

Major Ecosystem	Specific Conditions	Benthic Algae	Planktonic Algae
Standing waters – Lakes and ponds	Wide range of conditions	*Oedogonium*	*Scenedesmus, Pediastrum, Tetraedron, Chlamydomonas*
	Eutrophic to Hypertrophic	*Cladophora*	*Scenedesmus, Pediastrum.*
	Oligotrophic		Desmids
	High (H) and low (L) pH	*Chara, Nitella* (H) *Coleochaete* (L)	*Mougeotia* (L)
	Hard water – high (Ca) concentration	*Hydrodictyon*	
	Salt lakes		Small unicells, *Dunaliella*
Standing waters – Wetlands	Wide range of conditions		*Selenastrum Chlamydomonas*
	Low nutrient, low pH bogs		*Spirogyra, Mougeotia.* Desmids.
Running waters – Streams and rivers	Wide range of conditions		*Chlamydomonas Spirogyra*
	Hardness (Ca) concentration	*Hydrodictyon*	
Running waters – Estuaries, brackish water	Wide range of conditions		*Scenedesmus*
	High salinity		*Dunaliella*
Specialised microenvironments	Very high nutrients – e.g. high sewage level	*Prototheca*	*Euglena*
	Endosymbiont in invertebrates	*Chlorella*[a]	
	Associated with Calcareous deposits	*Oocardium stratum*	

[a] Also widely occurring as a free-living organism.

(*Stephanodiscus*) is replaced in early summer by populations of rapidly growing unicellular and small colonial green algae (*Scenedesmus* and *Pediastrum*), which continue to dominate for the rest of the annual cycle (Pechar *et al.*, 2002). Extensive growth of attached algae (periphyton) may also occur under high nutrient conditions. Branched thalli of *Cladophora* can form extended communities of shallow water vegetation along the shores of eutrophic lakes and streams, breaking off in storms to form loose biomass which can degrade to generate noxious odours. Another filamentous alga, *Mougeotia*, can form large mucilaginous subsurface growths in lakes that have been affected by acid rain.

Species preferences

Individual species of green algae differ considerably in their ecological preferences, ranging from broad spectrum organisms (*Spirogyra*, *Chlamydomonas*) to species with very restricted habitats.

Spirogyra occurs in a wide range of habitats, where it is typically attached to stable substratum (as periphyton) but also occurs as free-floating mats (Lembi *et al.*, 1988 – derived by detachment either from periphyton (vegetative propagation) or from benthic zygotes (sexual derivation). In surveys of North American sites, McCourt *et al.* (1986) recorded *Spirogyra* at nearly one-third of all locations, and Sheath and Cole (1992) detected *Spirogyra* in streams from a wide range of biomes – including desert chaparral, temperate and tropical rainforests, and tundra.

Other green algae show clear preferences for particular environmental conditions (Table 1.8) that include degree of water movement (lentic vs. lotic conditions), inorganic nutrient status (oligotrophic to eutrophic), pH, hardness (Ca concentration) and salinity. These conditions frequently occur in combination, with many moorland and mountain water bodies being low nutrient and neutral to acidic, while lowland sites in agricultural areas are typically eutrophic and alkaline.

Nutrient status   Many desmids (e.g. *Closterium*) are particularly characteristic of oligotrophic (low nutrient) lakes and ponds – often in conditions that are also slightly acidic (see above) and dystrophic (coloured water). Desmid diversity is particularly high in these conditions, sometimes with hundreds of species occurring together at the same site (Woelkerling, 1976). Desmids are also typical of nutrient-poor streams, where they are permanent residents of periphyton – making up to 10% of total community biomass. These desmids are associated with plants such as the bryophyte *Fontinalis*, achieving concentrations as high as 10^6 cells g^{-1} of substrate (Burkholder and Sheath, 1984). Desmids often constitute a significant proportion of algal biomass in peat wetlands (see Section 3.3.2), where they are also a major aspect of species diversity.

Some species of desmids – such as *Closterium aciculare* – are more typical of high-nutrient, slightly alkaline lakes and are indicators of eutrophic conditions. Periphytic algae typical of eutrophic waters include *Cladophora*.

Acid lakes   Although desmids are typical of neutral to slightly acid lakes, other green algae are adapted to more extreme conditions. These include the filamentous green alga, *Mougeotia*, which can form substantial sub-surface growths in acidified waters and is widely regarded as an early indicator of early environmental change (Turner *et al.*, 1991). Where acidification is the result of acid precipitation, experimentation (Webster *et al.*, 1992) or industrial pollution, the acid conditions also tend to be associated with

- Relatively clear waters, due to low levels of phytoplankton

- Increases in the concentration of metals such as Al and Zn

- Reduced levels of dissolved organic carbon, with derivation largely from external lake (allochthonous) rather than internal (autochthonous) sources.

- Food web changes, including a reduction in the number of herbivores.

Laboratory studies (Graham *et al.*, 1996a, b) showed that *Mougeotia* was physiologically adapted to such conditions, with the ability to photosynthesise over a wide range of irradiances (300–2300 µmol quanta m^{-2} s^{-1}) and a tolerance to a broad range of both pH (pH 3–9) conditions and metal concentrations. The physiological adaptations of this alga, coupled with reduced grazing from herbivores, probably accounts for the extensive growth and domination of the benthic environment in many acidic waters.

Alkaline and calcium-rich waters   Some planktonic (*Closterium aciculare*) and benthic algae (e.g. *Nitella* and *Chara*) are adapted to grow in alkaline habitats, many of which are also rich in calcium (hard waters). *Nitella* produces lush meadows on the bottoms of neutral to high pH lakes, and *Chara* can form dense, lime encrusted lawns in shallow alkaline waters. *Hydrodictyon* is also typical of hard-water conditions, occasionally blanketing the surface of ponds and small lakes, and occurring widely in larger alkaline standing waters and hard-water streams. Extensive growths of this alga can also be observed in agricultural situations such as irrigation ditches, rice paddy fields and fish farms – where there is some degree of eutrophication.

Some calcium-loving algae occur in very restricted environments. The unusual, slow-growing desmid *Oöcardium stratum*, for example, occurs in calcareous streams and waterfalls at the top of branched calcareous tubes, in association with deposits of 'travertine' and 'tufa'.

Saline waters   Various green algae are adapted to saline conditions in salt lakes and estuaries, where salt levels range from low (brackish waters) to high. With some widely occurring genera such as *Scenedesmus*, brackish conditions are at one end of the continuum of environmental tolerance. Other algae, such as *Dunaliella*, are specifically adapted to a more extreme situation. This organism occurs in the highly saline waters of commercial salt pans (salterns) and natural salt lakes such as the Great Salt Lake of Utah (USA). *Dunaliella* is able to tolerate external salt concentrations of $>5M$, by balancing the high external osmotic pressure (OP) with a high internal OP generated by photosynthetically produced glycerol.

Specialised environments In addition to the free-living occurrence of particular algae in lakes and rivers that have a range of characteristics (see above), other green algae are adapted to more specialised conditions. Some of these involve a change in the mode of nutrition from autotrophic to heterotrophic and include various species of the non-flagellate unicellular algae, *Chlorella* – a widespread endosymbiont of freshwater invertebrates. Other green algae retain their free-living existence, but are able to supplement photosynthesis (mixotrophy) by absorption of exogenous dissolved carbon (Tuchman, 1996) such as amino acids and sugars through their cell surface (osmotrophy). The ability to use external organic carbon may be important in two situations – when photosynthesis is limiting and when external soluble organic is in excess.

• Limiting photosynthesis. This may occur, for example, under conditions of chronic low light intensity (e.g. arctic lakes). It is also typical of many acid lakes, where low pH reduces the level of dissolved inorganic carbon to levels that are unable to saturate photosynthesis. Laboratory experiments on *Coleochaete*, an acidophilic alga, have demonstrated an ability to use dissolved organic carbon in the form of hexose sugars and sucrose (Graham *et al.*, 1994).

• High levels of external dissolved organic carbon. Some osmotrophic green algae are photoheterotrophic, only able to use organic carbon when light is present and when their photosynthesis is inhibited by the high availability of dissolved organic carbon (Lewitus and Kana, 1994). The colourless unicellular alga *Prototheca* has completely lost the ability to photosynthesise and obtains all its carbon from external sources present in soil and freshwater environments that are contaminated by sewage. The evolutionary relationship of this obligate heterotroph to green algae is indicated by the presence of starch-containing plastids.

1.4.4   Green algae as bioindicators

Contemporary algae

Habitat preferences of contemporary green algae (Table 1.8) are frequently useful in providing information on physicochemical characteristics of the aquatic environment. Filamentous green algae, for example, often dominate environments stressed by cultural eutrophication, acidification and metal contamination (Cattaneo *et al.*, 1995).

Fossil algae

Green algae do not typically produce resistant walls that persist in aquatic sediments and are much less useful than diatoms and chrysophytes as bioindicators in terms of the fossil record. There are some exceptions to this, where the cell wall contains fatty acid polymers known as algaenans that can withstand millions of years of burial (Gelin *et al.*, 1997). *Scenedesmus* and *Pediastrum* are among the most common members of green algae that have fossil records, containing algaenans and also some silicon. The desmid *Staurastrum* is also frequently seen as fossil remains, due to the impregnation of cell wall material with polyphenolic compounds which confer resistance to bacterial decay (Gunnison and Alexander, 1975).

1.5   Euglenoids

Euglenoid algae (Euglenophyta) are almost entirely unicellular organisms, with a total of 40 genera worldwide (about 900 species) – most of which are present in freshwater. Cells are typically motile, either via flagella or (in non-flagellate cells) by the ability of the body to change shape (referred to as 'metaboly').

About one-third of the euglenoids are photosynthetic and classed amongst the algae. The rest are colourless, being either heterotrophic or phagotrophic, and are usually placed in the Protozoa. In photosynthetic organisms, pigmentation is closely similar to that in green algae (Table 1.3), but the variable presence of carotenoid pigments means that these organisms can routinely vary in colour from fresh green (Fig. 4.51) to yellow-brown. In some situations, the accumulation of the carotenoid astaxanthin gives cells a bright red colouration. This is seen particularly well in organisms such as *Euglena sanguinea*, which forms localised blooms in ponds and ditches.

1.5.1 Cytology

Euglenoids are typically elongate, spindle-shaped organisms (Figs. 1.6, 4.51) and usually contain

several chloroplasts per cell, which vary in appearance from discoid to star-, plate- or ribbon-shaped. Cytological features that distinguish euglenoids (Table 1.3) include the following:

- The presence of an anterior flask-shaped depression or reservoir within which the flagella are inserted. Although two flagella are present only one emerges from the reservoir into the surrounding medium, the second being reduced and contained entirely within the reservoir. An eyespot is often present and is located close to the reservoir.

- The production of paramylon as storage reserve. This β-1,3-linked glucan does not stain blue-black with iodine solution and is found in both green and colourless forms.

- The presence of a surface coat or pellicle that gives the cell a striated appearance. This occurs just below the plasmalemma and is composed of interlocking protein strips that wind helically around the cell. In some the pellicle is flexible (allowing the cell to change its shape), while in others the pellicle is completely rigid giving a permanent outline to the cell. Genera such as *Trachelomonas* have cells surrounded by a lorica, the anterior end of which is a narrow, flask-shaped opening through which the flagellum emerges.

Reproduction is asexual by longitudinal division, with sexual reproduction being completely unknown in this group of organisms.

1.5.2 Morphological diversity

Almost all euglenoids are unicellular, with colonial morphology being restricted to just a few organisms where cells are interconnected by mucilaginous strands. One of these is *Colacium*, a stalked euglenoid that is widespread on aquatic substrates – but is particularly commonly attached to zooplankton such as *Daphnia*. In the sessile state, this organism does not have emergent flagella on the cells – which occur at the end of the branched mucilaginous attachment

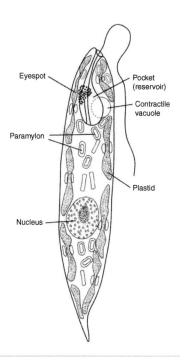

Figure 1.6 Diagrammatic view of *Euglena*, showing major cytological features. Graham & Wilcox, 2000. Reproduced with permission from Prentice Hall. See also Fig. 4.51 (live *Euglena*).

Eyespot

Pocket (reservoir)

Contractile vacuole

Paramylon

Plastid

Nucleus

stalks. Individual cells may produce flagella, however, swim away and form a new colony elsewhere.

With the relative absence of colonial form, diversity in this group is based mainly on variation in ultrastructural features – including feeding apparatus, flagella and pellicle structure (Simpson, 1997). A further aspect of diversity is the distinction between green and colourless (autotrophic/heterotrophic) cells, with approximately two-thirds of all species being heterotrophic. Some euglenoids (e.g. strains and species of *Euglena*) are facultative heterotrophs, which are able to carry out heterotrophic nutrition when photosynthesis is limiting or when surrounding concentrations of soluble organic materials are high. In the heterotrophic state, these organisms retain their plastids as colourless organelles and absorb soluble nutrients over their whole surface (osmotrophy). Other euglenoids (*Petalomonas, Astasia, Peranema*) are obligate heterotrophs and have lost their plastids completely. Many of these organisms consume particulate organic material (phagocytic) and have evolved a complex feeding apparatus.

1.5.3   Ecology

Euglenoids are generally found in environments where there is an abundance of decaying organic material. This is in line with the heterotrophic nature of many of these organisms and the ability to take up complex organic material either in the soluble or particulate state. Typical habitats include shallow lakes, farm ponds, wetlands, brackish sand and mudflats. Within these environments, euglenoids are particularly associated with interfaces such as sediment-water and air-water boundaries (Walne and Kivic, 1990) and should probably not be regarded as open water truly planktonic algae (Lackey, 1968).

Certain euglenoid algae are able to tolerate extreme environmental conditions. One of these, *Euglena mutabilis*, is able to grow in very low pH waters. This alga has an optimum pH of 3.0, can tolerate values below pH 1.0 and is typical of acidic metal-contaminated ponds and streams draining mines. Other euglenoids, found in brackish habitats, are able to tolerate a wide range of salinity (Walne and Kivic, 1990).

1.5.4   Euglenoids as bioindicators

Euglenoid algae are not particularly useful as environmental bioindicators, in terms of either contemporary populations or fossil records. Although present-day algae show some adaptations to specific environments (see previous section), there is no established environmental library and species may be difficult to identify. Within lake sediments, the lack of calcified or silicified structures that are resistant to decay means that hardly any remains have survived in the fossil record.

1.6   Yellow-green algae

Yellow-green algae (Xanthophyta) are non-motile, single-celled or colonial algae, with a distinctive pigmentation that gives the cells a yellow or fresh green appearance (e.g. Fig. 4.18 – *Tribonema*). Although there is a wide range in morphology (Table 1.9), this phylum contains relatively few species (compared to major groups such as green algae) and the algae tend to be ecologically restricted to small water bodies and damp soils.

1.6.1   Cytology

Yellow-green algae are most likely to be confused with green algae when observed in the fresh condition, but differ from them (Table 1.3) in a number of key features.

- Distinctive pigmentation, with chlorophylls (a, c_1 and c_2 – but not b), carotenoids (especially β-carotene) and three xanthophylls (diatoxanthin, vaucheriaxanthin and herteroxanthin).

- Carbohydrate storage as oil droplets or chrysolaminarin (usually referred to as leucosin) granules.

- Walls composed mainly of pectin or pectic acid (sometimes with associated cellulose or siliceous material). The walls often occur as two spliced and overlapping sections, breaking into 'H'-shaped pieces on dissociation of the filaments.

- When flagella are present they are of unequal length (heterokont) the longer one pointing forwards and the shorter (which can be more difficult to see using a light microscope) sometimes pointing forwards or backwards

- Chloroplast fine structure – four outer membranes (incorporating an endoplasmic reticulum) and thylakoids occurring in threes.

Cells contain two or more discoid and green to yellow-green chloroplasts, with associated pyrenoids rarely seen. An eyespot may be present.

1.6.2 Morphological diversity

Yellow-green algae show a range of morphology, from unicellular to colonial and filamentous (Table 1.9). Unicellular forms are non-flagellate in the mature (vegetative) state, with motility being restricted to biflagellate zoospores (asexual) or motile gametes (sexual) at reproductive stages. The zoospores have unequal and morphologically dissimilar flagella (heterokont algae – Fig. 1.7). Unicellular algae include simple free-floating forms (*Botrydiopsis*, *Tetraedriella*) or may have associated rhizoidal

systems or an attachment stalk. Some yellow-green algae form simple globular colonies, which may be mucilaginous and free floating (e.g. *Gloeobotrys*) or attached via a stalk (*Ophiocytium*). Filamentous forms may be unbranched (*Tribonema* – Fig. 4.18) or branched (*Heterococcus*) and reach their most massive state in the coenocytic siphonaceous genera *Botrydium* and *Vaucheria* (Fig. 4.3).

1.6.3 Ecology

Yellow-green algae are rather limited in their exploitation of aquatic habitats, tending to occur on damp mud (at the edge of ponds) and soil, but not occurring extensively in either lentic or lotic systems. Where planktonic forms do occur, they tend to be in ditches or small ponds. *Tribonema* species are often found as free-floating filaments in temporary waters, and *Botrydiopsis arrhiza* is found at the edge of ponds and in patches of water in sphagnum bogs, where it may form yellowish water blooms. Even *Gloeobotrys limneticus*, which has widespread occurrence as mucilaginous colonies in lake plankton, never forms extensive or dominant populations (as seen, e.g. by blue-green algae). Various yellow-green algae are epiphytic, including *Mischococcus*

Table 1.9   Range of Morphologies in Yellow-Green Algae.

Major Morphotype	Attached or Planktonic	Examples
Unicells		
Coccoid	Planktonic or benthic	*Botrydiopsis*
Pear-shaped, with rhizoidal system	Present on mud surface	*Botrydium*
Ovoid cells on stalk	Epiphytic on attached or planktonic algae	*Characiopsis*
Amoeboid	Benthic, often in hollow cells of *Sphagnum*.	*Chlamydomyxa*
Colonial		
Small to large globular colonies	Planktonic	*Gloeobotrys*
Branched colony on stalk	Planktonic or attached	*Ophiocytium*
Filaments		
Unbranched	Typically planktonic, occasionally attached when young	*Tribonema* (Fig. 4.18)
Branched	Attached, damp soils	*Heterococcus*
Coenocytic, (siphonaceous)	Attached	*Botrydium*, *Vaucheria* (Fig. 4.3).

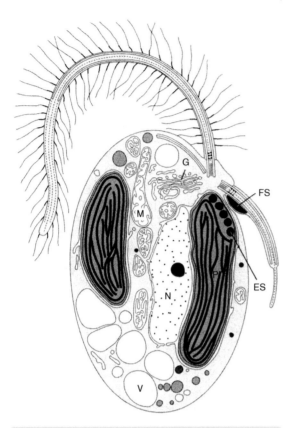

Figure 1.7   Line drawing of heterokont zoospore, typical of xanthophyte algae, with a long anterior flagellum bearing two rows of stiff hairs plus a short posterior flagellum. This is smooth and often bears a swelling (FS) that is part of the light-sensing system. ES, eyespot; G, Golgi body; M, mitochondrion; N, nucleus; P, plastid. Graham & Wilcox, 2000. Reproduced with permission from Prentice Hall.

and *Ophiocytium* (attached to filamentous algae) and *Chlorosaccus* (attached to macrophytes).

As with other algal groups, some yellow-green algae have developed alternative modes of nutrition. *Chlamydomyxa* is an amoeboid, naked form that has retained its photosynthetic capability (still has chloroplasts) but has also become holozoic, ingesting desmids, diatoms and other algae and digesting them within internal food vacuoles. *Chlamydomyxa* is typical of low-nutrient acid bogs, and the holozoic mode of nutrition may be important

for supplementing nitrogen uptake in such adverse conditions.

1.6.4   Yellow–green algae as bioindicators

Yellow-green algae have not been widely used as bioindicator organisms, partly because they are not a prominent group in the aquatic environment. Different species do have clear environmental preferences, however, that could be used to provide information on ambient conditions. These are discussed by John *et al.* (2002) and include algae that are prevalent in acid bogs (*Botrydiopsis* spp., *Centritractus* spp.), calcareous waters (*Mischococcus* spp., *Ophiocytium* spp.), humic waters (*Botrydiopsis* spp., *Tribonema minus*), organically rich conditions (*Chlorosaccus* spp.), inorganic nutrient-enriched waters (*Goniochloris fallax*) and brackish (partly marine) environments (*Vaucheria prolifera, Tetraedriella* spp.).

1.7   Dinoflagellates

Dinoflagellates (Dinophyta) are mostly biflagellate unicellular algae, although some (e.g. *Stylodinium*) are without flagella and are attached. They are predominantly found in the surface waters of marine systems (about 90% are marine), with only about 220 species present in freshwater environments. Dinoflagellates contain chlorophylls-*a* and -*c*, but are typically golden- or olive-brown (Fig. 1.3) in colour due to the major presence of carotene and the accessory xanthophyll peridinin. The chloroplasts can be plate-like (Fig. 4.57) in some species or elongate in others. Pyrenoids are present, and the main storage product is starch. Lipid droplet reserves may also be found, and some dinoflagellates possess an eyespot.

1.7.1   Cytology

Dinoflagellates have a number of distinctive cytological features.

- A large central nucleus (usually visible under the light microscope) containing chromosomes

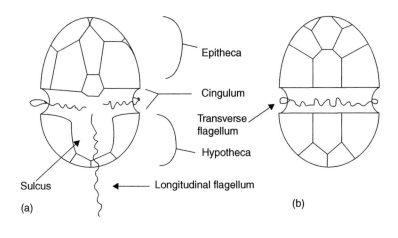

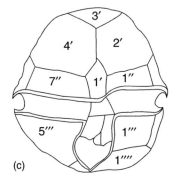

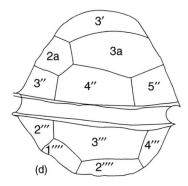

Figure 1.8 Dinoflagellate symmetry and plates. Top: Diagrammatic view of a typical dinoflagellate cell showing flagellar insertion and grooves within the plate system. (a) Ventral view; (b) Dorsal view. Bottom: Pattern of numbered thecal plates characteristic of the dinoflagellate *Peridinium*. (c) Ventral view; (d) Dorsal view. See also Fig. 4.57. Wehr & Sheath, 2003. Reproduced with permission from Elsevier.

that are typically condensed throughout the entire cell cycle. These have an unusual structure, with genetically active (transcriptional) DNA on the outside and genetically inactive (structural) DNA forming a central core (Sigee, 1984).

- Cell wall material lies beneath the cell membrane, in contrast to many other algae – where the cell wall, scales or extracellular matrix occurs on the outside of the plasmalemma. The cell wall in dinoflagellates is composed of cellulose (within subsurface membrane-bound vesicles), forming discrete discs (thecal plates) which give the dinoflagellate a distinctive armoured appearance. The presence of thecal plates is difficult to see in light microscope images of living cells, but is clear when chemically fixed cells are viewed

by scanning electron microscopy (Fig. 4.57). The number, shape and arrangement of thecal plates are taxonomically diagnostic (compare Figs. 1.8 and 1.9). A typical arrangement of plates can be seen in *Peridinium* (Fig. 1.8), with a division into two major groupings, forming an apical epitheca (or epicone) and a posterior hypotheca (or hypocone). Heavily armoured species tend to be more angular in outline, and plates may be extended in some species into elaborate projections. The presence of such projections or horns in *Ceratium,* for example (Fig. 1.9), increases the surface area to volume ratio of the cell and possibly reduces its sinking rate by increasing frictional resistance. Projections may also help in reducing predation. In contrast to armoured dinoflagellates, some species have very thin plates or they may be absent altogether (naked dinoflagellates).

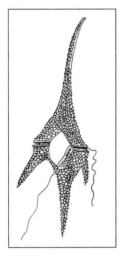
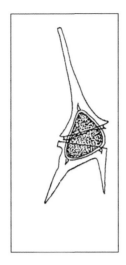

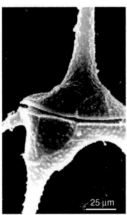

Figure 1.9 Morphology of *Ceratium*. Top: Diagrammatic views of a matured cell (left) and cyst formation (right). Bottom: Central region of a matured cell (SEM preparation), showing details of equatorial groove and thecal plates with pores. See also Figs. 2.7 (cyst), 2.16 (iodine-stained sample), 2.18 (cell death) and 4.56 (live cells).

1.7.2 Morphological diversity

These biflagellate unicellular organisms are of two main types, depending on the point of insertion of the flagella.

- Desmokont dinoflagellates, where the two flagella emerge at the cell apex (e.g. *Prorocentrum*).

Relatively few species occur in this group, which is also characterised by the presence of two large plates (valves) covering a major portion of the cell.

- Dinokont cells, where the flagella emerge from the middle of the cell (e.g. *Ceratium*, *Peridinium*). Dinokont cells have a distinct symmetry (Fig. 1.8) with dorsoventral division into epitheca and hypotheca. These are separated by an equatorial groove (cingulum), with a further small groove – the sulcus, extending posteriorly within the hypocone. The two are flagella inserted in the ventral region of the cell. Beating of the flagella – one of which is extended while the other is contained in the equatorial groove – gives the cell a distinctive rotatory swimming motion. The term 'dinoflagellate' is derived from the Greek word *dineo*, which means 'to whirl'.

Dinoflagellates do not show the range of morphology, from unicellular to colonial forms, seen in other algal groups. There are some unusual non-flagellate amoeboid, coccoid and filamentous forms, however, which reveal their phylogenetic relationship to mainstream dinoflagellates by the characteristic biflagellate structure of their reproductive cells – known as dinospores (zoospores).

1.7.3 Ecology

In freshwater environments, dinoflagellates are typically large-celled organisms such as *Ceratium*, *Peridinium* and *Peridiniopsis*. Their large size reflects the high nuclear DNA levels (large amounts of genetically inactive chromatin) and correlates with a long cell cycle and low rate of cell division. These features are typical of organisms that are K-strategists, dominating environments that contain high populations of organisms living under intense competition (Sigee, 2004).

Dinoflagellates are meroplanktonic algae, present in the surface waters of lakes and ponds at certain times of year. The annual cycle is characterised by two main phases.

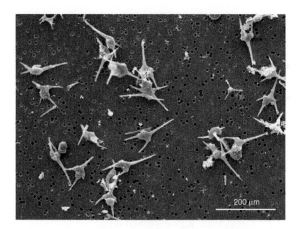

Figure 1.10 Autumn dinoflagellate bloom in a eutrophic lake. SEM view of epilimnion phytoplankton sample (September), showing almost complete dominance by *Ceratium hirundinella*. See also Figs. 2.8 (seasonal cycle) and 4.56 (live *Ceratium*).

- A midsummer to autumn bloom, when phytoplankton populations are very high and the surface waters are dominated by either dinoflagellates (Fig. 1.10) or colonial blue-green algae. At this point in the seasonal cycle, the surface water concentration of phosphorus is very low and dinoflagellates such as *Ceratium* survive by diurnal migration into the lower part of the lake – where P levels are higher. Dinoflagellates are adapted to this daily migratory activity by their strong swimming motion – which is coupled to an efficient phototactic capacity.

- An overwintering phase, where dinoflagellates survive on the sediments as resistant cysts. These nonflagellate cells (Fig. 2.7) lack an equatorial groove and initially form in surface waters at the end of the summer/autumn bloom before sinking to the bottom of the lake. Germination of cysts occurs in early summer, either on sediments or recently mixed waters, giving rise to vegetative cells which take up newly available phosphorus.

Ceratium and *Peridinium* are geographically widely dispersed, occurring particularly in waters that have high calcium-ion concentrations (hard waters) and low levels of surface inorganic nutrients (see above). Although these organisms are typically phototrophic, obtaining their nutrients in inorganic form, other dinoflagellates exhibit some degree of heterotrophy. Some are able to ingest food particles by engulfment of whole cells, others by formation of a feeding veil (pallium) or use of a feeding tube (phagopod). In some cases, these dinoflagellates are pigmented and mixotrophic, combining heterotrophic nutrition with photosynthesis. Other dinoflagellates such as *Peridiniopsis* are obligate heterotrophs and colourless. The common freshwater heterotrophic dinoflagellate *Peridiniopsis berolinensis* uses a fine cytoplasmic filament to make contact with suitable prey such as insect larvae and then ingests the contents.

Heterotrophic dinoflagellates are particularly adapted to conditions where photosynthesis is limiting. A number of colourless species, for example, are found under the ice of frozen lakes during the winter season, where photosynthesis is severely limited by low irradiance levels – which are about 1% of surface insolation.

1.8 Cryptomonads

Cryptomonads (Cryptophyta) are a group of relatively inconspicuous algae that are found in both marine and freshwater environments. They are generally small- to medium-sized unicells, and in many standing waters are a relatively minor part of the phytoplankton assemblage – both in terms of cell numbers and biomass.

1.8.1 Cytology

The fine structure of cryptomonads is illustrated by the light microscope images of *Rhodomonas* (Fig. 4.52) and *Cryptomonas* (Fig. 4.53), and by the line drawing shown in Fig. 1.11. Each cell bears two unequal flagella, which are about the same length as the cell and slightly unequal. One of these flagella propels the cell, while the other is stiff and nonmotile. The flagella are inserted near to an anterior ventral depression, the vestibulum. The cells tend to be more convex on one side than the other and

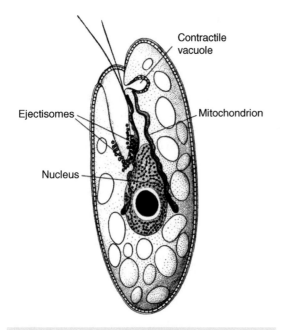

Figure 1.11 Cryptomonad morphology. Diagram of a non-photosynthetic cryptomonad with two flagella emerging from a depression. Graham & Wilcox, 2000. Reproduced with permission from Prentice Hall.

the front end, where the flagella arise, tends to be obliquely truncated. On this side arises a gullet which is not always clearly visible using a light microscope. Two chloroplasts are present each elongated in shape and lying along the length of the cell (Fig. 4.53) and on either side of the mid-line. The colour of the chloroplasts may be blue, blue-green, reddish, olive green, brown or yellow-brown depending upon the accessory pigments present. These may include Phycocyanin, phycoerythrin, carotene and xanthophylls. A pyrenoid is present and starch is stored. Further distinctive cytological features include:

- A prominent contractile vacuole

- Large ejectile organelles (ejectisomes), which discharge their content into the vestibular cavity and probably act as a defence response to herbivorous zooplankton

- The presence of an internal 'cell wall' (the periplast) below the plasmalemma that is composed of round, oval, rectangular or hexagonal organic plates

- A complex flagellar root apparatus

- distinctive plastids, with the associated presence of a much reduced nucleus – referred to as a nucleomorph. This phylogenetic survival is thought to be derived from the original (red algal) symbiont that lead to the establishment of plastids in cryptomonads.

1.8.2   Comparison with euglenoid algae

Cryptomonads share a number of similarities with euglenoid algae. In both cases:

- Algae are fundamentally biflagellate unicells, with flagella emerging from an apical depression.

- Cells are essentially naked, with a rigid covering of the cell occurring inside the plasmalemma.

- Species able to produce thick-walled cysts that can survive adverse conditions.

- Different species occur in a variety of aquatic environments, both freshwater and marine.

- Heterotrophic (colourless) algae are well-represented alongside autotrophic (green) species.

In spite of these similarities, there are a number of major differences between the algal groups, indicating a distinct phylogenetic origin. Key biochemical differences include the presence of starch (stains blue-black with iodine) in cryptomonads, with photosynthetic pigments which include phycobilins and show some similarity to red algae. Ultrastructural differences also occur in terms of flagellar ornamentation, and flagellar roots are different from euglenoids. The environmental biology of cryptomonads is also quite distinct from euglenoids, indicating a different ecological role for these organisms. Cryptomonads are genuinely planktonic algae, widespread amongst the plankton of a range

Table 1.10 Variation in Plastid Number and Coloration in Cryptomonads.

Trophic State	Plastid Characteristics
Autotrophic	*Rhodomonas* – a single boat-shaped, red-coloured plastid (Fig. 4.52) *Chroomonas* – a single H-shaped blue-green plastid with a pyrenoid on the bridge. *Cryptomonas* – two plastids, each with a pyrenoid (Fig. 4.53).
Heterotropohic	*Chilomonas* – a colourless cell that contains a plastid (leucoplast) lacking pigmentation. No pyrenoid.

of water bodies – in contrast to euglenoids, which tend to occupy mud, sand and water interfaces.

1.8.3 Biodiversity

Cryptomonads are a relatively small group of algae, with a total of about 12 genera (Hoek *et al*., 1995) containing approximately 100 known freshwater species and 100 known marine species.

The low diversity in terms of number of species is reflected in the uniformity of morphology. There are no clear colonial forms, though some species are able to form irregular masses embedded in mucilage (palmelloid stage) under adverse conditions. Differences do occur in relation to plastids – including presence or absence, number and pigmentation (Table 1.10). Variations also occur in other key cytological features such as the presence or absence of nucleomorphs, occurrence of ejectisomes and structure of the periplast – providing a basis for the classification of these organisms (Novarino and Lucas, 1995).

1.8.4 Ecology

Cryptomonads are typical of temperate, high latitude standing waters that are meso- to oligotrophic. They appear to be particularly prominent in colder waters, becoming abundant in many lakes in early spring, when they may commence growth under ice. Algae

such as *Rhodomonas* may form red blooms when the ice melts in early spring. In other (Antarctic) lakes, where ice persists, cryptomonad species such as *Cryptomonas* sp. and *Chroomonas lacustris* may contribute up to 70% of the phytoplankton biomass, dominating the algal flora for the whole of the limited summer period (Spaulding *et al*., 1994).

Cryptomonads are also typical of the surface waters of temperate lakes during the clear-water phase of the seasonal cycle, between the diatom spring bloom and the beginning of the mixed summer bloom. During this period, when algal populations are severely predated by zooplankton populations, the limited competition between algal cells favours rapidly growing r-selected organisms such as cryptomonads (Sigee, 2004).

In oligotrophic freshwater lakes, cryptomonads often form large populations about 15–20 m deep within the lake, where oxygenated surface waters interface with the anoxic lower part of the water column. In this location, these algae form deep-water accumulations of photosynthetic organisms, referred to as 'deep-chlorophyll maxima'. Studies on cultures of *Cryptomonas phaseolus* and *Cryptomonas undulata* isolated from deep chlorophyll maxima (Gervais, 1997) showed that these organisms had optimal growth under light-limiting conditions, suggesting photosynthetic adaptation to low light intensities. Other factors which may be important in the ability of these organisms to form large depth populations include the ability for heterotrophic nutrition, relative absence of predation by zooplankton, access to inorganic nutrients regenerated by benthic decomposition and tolerance of sulphide generated in the reducing conditions of the hypolimnion.

The ability of cryptomonads to carry out heterotrophic nutrition, using ammonium and organic sources as a supply of carbon and nitrogen, is ecologically important. Evidence for uptake of particulate organic material comes from direct observation of phagocytosis and has been documented in both pigmented (e.g. *Cryptomonas ovata*) and various colourless species (Gillott, 1990). Ultrastructural studies also provide evidence that many pigmented cryptomonads are mixotrophic, capable of both autotrophic and heterotrophic nutrition. The blue-green cryptomonad, *Chroomonas pochmanni*, has a specialised vacuole for capturing and containing bacterial cells. Bacteria are drawn into this through a small pore at the cell surface, and subsequently pass into digestive vesicles within the cytoplasm.

1.8.5 Cryptomonads as bioindicators

Although they are common in oligo- and mesotrophic conditions, cryptomonads also occur in eutrophic lakes (Fig. 2.16) and have limited use as bioindicators. Individual species do not provide the contemporary diagnostic range for different habitats seen in other algal groups, and cryptomonads do not survive as clearly identifiable remains within lake sediments.

1.9 Chrysophytes

Chrysophytes (Chrysophyta), commonly referred to as the 'Golden Algae', are a group of microscopic algae that are most readily recognised by their golden-brown colour (Figs. 1.3 and 4.37). This is due to the presence of the accessory pigment fucoxanthin within the chloroplasts, masking chlorophylls-*a* and -*c* (Table 1.3). There are around 200 genera and 1000 species of chrysophytes, present mainly in freshwater although some may be found in brackish and salt waters.

1.9.1 Cytology

Distinguishing cytological features (Table 1.3) include the presence of two dissimilar (long and short) flagella (heterokont condition) in motile organisms, plastids with four outer membranes plus triplet thylakoids, chrysolaminarin as the main storage product (present in special vacuoles) and cell walls composed of pectin – covered in some species with small silica spines or scales. Lipid droplets also frequently occur, and an eyespot may be visible in some species, associated with the chloroplast. In order to survive adverse conditions, round-walled cysts (stomatocysts) may be produced, with silica scales present in some species. Although most chrysophyte species are photosynthetic, some may be partially heterotrophic or even fully phagotrophic (Pfandl *et al.*, 2009). Many

species are small and are important members of the nanoplankton. They are widely distributed throughout freshwater systems and are of particular interest in relation to their morphological diversity, ecology and as indicator organisms. Details of taxonomy and ultrastructure are given by Kristiansen (2005).

1.9.2 Morphological diversity

Chrysophytes exhibit considerable diversity in their general organisation, ranging from unicellular to spherical (e.g. *Synura* – Fig. 4.37) and branching (e.g. *Dinobryon* – Fig. 4.2) colonial types. The great morphological diversity shown by these organisms (Table 1.11) follows a similar diversity seen in other algal phyla and includes flagellates, amoeboid forms, cells in jelly (capsulate), filamentous algae and plate-like (thalloid) organisms. Each of these types, however, can transform into another phase (e.g. flagellates can become amoeboid) – so that grouping in relation to morphology is dependent on the stage of life cycle. The occurrence of a resistant stage (stomatocyst) phase (see previous section) is also a key part of the life cycle.

1.9.3 Ecology

Chrysophytes are ecologically important in a number of ways (Kristiansen, 2005).

- In many ecosystems, they play an important role as primary producers. This is particularly the case for adverse conditions – low nutrient and acid lakes (e.g. Nedbalova *et al.*, 2006).

- They have a versatile nutrition, with many members (e.g. *Dinobryon*) being mixotrophic. These organisms are able, in addition to photosynthesis, to obtain their carbon from organic sources – either by surface absorption of organic compounds or by phagocytosis of particulate organic matter.

- They can be nuisance algae, giving a fishy smell to drinking water reservoirs when they reach high population levels.

Table 1.11   Range of Body Structure and Form in Chrysophytes.

General Structure	Colony Status	Examples
Flagellate cells (monads)		
Simple cell wall	Single cells:	*Chromulina, Ochromonas*
	Colonies	*Uroglena*
Surrounded by envelope (lorica)	Colony	*Dinobryon* (Fig. 4.2)
cells covered in silica scales	Single cells	*Mallomonas*
	Colony	*Synura* (Fig. 4.37)
Non-flagellate unicells		
Motile by pseudopod	Golden amoebae	*Chrysamoeba*
Non-motile	Cells surrounded by a jelly	*Chrysocapsa*
	Cells surrounded by solid cell wall	*Stichogloea*
Colonies with distinctive morphology		
	Simple branched filaments	*Sphaeridiothrix*
		Phaeothamnion
	Two-dimensional cell plates	*Phaeoplaca*

1.9.4   Chrysophytes as bioindicators

Their diverse ecological preferences (Table 1.12) make chrysophyte species potentially useful as environmental indicator organisms – both in terms of contemporary water assessment and sediment analysis.

Contemporary water quality

Although potentially very useful, chrysophytes are rarely used in monitoring projects. The main reason for this is that the best ecologically studied species (with silica scales) require specialised light microscopy (Cumming *et al.*, 1992) or scanning electron microscopy – neither of which may be available for quality assessment.

Although chrysophytes have traditionally been held to indicate oligotrophic conditions, the situation is more complex. Studies by Kristiansen (2005) have shown that the presence of a few species such as *Uroglena* and *Dinobryon* may indicate oligotrophy, but greater chrysophyte species diversity at lower overall biomass is more indicative of eutrophic conditions. The value of individual chrysophyte species as ecological indicators varies considerably (Kristiansen, 2005). *Synura petersenii*, with a broad ecological range, is of limited use – but other narrow-range species (Table 1.12) can be used particularly in relation to pH and salinity.

Table 1.12   Chrysophyte Species as Bioindicators.

Strongly acid, often humic conditions	*Dinobryon pediforme, Synura sphagnicola, Mallomonas paludosa, Mallomonas hindonii, Mallomonas canina*
Weakly buffered, slightly acid clear water lakes	*Mallomonas hamata, Synura echinulata, Dinobryon bavaricum* var. *vanhoeffenii*
Alkaline conditions	*Mallomonas punctifera, Synura uvella*
Alkaline/saline conditions	*Mallomonas tonsurata, Mallomonas tolerans*

Sediment analysis

The main application of chrysophytes as environmental bioindicators is in paleoecology. Their major advantage for this (as with diatoms) is the presence of non-degradable silica cell wall material, so that they remain undamaged in the lake sediments. The use of these algae in paleoecology depends on accurate means of classification and identification, undisturbed sediment layers and absolute dating of the layers – for example by the lead isotope Pb[210]. Techniques for the extraction and counting of chrysophyte remains from sediment cores, with determination of concentrations using microspheres, are described by Laird *et al.* (2013).

Two main types of chrysophyte remains are useful in freshwater sediments – silica scales (from vegetative planktonic cells) and stomatocysts (resistant spores). Stomatocysts have the advantage of better preservation compared to scales, but the disadvantage that (with a few exceptions) they cannot be referred to particular species. Although stomatocysts cannot be identified in terms of species, they can be recorded as distinct morphological types (morphotypes). These can be correlated with other microfossils (e.g. diatoms) and with pollen to obtain environmental indicator values. Quantitative assessment of past aquatic environments in relation to fossil chrysophytes (scales, stomatocysts) parallels that of diatoms (see Section 3.2.2) and can be obtained in relation to species ratios (environmental indices), transfer functions (Facher and Schmidt, 1996), group analysis and total planktonic chrysophyte populations (flux).

Species ratios   The ratio of chrysophytes to other algal groups may provide a useful index to assess environmental change. Studies by Smol (1985) on the surface sediments of Sunfish Lake (Canada) demonstrated a marked increase in the ratio of diatoms to stomatocysts, coincident with the arrival of settlers to the lake catchment area. The diatom/stomatocyst ratio is a trophic index, signalling an increase in eutrophic status of the lake at the time of human settlement. This ratio change resulted from a population decrease in the chrysophyte *Mallomonas* (indicating oligotrophic conditions) and an increase in eutrophic diatoms. The increase in trophic ratio parallels an increase in the pollen count of ragweed, an indicator of forest clearance and farming activity.

Transfer functions   The environmental properties of stomatocyst morphotypes can be assigned as numerical descriptors or 'transfer functions'. These define each morphotype in relation to a range of environmental data (pH, conductivity, phosphorus contents and maximal depth of the lake).

The transfer functions have been derived by multivariate analysis of sediment calibration sets, obtained from a series of North American lakes in which both environmental data and stomatocyst assemblages were available (Charles and Smol, 1994).

Group analysis   The various environmental factors which influence correlations between morphotypes can be identified by statistical analysis – particularly principle component analysis (PCA). Studies by Duff and Smol (1995) on 181 morphotypes obtained from 71 different lakes in the United States showed that differences in pH and lake morphometry had the greatest influence in determining statistical associations. PCA separated morphotypes into three main groups relating primarily to acid environments, neutral/alkaline environments and lake morphometry (e.g. lake depth).

Total Chrysophyte populations   Assessment of overall changes in planktonic chrysophytes can be determined from sediments by multiplying the concentration of chrysophyte scales by the annual sediment rate. Laird *et al.* (2013) used this approach in their analysis of limnological changes of Saskatchewan (USA) lakes, showing that an increased flux of chrysophyte scales was consistent with recent climatic changes deduced from changes in diatom assemblages.

1.10   Diatoms

Diatoms (Bacillariophyta) are a very distinct group of algae, identifiable under the light microscope by their yellow-brown coloration (Fig. 1.13) and by the presence of a typically thick silica cell wall. This normally appears highly refractive under the light microscope, giving the cell a well-defined shape. Removal of surface organic material by chemical oxidation (see Section 2.5.2) often reveals a complex cell-wall ornamentation – as illustrated in the light microscope images comparing living and digested cells of *Stephanodiscus* (Fig. 4.58). Cell-wall ornamentation can be seen even more clearly with the higher resolution of the scanning electron microscope and is important in species identification.

Diatoms occur as non-flagellate single cells, simple colonies or chains of cells, and are widely distributed in both marine and freshwater environments. Their success in colonising and dominating a wide range of aquatic habitats is matched by their genetic diversity – with a worldwide total of 285 recorded genera, encompassing 10,000–12,000 species (Round *et al.*, 1990; Norton *et al.*, 1996). Diatoms are also very abundant in both planktonic and benthic freshwater environments, where they form a major part of the algal biomass, and are a major contributor to primary productivity.

1.10.1 Cytology

Distinctive cytological features (Table 1.3) include:

- Plastids with periplasmic endoplasmic reticulum and the presence of girdle lamellae.

- Chrysolaminarin and lipid food reserves, present outside the plastid.

- A distinctive cell wall, the frustule, composed of opaline silicon dioxide (silica) together with organic coatings.

Silica cell wall

The presence of silica in the diatom cell wall can be detected by cold digestion in a strongly oxidising acid to remove organic matter (see Section 2.5.2). If the cell wall resists this treatment it is probably silica. The cell wall of diatoms differs from that of other algae in being almost entirely inorganic in composition. It has evolved as an energy-efficient structure, requiring significantly less energy to manufacture than the cellulose, protein and mucopeptide cell walls of other algae (Falkowski and Raven, 1997). This gives diatoms a major ecological advantage at times when photosynthesis is limited (e.g. early in the seasonal cycle), but has a number of potential disadvantages.

- The cell wall of diatoms is very dense. These organisms are only able to stay in suspension in turbulent conditions, limiting the development of extensive planktonic diatom populations to well mixed unstratified waters.

- The formation of the diatom cell wall is dependent on an adequate supply of soluble silica (silicic acid) in the surrounding water. The diatom spring bloom of temperate lakes strips out large quantities of silica from the water, reducing concentrations to a level that becomes limiting.

- Unlike other types of cell wall material, silica is rigid and unable to expand. This means that daughter cells are unable to enlarge and progressive cell divisions result in a gradual decrease in cell size. Ultimately this decrease reaches a critical level, at which point sexual reproduction is required to completely shed the original cell wall and form new, large daughter cells.

Cell wall structure

The frustule is composed of two distinct halves – the epitheca and hypotheca – which fit together rather like the lid and base of a pill-box (Fig. 1.12). At cell division, two new walls are formed internally at the cell equator 'back to back', and become the hypothecae of the two daughter cells. The original epitheca and hypotheca of the parent diatom become the epithecae of the two daughter cells, so the epitheca is always the oldest part of the frustule.

In both centric and pennate diatoms the epitheca consists of two main parts – a circular disc (the epivalve) with a rim (the mantle) plus an attached band or girdle (the epicingulum). The hypotheca correspondingly consists of a hypovalve, hypovalve mantle and hypocingulum. When the epitheca and hypotheca fit together to form the complete frustule, the hypocingulum fits inside the epicingulum to give an overlap.

Because of frustule morphology (Fig. 1.12), diatoms can be observed from either a girdle (side) view or a valve (face) aspect – with girdle and valve views of live diatoms being clearly distinguishable

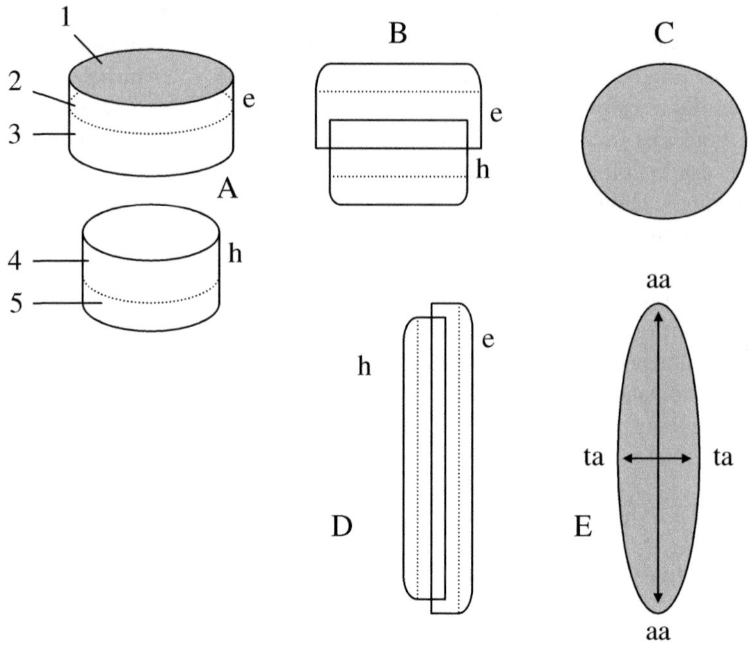

Figure 1.12 Diatom frustule structure – comparison of centric and pennate diatoms. **Centric diatom**: A. Separate views of epitheca (e: 1. Epivalve, 2. Mantle of epivalve, 3. Epicingulum) and hypotheca (h: 4. Hypocingulum, 5. Hypovalve). B. Complete frustule – girdle or side view (showing overlap of cingula). C. Valve or face view of epivalve. **Pennate diatom**: D. Complete frustule – girdle view. E. Valve view, showing apical or longitudinal axis (aa) and transapical or transverse axis (ta).

under the light microscope (Fig. 1.13). The two possible valve views, of the top (epivalve) or bottom (hypovalve) of the diatom, are not distinguishable in many genera – but do differ in some cases (e.g. *Cocconeis*, *Achnanthes*). Axes of symmetry in pennate diatoms are shown in Figs. 1.12 and 1.13.

Frustule markings

A wide range of surface markings can be seen on the face (epivalve and hypovalve) of diatoms. These have been recorded in considerable detail (Barber and Haworth, 1981; Round *et al.*, 1990; Wehr and Sheath, 2003) and form the basis for the classification and identification of these organisms (see Chapter 4).

Clear visualisation of the frustule markings requires the high resolution of either oil immersion (light microscopy) or scanning electron microscopy, and is normally carried out after the removal of surface organic matter by chemical (acid digestion) cleaning. Chemical fixation for scanning electron microscopy may also strip away some of the

overlying organic material to reveal frustule surface structure.

The terminology of diatom morphology (see Glossary, Chapter 4) includes various descriptors of frustule markings – including eye-shaped structures (ocelli), small pores (punctae) and fine lines (striae). Illustrations of diatom surface markings are shown diagrammatically in Figs. 1.13–1.14, and in various figures and plates in Chapter 4. Although the biological significance of much of this surface detail is obscure, the presence of one major surface structure – the raphe – is clearly associated with locomotion. The secretion of mucus from this channel or canal promotes movement on solid surfaces. In some diatoms such as *Nitzschia* (Fig. 4.70a,b), the raphe is elevated from the main diatom surface as a keel, allowing more intimate contact between the raphe and substrata. Such keeled diatoms are able to move particularly well on fine sediments, and reach their maximum abundance in the epipelon of pools and slowly flowing streams (Lowe, 2003). A raphe is not always present in pennate diatoms and is never seen in centric diatoms.

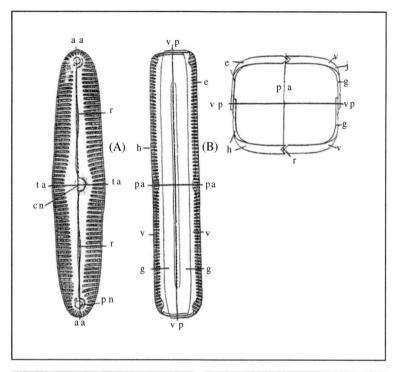

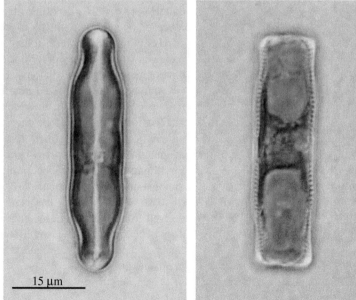

Figure 1.13 Details of frustule structure in a pennate diatom (*Pinnularia*). Top: Diagrams of cell wall structure, with principal axes. Top left: Valve view. aa, apical axis; ta, transapical axis; pn, polar nodule; r, raphe. Top middle: Girdle view. pa, pervalvular axis; vp, valvar plane; h, hypotheca; e, epitheca; g, girdle. Top right: Transverse section. Fritsch, 1956. Reproduced with permission from CUP. Bottom: Light microscope images of fresh (unfixed) cells in valve (left) and girdle (right) view.

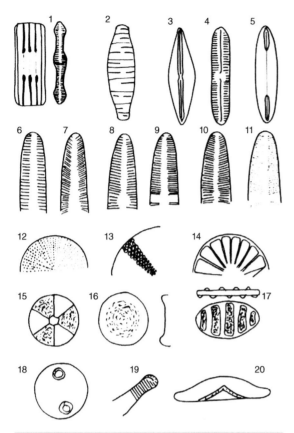

Figure 1.14　Diatom valve markings: (1) internal septa; (2) transapical costae; (3) raphe in thickened ribs; (4) normal raphe; (5) shortened raphe in ribs; (6) parallel striae; (7) radiate striae; (8) central area round; (9) central area transverse; (10) central area small; (11) central area acute angled; (12) punctae in radial rows interspersed with subradial rows; (13) coarse areolae; (14) areolae grouped in segments; (15) sections of valve alternately raised and depressed giving shaded appearance; (16) central region raised or depressed; (17) pennate with raised or depressed areas; (18) ocelli on surface; (19) rudimentary raphe, as in *Eunotia*; (20) raphe curved to point at centre, as in *Epithemia*.

1.10.2　Morphological diversity

The morphological diversity of diatoms can be considered in relation to two main aspects – the distinction between centric/pennate diatoms and the range of unicellular to colonial forms.

Centric and pennate diatoms

Diatoms can be separated into two major groupings – centric and pennate diatoms, based primarily on cell shape and frustule morphology (Fig. 1.12). Centric diatoms typically have a discoid or cylindrical shape, with a radial symmetry when seen in face or 'valve' view. Pennate diatoms have an elongate bilateral symmetry, with longitudinal and transverse axes, and often have the appearance of a 'feathery' (hence 'pennate') ornamentation when seen in surface view (Fig. 1.13). Pennate diatoms have a range of shapes (Barber and Haworth, 1981), with broad division (Fig. 1.15) into isopolar (ends of valve similar in size and shape), heteropolar (ends of valve differing in size and shape), asymmetrical (outline dissimilar either side of the longitudinal axis) and symmetrical (outline of valve similar either side of the longitudinal axis).

The fundamental differences between centric and pennate diatoms in relation to structure and symmetry reflect a range of other distinguishing features (Table 1.13) – including motility, number and size of plastids, sexual reproduction and ecology. Centric diatoms are mainly planktonic algae, and the occurrence of oogamy – with the production of a large number of motile sperms, is regarded as a strategy for increasing the efficiency of fertilisation in open-water environments. Pennate diatoms are almost always isogamous, with equal-sized non-flagellate gametes (two per parent cell). In these algae, efficiency of fertilisation is promoted by the pairing of gamete parental cells prior to gamete formation, a possible adaptation to more restricted benthic environments. Fossil and molecular evidence suggests that centric diatoms arose prior to pennate ones, implying that in these algae isogamy has been phylogenetically derived from oogamy (Edlund and Stoermer, 1997). This is in contrast to the evolutionary sequence normally accepted for other algal groups – particularly green and brown algae.

Unicellular and colonial diatoms

The genetic diversity of diatoms is not matched by the complexity of their morphological associations – which are limited to small chains and groups of cells.

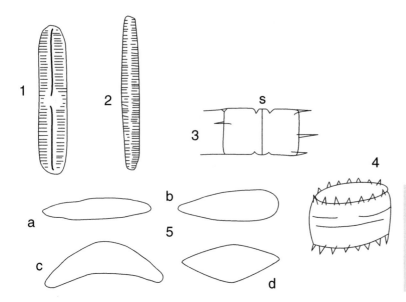

Figure 1.15 Diatom morphology: (1) pennate diatom with raphe; (2) pennate diatom with pseudoraphe; (3) sulcus (s) on *Melosira*; (4) spines on margin of *Stephanodiscus*; (5) valve shapes: (a) isopolar, (b) heteropolar, (c) asymmetrical, (d) symmetrical.

In planktonic diatoms, simple filamentous (e.g. *Aulacoseira* – Fig. 4.9) and radial (e.g. *Asterionella* – Figs. 1.16 and 4.42) colonies have evolved, but there is no formation of spherical globular colonies that are a key feature of other algal groups.

Colonial forms of diatoms arise by means of simple linkages, which are of three main types.

- Mucilage pads. These occur at the ends of cells to join diatom cells in various ways. *Asterionella*

Table 1.13 Centric and Pennate Diatoms.

Characteristic	Centric Diatoms	Pennate Diatoms
Symmetry	Radial	Bilateral
Examples	*Stephanodiscus* (Fig. 4.58)	*Pinnularia* (Figs. 1.13 and 4.71)
	Cyclotella (Fig. 4.62)	*Navicula* (Fig. 4.73)
Gliding Motility	Non-motile	Some (raphid) diatoms are actively motile.
Plastids	Many discoid plastids	Two large plate-like plastids
Sexual reproduction (Hasle and Syvertsen, 1997)		
Gamete formation	Independent formation of gametes from parent cells	Pairing of parental cells prior to gamete production
Egg cells	Oogamous – production of one or two eggs per parent cell	Isogamous
Sperm cells	4–128 sperms per parent cell. Each with a single flagellum bearing two rows of mastigonemes.	Few amoeboid, non-flagellate sperm cells.
Ecology	Mainly planktonic, typical of open water	Planktonic, epiphytic and benthic forms

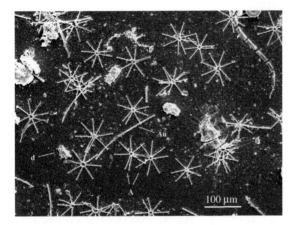

Figure 1.16 Spring diatom bloom in a eutrophic lake. SEM view of lake surface phytoplankton sample (March), showing numerous stellate colonies of *Asterionella* (A) with some filamentous *Aulacoseira* (Au). The large amount of debris (d) occurs due to the complete mixing of the water column at this time of year. See also Figs. 2.8 (seasonal cycle), 4.9 (live *Aulacoseira*) and 4.42 (live *Asterionella*).

has elongate cells which are joined at their inner ends to form stellate colonies. In *Tabellaria*, cells are linked valve to valve as short stacks, with connection of the stacks at frustule edges by mucilage pads to form a zig-zag pattern. Mucilage pads are seen particularly clearly in Fig. 2.28, where chains of *Tabellaria* are attached to *Cladophora*.

- Interlinking spines. In diatoms such as *Aulacoseira*, chains of cells are joined valve to valve by spines. The valves are of two main types (Davey and Crawford, 1986):

  - Separation valves, with long tapering spines and straight rows of pores

  - Linking valves, with short spines and curved rows of pores

Cells are strongly linked into chains via the short spines on the linking valves, but weakly linked via the long spines of the separation valves – where filament breakage easily occurs (Fig. 4.9). The length of filaments within a population of *Aulacoseira*

depends on the proportion of separation valves to linking valves.

- Gelatinous stalks. Biofilm diatoms such as *Rhoicosphenia* (Fig. 4.67) and *Gomphonema* (Fig. 2.29) have gelatinous stalks, which attach the diatom to the substratum and join small groups of cells together as a single colony. In these diatoms, the stalk material is secreted via apical pores. Some species of *Cymbella* also produce a gelatinous tube-like filament.

1.10.3 Ecology

Diatoms are ubiquitous in both standing and running freshwaters – occurring as planktonic, benthic, epiphytic (on higher plants and other algae) and epizoic (on animals such as zooplankton) organisms. They are equally successful as free-floating and attached forms, where they may be respectively important in plankton bloom formation and biofilm development.

Planktonic diatoms: bloom formation

In many temperate lakes, diatoms such as *Asterionella* and *Tabellaria* dominate the phytoplankton population in Spring and early Summer (Fig. 1.16), at a time when inorganic nutrients (N, P, Si) are at high concentration, light and temperature levels are rising and lake turbulence is maintained by moderate wind action. At this point in the seasonal cycle, diatoms are able to out-compete other microalgae due to their tolerance of low temperature and low light conditions, coupled with their ability to grow in turbulent water. Autumn blooms of *Fragilaria* and *Asterionella* are also a common feature of many lakes, with the diatoms part of a mixed phytoplankton population.

Diatoms such as *Aulacoseira italica* are regarded as meroplanktonic, with major growth in very early Spring, but spending the rest of the year as resistant cells on lake sediments. Other diatoms, such as *Asterionella* and *Tabellaria* also form major blooms in Spring, but are additionally present as minor constituents within the mixed phytoplankton population over much of the annual cycle – and are regarded as holoplanktonic.

Attached diatoms: biofilm formation

Diatoms are major components of biofilms, where they are present early in the colonisation sequence and also within the mature periphyton community. See later for examples of diatoms in river biofilm (Fig. 2.23), reed biofilm (Fig. 2.29) and as attached epiphytes (Fig. 2.28).

1.10.4   Diatoms as bioindicators

The major use of diatoms as bioindicators of water quality (both lotic and lentic systems) is detailed in Chapter 3. These algae can be isolated from existing live populations (contemporary analysis) or from sediments (fossil diatoms), and quantitative information on water quality from species counts can be obtained from taxonomic indices, multivariate analysis, transfer functions and species assemblage analysis.

Contemporary analysis

Many diatom species have distinct ecological preferences and tolerances, making them useful indicators of contemporary ecological conditions. These preferences relate to degree of water turbulence, inorganic nutrient concentrations, organic pollution, salinity, pH and tolerance of heavy metals. Round (1993) also lists a number of other advantages in using diatoms as indicators – including easy field sampling, high sensitivity to water quality but relative insensitivity to physical parameters in the environment and easy cell counts. Diatoms are also the most diverse group of algae present in freshwaters, making them the most ideal assemblage for calculation of bioindices (Section 3.4.5).

Diatoms, once cleaned and mounted for identification, make excellent permanent slides which can form an important historical record for a location. Although there are hundreds of diatom species that may be present, only dominant species need to be used in an assessment. Benthic diatoms have been particularly useful in assessing water quality of rivers,

and the wide range of indices that have been devised is discussed in Chapter 3.

Lake sediments

Diatoms, more than any other group of algae, are used to monitor historical ecological conditions from sediment analysis – as discussed in Section 3.2.2. The major role of diatoms in this lies in the resistance of the diatom frustule to biodegradation, coupled with ease of identification from frustule morphology and the wide range of species with clear ecological preferences seen in freshwater environments.

The importance of diatoms in environmental analyses is indicated by the European Diatom Database Initiative (EDDI). This key web-based site includes electronic images and data handling software and is particularly useful for lake sediment analysis (see Section 3.2.2). Other databases, such as the stream benthic database of Gosselain *et al.* (2005), are more applicable to contemporary environmental analysis. Techniques for the collection of sediment cores, preparation of diatom samples and procedures for making counts are widely described (e.g. Laird *et al.*, 2013). Methods for cleaning diatoms in preparation for microscopy are given in Section 2.5.2.

1.11   Red algae

Red algae (Rhodophyta) are predominantly marine in distribution, with only 3% of over 5000 species worldwide occurring in true freshwater habitats (Wehr and Sheath, 2003). Although termed red-algae, the level of accessory pigments (phycoerythrin and phycocyanin) may not be sufficient to mask the chlorophyll – resulting in an olive-green to blue (rather than red) coloration. The relatively common river alga *Batrachospermum*, for example, typically appears bluish-green in colour (Fig. 4.4) – giving no clue to its rhodophyte affinity.

Major cytological features of red algae (Table 1.3) include the absence of flagella, presence of floridean starch as the major food reserve, characteristic photosynthetic pigments (chlorophyll-*a* only, presence of phycobilins) and distinctive plastid structure (unstacked thylakoids, no external endoplasmic reticulum).

Many of the freshwater species (which occur mainly in streams and rivers) are quite large, being visible to the naked eye when occurring in reasonable numbers (Sheath and Hambrook, 1990). The range of morphologies includes gelatinous filaments (e.g. *Batrachospermum*), pseudoparenchymatous forms (e.g. *Lemanea*) and a flat thallus of tiered cells (e.g. *Hildenbandia*). Many of these shapes can be an advantage in resisting the forces exerted in swiftly flowing waters.

Although freshwater red algae (such as the large filamentous alga, *Batrachospermum*) are largely found in streams and rivers, Rhodophyta may also occur as marine invaders of lakes and brackish environments. Certain freshwater red algae in the littoral zones of the Great Lakes Basin (USA), for example, appear to be originally marine and to have lost the capacity for sexual reproduction. These include the filamentous red alga *Bangia atropurpurea* (Lin and Blum, 1977), which reproduces only by asexual monospores – in contrast to marine species which undergo alternation of generations and carry out sexual reproduction. Attached red algae (e.g. *Chroodactylon*

ramosum) also contribute to the epiphytic flora of lake periphyton.

1.12 Brown algae

As with red algae, brown algae (Phaeophyta) are almost entirely marine – with less than 1% of species present in freshwater habitats (Wehr and Sheath, 2003). These species are entirely benthic, either in lakes or rivers, and have a very scattered distribution.

Cytological features of this group (Table 1.3) include heterokont flagella (reproductive cells only), presence of laminarin as the major food reserve, characteristic photosynthetic pigments (chlorophyll-*a* and -*c*, β- and ε-carotenes) and distinctive plastid structure (triple thylakoids, enclosing endoplasmic reticulum). Freshwater brown algae include genera such *Pleurocladia* and *Heribaudiella* and are the least diverse of all freshwater algae. Their morphologies are based on a relatively simple filamentous structure (tufts or crusts), and they lack the complex macro-morphology typical of the brown seaweeds. The freshwater brown-algae have not been fully studied and their ecological characteristics are not well known. All are benthic and may be recognized by their possession of large sporangia (Wehr, 2002).

2

Sampling, Biomass Estimation and Counts of Freshwater Algae

This chapter concentrates on techniques of collection, biomass estimation and numerical assessment (microscope counts) of freshwater algae. Aspects of algal identification are dealt with in Chapter 1 (brief account of molecular analysis) and Chapter 4 (algal key based on morphological identification). The reader is also referred to other texts on limnological analyses, including works by Wetzel and Likens (1991) and Eaton *et al.* (2005).

Algal populations can be sampled in a number of different ways, depending on the nature of the habi-

tat and the location of the algae within the water body. Some indications of the ecological diversity of freshwater algae are shown in Fig. 2.1, where algae within the water column (plankton) are in dynamic equilibrium with both attached and unattached forms associated with plant surfaces and sediments at the bottom of the water column. Techniques for sampling and enumeration can be conveniently separated in relation to these two main groups of algae – planktonic and non-planktonic (substrate-associated) organisms.

A. PLANKTONIC ALGAE

Lake and river phytoplankton include all those photosynthetic free-floating organisms that are present within the main water body, and which generate oxygen by photosynthesis. This includes prokaryotic blue-green algae but excludes photosynthetic bacteria – which do not evolve oxygen and which do not normally make a major contribution to overall biomass formation.

The schedule for sampling and assessment of phytoplankton involves a sequence of stages (Fig. 2.2), which are essentially the same for lakes and rivers. These involve protocol for collection, mode of collection and laboratory analysis of the sample. It is

essential to clarify the purpose of the study and exactly what information is required before planning the fieldwork, sampling and algal analysis.

2.1 Protocol for collection

As an initial step, information should be obtained on the morphology and hydrological characteristics of the water body to be studied. For standing waters, these include surface area, mean/maximum depths, water retention time plus inflow/outflow rates and for rivers – width, transverse depth profiles, flow rates

Freshwater Algae: Identification, Enumeration and Use as Bioindicators, Second Edition. Edward G. Bellinger and David C. Sigee.
© 2015 John Wiley & Sons, Ltd. Published 2015 by John Wiley & Sons, Ltd.

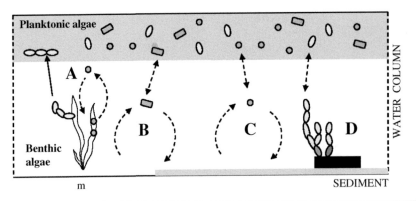

Figure 2.1 General occurrence of algae in the freshwater environment (not to scale). Freely suspended planktonic algae present in surface waters (shaded area) are in equilibrium (dynamic interchange) with a range of sediment- or macrophyte-associated organisms. (A) Epiphytic algae present on macrophytes (m) include filamentous and unicellular forms. Filamentous algae may detach (→) to form planktonic filament masses (metaphyton). (B) Unattached algae on the substratum include actively photosynthetic organisms (shallow waters) and dormant overwintering cells and colonies. (C) Algae present within a biofilm. Along with other microorganisms, the algae may be enclosed within a mucilaginous matrix (□). (d) Extended growths of filamentous algae (with associated microalgae) form a dense mat (periphyton). The filamentous algae are typically attached to solid surfaces such as rocks and stones (■).

and substrate mobility. In both cases, human activities – inflow discharges, recreational activities, water removal and surrounding agriculture – may also be important. As a final step, the past aquatic history of the water body may also be of relevance – including previous contemporary studies and analysis of the sediments (see Section 3.2.2).

Ecological interpretation of environmental samples is only as meaningful as the samples themselves, so great care must be taken in planning how and where the samples are collected. Strategies for phytoplankton collection will also vary with the dynamics of the aquatic system, which is clearly different for standing waters (lakes and wetlands) compared to flowing waters (rivers and estuaries).

2.1.1 Standing water phytoplankton

Phytoplankton is normally collected at various sites in the water body. These will normally be located in a deep part of the lake and away from the shore, to avoid contamination by sediment and littoral organisms, respectively. The strategy for phytoplankton sampling in standing waters varies in relation to number and location of sites across the water body, whether these are taken at specific points or along a line (trawl net), frequency of sampling and depth within water column. Phytoplankton samples should be taken at the same (or close to) sites that are used for obtaining physicochemical lake data and may be selected on a random basis in relation to a grid network or transect lines (Eaton *et al.*, 2005) or on the basis of specific features of the water body – for example near the inlet or outlet of a reservoir.

The exact sampling schedule employed depends on the overall aims and objectives of the project and can provide information on the occurrence of planktonic algae at various ecological levels (Fig. 2.3) – including general phytoplankton characteristics, diversity within the water column and small-scale patchiness of microalgal distribution.

General phytoplankton characteristics

In many cases, phytoplankton sampling is carried out simply to keep a broad ecological record of phytoplankton populations (total biomass, species composition, seasonal changes) present. These parameters can provide useful information on lake

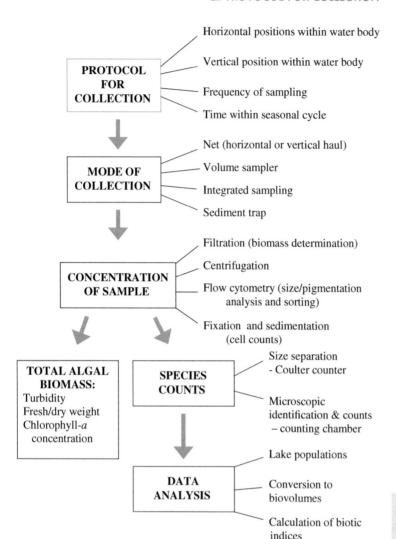

Figure 2.2 Sampling and assessment of lake phytoplankton populations.

trophic status (see Section 3.2.3), algal productivity and the development of major algal blooms within the water body.

Broad ecological monitoring normally involves taking samples from at least three separate points across the lake (Fig. 2.3a) on any particular date. Collection at more than one site is necessary because of uneven horizontal distribution (patchiness), which can be a source of sampling error. Samples can either be combined together to make a composite sample or be analysed separately – allowing statistical evaluation of biomass and species composition (means $\pm$ SD) across the water body.

Sampling is normally carried out within the epilimnion or to just below the thermocline, corresponding approximately to the main depth of light penetration – the photic zone. This can be defined (Sigee, 2004) as the vertical distance between the lake surface and the light compensation point (depth at which photosynthesis balances respiration). The photic zone can be roughly estimated as twice the Secchi disk reading (see Section 2.3.1) and is the major zone of phytoplankton population and productivity within the water column.

Frequency of sampling varies from once or twice yearly (general checks on lake status – summer

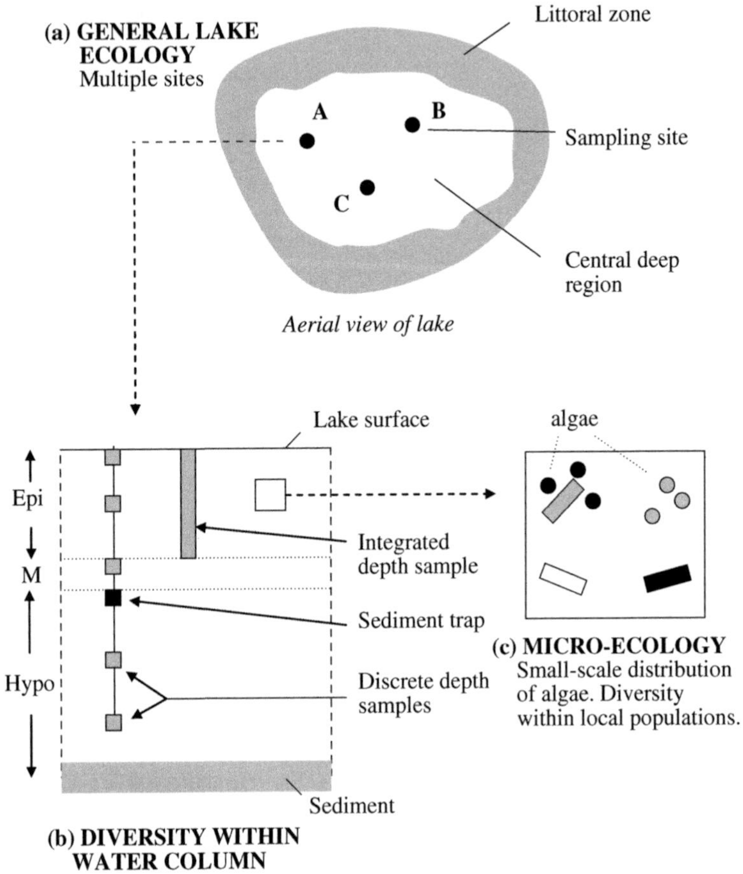

Figure 2.3 Strategies for sampling lake phytoplankton. Different strategies provide information on: (a) General lake ecology – samples from different pelagic sampling sites (A, B, C) and littoral zone. (b) Diversity within the water column – depth samples, sedimenting algae and integrated samples. Vertical zonation – Epi (epilimnion), M (metalimnion) and Hypo (hypolimnion). (c) Microecology of phytoplankton, looking at small scale distribution, patchiness and associations. Analysis of microsamples and high-resolution *in situ* monitoring of phytoplankton biomass.

samples), to monthly (general seasonal progression) and higher frequencies (monitoring rapidly occurring lake changes). An intensive sampling programme, involving collection once or twice a week (Tittel *et al.*, 1998), is necessary to capture detailed population dynamics and to quantify short-term fluctuations in phytoplankton abundance and biomass.

Diversity within the water column

Vertical separation of algal biomass into actively growing photosynthetic populations (high light: epilimnion) and non-photosynthetic cells (low light: hypolimnion) represents one of the broad ecological sub-divisions within the pelagic ecosystem. Cells within the hypolimnion may be sinking in the water column by passive sedimentation – including cells undergoing senescence (see Section 2.6.2) and cells

entering an overwintering phase (see Section 2.7.1). Other algae, particularly colonial blue-greens and dinoflagellates, periodically enter the hypolimnion as part of their diurnal cycle, taking up inorganic nutrients at a time of year (in many lakes) when the epilimnion has been largely stripped of soluble nitrates and phosphates. Collection of algal samples at specific points in the water column (Fig. 2.3b) can provide dynamic information on all of these processes and may involve special collecting devices such as depth samplers (Fig. 2.5) and sedimentation chambers.

Small-scale characteristics and associations of phytoplankton

Localised distribution and associations of phytoplankton cells/colonies within the water column

(Fig. 2.3c) are important aspects of lake ecology since all areas of plankton biology ultimately occur at the microlevel. The spatial scale at which sampling should be carried out relates particularly to organism size – including both single organisms/colonies and aggregates (DeAngelis, 2004). In the case of phytoplankton, the size factor ranges from population aggregates (aspects of horizontal and vertical patchiness, over distances of 0.1–5 m) down to single cells and colonies, with analysis of nanoscale heterogeneity at the microscopic level.

Patchiness of phytoplankton distribution

In both marine and freshwater systems, observations of patchiness have stimulated new approaches to phytoplankton sampling and recording. Conventional sampling procedures, using phytoplankton nets or volume samplers (see Section 2.2), provide little information at the microlevel, since quite large volumes are trawled or collected, and are assumed to be homogeneous for the purpose of measuring aquatic parameters. Studies by Wolk *et al.* (2004), for example, taking 20-ml subsamples of lake water from a 5-l Niskin volume sampler, showed considerable heterogeneity of algal biomass (chlorophyll-*a* concentration) within the collected sample. This information would normally be lost, since the water samples were normally split up into different aliquots for bulk analysis of inorganic nutrients, dissolved organic matter, phytoplankton (biomass), phytoplankton production and particulate organic matter.

The development of new *in situ* instrumentation (see also Section 2.3.3) has been particularly useful in resolving small-scale phytoplankton distribution in the lake environment. Wolk *et al.* (2004) used a high-resolution bio-optical sensor (fluorescence and turbidity) to monitor chlorophyll-*a* distribution in Lake Biwa (Japan), revealing vertical patterns of phytoplankton distribution over spatial scales of <1 cm. Cowles (2004) has emphasised the importance of small-scale plankton distribution in terms of 'pattern and process' – particularly in relation to rates of biological production and transfer of material through the food web. The importance of local phytoplankton distribution in relation to population growth includes aspects such as nutrient patchiness, local competition with other microorganisms and critical algal density for fungal infection and zooplankton grazing (Sigee,

2004). Localised phytoplankton distribution is also affected by wind-induced turbulence and mixing, plus directional drift in surface waters. In some cases, local plankton populations may be limited to a particular microenvironment – such as photosynthetic bacterial restriction in some mesotrophic and oligotrophic lakes at the bottom of the photic zone, where there is adequate light under anaerobic conditions (Sigee, 2004).

Heterogeneity at the microlevel

Looking at the nanoscale patchiness of algae and associated microorganisms within lake water (over distances of 10–100 μm) is potentially useful for studying intercellular dynamics. Defined associations between algae and other organisms are important in the air–water surface biofilm, and in the algal phycosphere (Sigee, 2004), and other associations may also be important. These can only be analysed via discrete samples if the scale of collection matches the scale of aggregation. Another approach proposed by Krembs *et al.* (1998) is to collect small volumes of water containing algae and other particulate material (without destroying their relative spatial distribution), rapidly freezing the sample, then analysing the distribution in two dimensions as a projection on a microscope slide.

Variation within species populations is also an important aspect of algal microecology. The microscopical techniques outlined in Section 2.6.2 involve observation and analysis of phytoplankton samples directly isolated from the lake and processed within a short period of time.

2.1.2 River phytoplankton

The phytoplankton biomass in rivers is typically (but not always) lower than standing waters since the transport of suspended algae by water current (potentially short residence time) limits the development of established populations. The phytoplankton community of rivers is also dominated by a smaller number of taxa compared to lakes, with fast-growing algae such as centric diatoms being particularly important. In shallow rivers, light penetration to the sediments means that sampling of benthic algae may also be important in assessing total algal population (see Sections 2.8–2.10).

Entry of phytoplankton into rivers can occur from standing waters or stream inlets, and assessment of river phytoplankton dynamics needs to take into account both irregularities of entry and displacement by current. Studies by Welker and Walz (1998) on grazing of phytoplankton in the River Spree (Germany), for example, took both of these factors into account by adopting a Lagrangian sampling strategy. This involved timed collection of phytoplankton in the middle of the river to achieve successive sampling of the same water parcel as it travelled downstream. Using this approach, these authors demonstrated a major removal of phytoplankton by freshwater mussels (unionid bivalves) after discharge of algae into the river from a highly productive shallow lake.

2.2   Mode of collection

Phytoplankton samples can be directly collected from the water body in three main ways – via a phytoplankton trawl net, volume sampler or as an integrated sample. In addition to this, samples of sedimenting phytoplankton can be passively collected over a period of time using sediment traps. In all cases, material for live examination should be brought back to the laboratory as quickly as possible and examined straightway (see Section 2.5.1). For chemically preserved material, the fixative should be added immediately (on the boat) after collection.

2.2.1   Phytoplankton trawl net

Plankton nets vary in design from basic tow-nets (conical or with truncated neck) to more complex devices fitted with flow meters and special opening/closing mechanisms (Eaton et al., 2005). In their simplest form (Fig. 2.4), they consist of long cones with a circular opening at their mouth and some form of collecting chamber at the narrow end.

Mesh size

The type of netting and mesh size determine filtration efficiency (extent of removal of different particle sizes), clogging tendencies, velocity of water flow within the net and condition of sample after collection. Trawl nets of different mesh sizes can be classified in relation to the minimum size range of organisms collected (Fig. 2.4b) and vary from largest zooplankton/ichthyoplankton nets (mesh aperture 1024 μm) to phytoplankton (53–76 μm) and nanoplankton (<50 μm) nets.

A typical phytoplankton net mesh size of about 50 μm will collect the majority of algae in the microplankton and macroplankton range (Table 1.1) – referred to collectively as 'net phytoplankton'. Small unicellular algae (nanoplankton) can be collected by smaller mesh sizes (commercially available nets go down to 10 μm), but problems with clogging and reduced water flow mean that these smaller organisms are best collected as bulk samples (see Section 2.2.2) followed by sedimentation or centrifugation (see Section 2.5.1).

In practice, routine collection of net zooplankton and phytoplankton involves using separate zooplankton and phytoplankton nets (mesh sizes about 250 μm and 50 μm, respectively). Using the two nets in combination has the advantages that both samples are collected from the same volume of lake water, and that zooplankton is largely removed from the phytoplankton sample – reducing sample contamination and algal ingestion during transport to the laboratory.

Application and quantitation

The phytoplankton net can be hauled horizontally, vertically or obliquely depending upon the type of sample required. A vertical haul will collect a composite community within all or part of the water column, whereas a horizontal haul through surface waters will collect a composite sample across an area of lake.

The volume of water passing through the net can be very approximately estimated from the area of the net aperture and the distance (vertical or horizontal) travelled. With horizontal trawls, more precise volumes can be determined using a flow meter (Eaton et al., 2005). The speed of the horizontal trawl should be sufficiently slow to allow the net to sink to a depth of about 0.5–1 m below the water surface and also to prevent the formation of a bow wave that reduces the efficiency of collection. Problems may arise with net clogging in the case of small mesh-size nets, and also

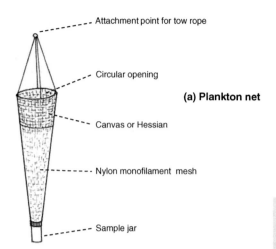

(a) Plankton net

- Attachment point for tow rope
- Circular opening
- Canvas or Hessian
- Nylon monofilament mesh
- Sample jar

(b) Net mesh sizes and classification

Size of Aperture (µm)	Approximate Open area (%)	Classification
1024	58	Largest zooplankton and ichthyoplankton
752	54	Larger zooplankton and ichthyoplankton
569	50	Large zooplankton and ichthyoplankton
366	46	Large microcrustacea
239	44	Zooplankton - Microcrustacea
158	45	Zooplankton, Microcrustacea, Rotifers
76	45	Net phytoplankton – Macroplankton and microplankton
64	33	
53	-	
10		Nanophytoplankton

Figure 2.4 Trawl net collection of phytoplankton. (a) Simple conical phytoplankton net, with collection jar. (b) General table of commercially available net mesh sizes and their classification. The phytoplankton net would have mesh sizes in the range 53–76 µm. Larger mesh sizes would collect colonial blue-greens and other large algae but not the major part of the net phytoplankton. Each mesh size will collect the specified group of organisms, plus all larger size groups, unless they are prevented from entry into the net by pre-filtration. Adapted from Easton *et al.*, 2005.

where there are dense populations of zooplankton and phytoplankton (particularly large globular algae such as *Microcystis*, and filamentous forms such as *Aulacoseira*).

After trawling, the sample is obtained by unscrewing the end fitting and flushing into a container. Using a standard phytoplankton net mesh size of 53 µm means that all particulate materials over this size (both phytoplankton and zooplankton) would be included. To avoid collecting zooplankton and other large particulate materials, a zooplankton net is typically placed within the phytoplankton net prior to trawling – though large-sized algae may also be retained by the inner net.

Advantages and disadvantages

The major advantage of trawl net collection is that a large sample of phytoplankton biomass can be obtained within a short period of time and that it is collected as a concentrated sample. This is particularly useful in oligotrophic waters and during winter in temperate mesotrophic/eutrophic lakes, where phytoplankton numbers may be very low.

Trawl net samples can be used for bulk analysis of phytoplankton (e.g. nitrogen, phosphorus and lipid content), microscopic analysis of the range of organisms present (e.g. light and electron microscopy, X-ray and infra-red microspectroscopy) or simply where a rapid collection procedure is required to give an idea of the main large-celled and colonial algae present in the lake.

The major disadvantages of trawl net sampling are:

- *Loss of small algae.* Algae below the mesh size are largely excluded from the sample. Exclusion is not absolute, since the mesh becomes partly occluded by large-celled and colonial algae, so some smaller organisms are retained.

- *Sample contamination.* The phytoplankton sample tends to become contaminated by zooplankton, which selectively consume algae and contribute to overall biomass. Pre-filtration using the zooplankton net removes larger zooplankton such as *Daphnia*, but smaller zooplankton such as rotifers and protozoa pass through. Small particulate inorganic and organic debris may also be retained with the phytoplankton, some of which (decomposing matter) may have an associated bacterial population. In stratified temperate lakes, debris contamination is particularly important during winter months, when the lake is vertically mixed and sediment at the bottom of the water column may become resuspended. In streams, contamination of the net sample by non-algal particulate material is most likely to occur at night-time (sediment disturbance by the nocturnal benthic community) and during winter months (Richardson *et al.*, 2009).

- *Sample deterioration.* Because of the concentrated nature of the sample, adverse changes in the water (e.g. oxygen depletion) during transport may lead to death of some of the more delicate algae.

- *Problems in quantitation.* For various reasons, net samples cannot be used to make a quantitative assessment of phytoplankton populations (per unit volume of water) within the lake. During sample collection, the vertical position of the net varies within the surface waters, so the sample is not truly representative of the entire epilimnion. Even when trawling is carried with a flow meter, complexities of water circulation around and through the net (see above) make it difficult to know the actual volume of water that has been filtered.

2.2.2   Volume samplers

Collecting phytoplankton samples within a volume of lake water has the advantages that biomass or species populations can readily be determined per unit volume of medium, and that all sizes of phytoplankton are obtained within the sample. A variety of collecting vessels are available – including the Kemmerer,

Van Dorn, Niskin and Nansen samplers (Eaton *et al.*, 2005).

Phytoplankton samples at vertical points in the water column (Fig. 2.3b) can be obtained by lowering a depth sampler such as the Van Dorn model (Fig. 2.5)

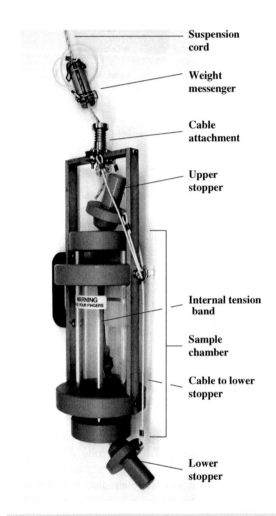

Suspension cord

Weight messenger

Cable attachment

Upper stopper

Internal tension band

Sample chamber

Cable to lower stopper

Lower stopper

Figure 2.5 Van Dorn Volume Sampler – shown in the open (sprung) position. The sampler is lowered to the required depth in the water column, then the weight messenger released down the suspension cord. This strikes the cable attachment, which releases the cables to the upper and low stoppers. These are then jerked into position at either end of the chamber by the internal tension band, creating a water-tight seal, which encloses the water sample. The sampler is subsequently pulled to the surface and the water sample collected.

to the required depth, then dropping a weight messenger to trigger closure. The volume of sample (usually 1–2 l) can be brought to the surface and analysed. Care must be taken to ensure that depth samplers are completely watertight at the depth being sampled, and that surfaces in contact with the water sample are completely inert.

Samples collected from volume samplers are typically divided into separate aliquots for determination of a range of parameters – including inorganic nutrient concentrations, soluble organics, phytoplankton and zooplankton biomass. These water samples can be very heterogeneous (Wolk *et al.*, 2004), however, and care must be taken that they are well-mixed (by pouring into a large container and stirring) before subdivision.

2.2.3 Integrated sampling

Analysis of the average community structure of the top part of the water column (epilimnion) of a stratified lake (Fig. 2.3b) is normally obtained from an integrated (composite) sample, using a collecting tube that is lowered from the boat. In the case of shallow unstratified lakes, the whole of the water column can be sampled.

The sampler is typically a flexible plastic tube, internal diameter approximately 2.5 cm and about 1 m longer than the depth through which the sample is to be collected, and is weighted at its end (Fig. 2.6). A long piece of cord is attached to the

weighted end and the tube lowered slowly and vertically into the water. This process has to be carried out with care in order to collect a column of water in the tube as undisturbed as possible, representing the water column of the lake. To ensure a vertical sample (tube not displaced by current), the tube should be sufficiently weighted and the boat should be stationary (attached to a buoy). The top of the tube is then closed by inserting a rubber bung. With the top still, the weighted bottom end is raised out of the water so the tube forms a loop. The weighted end is placed in a suitable large receptacle and the rubber bung removed from the other end. The trapped column of water now flows into the container and totally mixes. The integrated sample of the lake surface (top 5 m) can then be separated into subsamples to analyse water chemistry, chlorophyll content and species composition.

The use of integrated sampling techniques provides information on the overall phytoplankton population over a part of the water column (stratified lake) – but loses detail such as phytoplankton gradients and strata of particular algal species. Comparison of vertical trawls from different sites on the lake is particularly useful where horizontal rather than vertical variation in the lake is being investigated.

Once the lake water sample has been collected, the phytoplankton can then be analysed in various ways – including total biomass, species populations, biological interactions (e.g. fungal infection), physiological state, chemical composition and molecular characteristics.

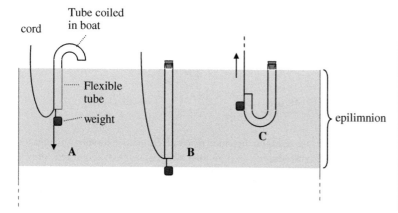

Figure 2.6 Collection of integrated water sample at lake surface (not drawn to scale). (A) Weighted flexible tube is lowered vertically into water. (B) Tube closed by bung when fully extended in epilimnion. (C) Weighted end of tube is raised by attachment cord. The whole tube is then taken into the boat, the bung removed and the lake water poured as a mixed sample into a collecting vessel.

2.2.4 Sediment traps

Sediment traps have been widely used in fresh water biology to monitor particle fluxes at the top of the water column of standing waters. The traps are typically suspended from a fixed buoy and are positioned in the lower part (or just below) of the epilimnion for a fixed length of time to collect algae that are sedimenting out of the major productive region (euphotic zone) of the lake. In addition to sedimenting phytoplankton, the traps will also collect organic and inorganic particulates deposited onto the lake's surface and sediment that has become re-suspended during a period of lake turbulence.

The traps typically consist of an open bottle, with a top funnel that directs sedimenting phytoplankton and other material into the bottle chamber. In some aquatic studies (particularly oceanographic projects) sediment traps may be left for a long period of time (weeks to months). These traps may have multiple collectors that cycle through time to record seasonal changes in algae. Over long time periods, where algal populations are simply being enumerated, chemical preservation of material is necessary to prevent algal decomposition and zooplankton ingestion of material within the trap. Chemical preservation cannot be used where nutrient cycling is being monitored, however, and in this situation the impact of zooplankton activity, solubilisation of collected material as well as problems with hydrodynamics (water flow across the trap mouth) all need to be considered if the sediment traps are to be used quantitatively (Buesseler et al., 2007).

Sediment traps have been used in lakes for both short-term and long-term studies. Sigee et al. (2007) employed traps over a relatively short (9 days) period in their studies on cell death during late-season bloom decline in a eutrophic lake. Sedimenting phytoplankton was sampled in sediment traps (1-l plastic bottles with entry funnels) at a depth of 4 m. Bottles were filled with sterile water before gently lowering to the required depth to prevent filling with lake water and contamination with suspended phytoplankton. Sediment traps were left in the water column for successive periods of 3 days during mid-November. At this late stage in the season, the sedimenting phytoplankton is dominated by resistant spores of algae such as Ceratium and Anabaena, with a lot of organic debris derived from decomposing vegetative cells (Fig. 2.7). Longer-term studies have been carried out to investigate seasonal changes in lake sedimentation. Livingstone and Reynolds (1981), for example, monitored seasonal changes over a 1.5-year period, emptying traps at approximately three

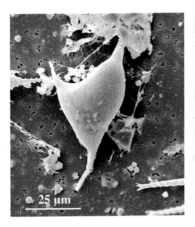
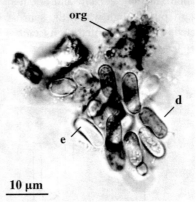

Figure 2.7   Lake sediment trap collection. Examples of late-season sedimented material, dominated by spores of algae such as (left) *Ceratium hirundinella* (SEM image) and (right) *Anabaena* (light microscope image). A group of spores (akinetes) of *Anabaena* are associated with organic debris (org) and appear to be either viable (d, dense contents) or non-viable (e, empty). The spores of these algae can be compared with actively growing vegetative cells – Figs. 4.56 and 4.24, respectively.

monthly intervals. Algal remains recovered from sediment traps broadly reflected phytoplankton periodicity and showed that turbulent resuspension of sedimented material occurred infrequently.

The use of traps to collect phytoplankton (sedimenting algae) can be compared with recruitment traps to collect benthic algae (rising from the sediments) – Fig. 2.22, Section 2.7.1. In both cases, the traps are emptied at regular intervals and there are potential problems with algal decomposition and the dynamics of collection.

2.3 Phytoplankton biomass

Determination of phytoplankton biomass is important since it provides information on the primary productivity of the aquatic system, and the amount of organic material that is available for consumption by zooplankton and the rest of the food chain.

Algal biomass can be determined as two major parameters.

- Total biomass. Where the overall value for the entire phytoplankton population is being evaluated by direct measurements from the lake water.

- Species and group biomass. Where the biomass for individual species and higher taxonomic groups is being indirectly estimated from population counts and biovolumes (see Section 2.5.4).

Total phytoplankton biomass can be directly estimated in three main ways – turbidity measurements, determination of fresh/dry weight and analysis of aquatic pigment concentration. Some of these measurements can be carried out from the boat and are part of the on-site collection of ecological data. These *in situ* methods include direct measurement of water turbidity (Secchi depth), use of fluorimeters to determine pigment/particulate concentrations and colorimetric assessment of chlorophyll concentration. Other techniques for determining pigment concentration and fresh/dry weight involve collection of water samples with subsequent laboratory analysis.

2.3.1 Turbidity

Turbidity is caused by suspended particulate and colloidal matter in water, including both inorganic (e.g. dispersed sediment particles) and organic (e.g. phytoplankton, zooplankton) material. Turbidity can be monitored manually using a Secchi disk or electronically via a fluorimeter probe as a nephelometer.

Secchi disk

Turbidity is frequently monitored as Secchi depth, which involves lowering a black/white sectored Secchi disk from the side of the boat until it can no longer be seen, then raising the disk until it is just visible. The distance from disk to surface is the 'Secchi depth'. For consistency over a period of time, the operation should be carried out on the sun-facing side of the boat and by the same operator. Although Secchi depth appears to be rather a crude estimate of turbidity, and includes both phytoplankton and non-phytoplankton particulates, there is a good inverse correlation with chlorophyll-*a* measurements (Fig. 2.8). In many standing waters, determination of Secchi depth has been found to be a simple and reliable approach to monitoring changes in seasonal phytoplankton biomass.

Nephelometer

Fluorimeter probes can function as nephelometers, collecting reflected light from suspended matter to determine the particulate concentration of phytoplankton and other fine materials in the water column. Nephelometer operation, calibration and interpretation of data are detailed by Eaton *et al.* (2005). These instruments have the advantage that they can operate as part of a remote-sensing system (Fig. 2.9), allowing turbidity to be measured on a continuous long-term basis. Fluorimeters are normally used to monitor chlorophyll-*a* concentration and turbidity simultaneously (Fig. 2.10) and are described in more detail in Section 2.3.3.

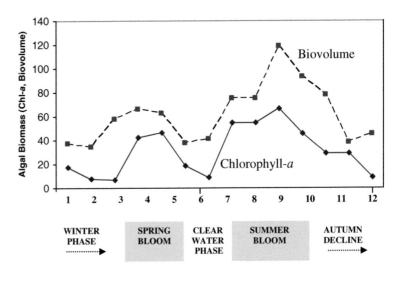

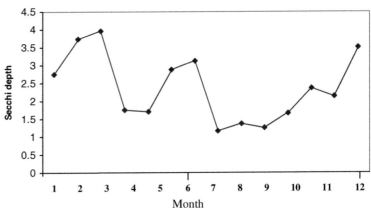

Figure 2.8 Monitoring phytoplankton biomass in lake water – relationship between chlorophyll-*a* concentration, total biovolume and Secchi depth. Seasonal changes in estimated biovolume (expressed as $\mu m^3 \; l^{-1} \times 10^4$) closely follow chlorophyll-*a* concentration ($\mu g \; l^{-1}$). Water turbidity, measured as Secchi depth (m), is inversely related to the above parameters. Reproduced with permission from Qari, 2006.

2.3.2 Dry weight and ash–free dry weight

One of the most direct ways to measure algal biomass is to simply collect phytoplankton in a net or using a membrane (glass fibre or Millipore© fibre) filter from a water volume sample and obtain values for wet weight (inorganic and organic biomass, plus water), dry weight (inorganic and organic biomass) or ash-free dry weight (organic biomass only). Wet weight is not normally determined because of the variability in removing free water from the sample.

In practice, dry weight or ash-free dry weight is determined from at least three replicate water samples. Dry weight can be measured by drying the

sample in a thermostatically controlled oven at 105°C to constant weight (normally taking about 24 h). To obtain ash-free dry weight, the sample is heated in a thermostatically controlled muffle furnace for 1 h (or to constant weight) at 300–500°C, and the resulting ash-weight deducted from the dry weight to obtain the ash-free value.

Limitations in the direct determination of biomass as dry weight are as follows:

• The phytoplankton sample is liable to contain contaminant non-algal material, such as particulate debris and zooplankton – both of which contribute to dry weight. The presence of organic and inorganic particulate material (from sediments) is

Figure 2.9 Monitoring buoy. Remote-sensing station used for continuous recording of lake parameters on Rostherne Mere, UK. Aerial sensors record air temperature, wind speed/direction, solar radiation and incident photon flux density. Underwater probes monitor conductivity, water temperature, chlorophyll-*a* concentration and turbidity (see Fig. 2.10) to a depth of 25 m.

particularly high during winter, when the lake may be fully mixed. This occurs at a time when the phytoplankton population in the epilimnion is seasonally low, and both of these factors lead to a high proportion of particulate debris in phytoplankton samples at this time of year. Wetzel (1983) has estimated that this particulate material may contribute over 80% of the total seston in certain situations.

- The drying process can result in appreciable loss of volatile organic compounds, leading to an underestimate of dry weight. This effect can be reduced by drying at lower temperature (80–90°C).

Because of the above limitations, phytoplankton biomass is normally measured in relation to pigment concentration rather than dry weight. Contaminant non-algal material, the major problem with dry weight determinations, is not recorded in chlorophyll-*a* estimations of biomass.

2.3.3 Pigment concentrations

Another approach for estimating phytoplankton biomass is to determine the concentration of a biomass-related constituent such as pigment (e.g. chlorophyll-*a*) or organic carbon, expressed per unit volume or per unit surface area of water. Using chlorophyll-*a*, the level of phytoplankton can either be monitored directly as aquatic pigment concentration (e.g. Fig. 2.8) or be converted to estimated biomass, with an appropriate conversion factor.

There are various limitations in the use of pigment concentrations to assess biomass.

- There is no precise relationship between pigment concentration and biomass. Pigment concentrations in algae vary between species and also within species in relation to external (light, temperature, nutrient availability) and internal (algal physiology) parameters. In general, chlorophyll-*a* content varies from 0.9% to 3.9% ash-free dry weight (Reynolds, 1990), and different authors have used different values to estimate biomass. Assuming a mean value of 1.5% ash-free dry weight, algal biomass can be estimated by multiplying the chlorophyll-*a* content by a factor of 67 (Eaton *et al.*, 2005). Chlorophyll-*a* can also be related to carbon content. Welker & Walz (1998), for example, converted their river chlorophyll concentrations to phytoplankton biomass assuming a carbon/chlorophyll-*a* ratio of 20 and a carbon content of 50% dry weight.

- Pigments of photosynthetic bacteria may also contribute to the total chlorophyll estimation, so that the final concentration will not only relate to algae.

- Any contaminant zooplankton in the sample may also contain ingested algae, which will also

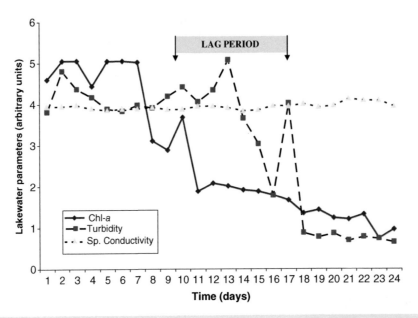

Figure 2.10 Fluorimetric monitoring of lake water parameters. Fluorimetric traces of mean lake water chlorophyll-*a* concentration and turbidity (both arbitrary units) obtained over a 24-day period (August 15–September 7, 2006) from the remote-sensing buoy shown in Fig. 2.9. Values were recorded at mid-day, depth 1 m. The decrease in chlorophyll-*a* corresponds to a decline in live algal biomass, with a lag period before the related decline in phytoplankton turbidity due to persistence of dead cells (no chlorophyll) in the water column. No significant change occurs in the mean specific conductivity. Fluorimeter parameters are given in Table 2.1 and text.

contribute to the pigment assessment of algal biomass. Normally this effect would be minimal.

- Chlorophyll pigments may degrade to phaeophytin products, which are relatively stable and which interfere with fluorimetric or spectrophotometric determination of chlorophyll. This degradation can be compensated for, however, since phaeophytin concentrations can be estimated separately on the same samples from which the chlorophyll was determined (Wetzel and Likens, 1991).

Techniques for the determination of chlorophyll-*a* concentration have been described in various standard texts (Wetzel and Likens, 1990; Eaton et al., 2005) and involve either direct on-site (lake water) measurement or subsequent analysis of aquatic samples back in the laboratory.

On-site measurement of chlorophyll concentrations

This can be carried out using fluorimeter probes or by colorimetric analysis.

Fluorimeter probes Fluorimeters measure the intensity and the wavelength distribution of light emitted as fluorescence from molecules that are excited at specific wavelengths. Using different excitation wavelengths and collecting light at different wavelength emissions (Table 2.1), they can be used to measure aquatic concentrations of chlorophyll-*a* (index of total phytoplankton biomass) and other algal pigments (assessment of different algal groups) within the water column.

Total chlorophyll Fluorimeter probes can be used for direct readings by manually lowering into the

Table 2.1 Fluorimeter Detection of Algal Pigment Concentrations and Aquatic Turbidity.

Measured Parameter	Total Algal Biomass	Cyanophyta	Cryptophyta	Turbidity
Pigment/ turbidity	Chlorophylls-*a* and -*b*	Phycocyanin	Phycoerythrin	Particle concentration
Excitation wavelength[a]	470 (30) nm	590 (30) nm	530 (30) nm	470 (30) FTU
Emission wavelength[a]	685 (30) nm	645 (35) nm	580 (30) nm	470 (30) FTU
Concentration range	0.03–100 µg l^{-1}	0.03–100 µg l^{-1}	0.03–100 µg l^{-1}	0.04–100 FTU

[a]Wavelength values are given as centre wavelength and bandwidth.
FTU, fluorimeter turbidity unit.

water column or as part of an automated remote recording system. The monitoring buoy illustrated in Fig. 2.9 is permanently positioned at a deep water site within a eutrophic lake and makes periodic depth recordings of chlorophyll-*a*, turbidity and water temperature. Equipment should be calibrated against known chlorophyll-*a* concentrations, and periodically checked. The sensitivity scale should also be adjusted so that low phytoplankton levels (few cells per litre) can be detected as well as blooms.

Such *in situ* profiling of lake parameters from a monitoring buoy provides a long-term record of phytoplankton biomass in relation to both short-term (e.g. diurnal fluctuations) and seasonal (relation to lake stratification, peaks of algal population) changes. The development of new high-resolution bio-optical sensors (Wolk *et al.*, 2004) also permits highly detailed phytoplankton profiles to be obtained, giving information on small-scale patterns of distribution (see Section 2.1.1). Environmental information collected by the remote-monitoring buoy is recorded and stored within a data logger and can be accessed via the GSM network.

Simultaneous recording of chlorophyll-*a* concentration and turbidity (Fig. 2.9) provides a useful comparison of these two parameters for measuring phytoplankton biomass. In this capacity the instrument is acting both as a fluorimeter (measuring chlorophyll-*a* concentration) and as a nephelometer (measuring particulate concentration) – where the generation and collection of scattered light at the same wavelength (470 nm) provide information on aquatic turbidity. During the period of phytoplankton decline (Fig. 2.10), the decrease in chlorophyll-*a* concentration (approximately 5–1 arbitrary units) is accompanied by a corresponding decrease in turbidity. There is a distinct lag phase between the two sets of data, however, consistent with a period of phytoplankton senescence (see Section 2.6.2) in which cells are losing their chlorophyll but remaining in suspension prior to sedimentation from the surface waters.

Fluorimetric analysis of phytoplankton within the water column is also relevant to river systems. Sherman *et al.* (1998), for example, used this approach to monitor biomass changes in a turbid-river weir pool that was dominated by the diatom *Aulacoseira*. Mean vertical water-column chlorophyll concentrations were computed at successive sites in the river, over a time period that coincided with the downstream transport of a parcel of water between sites. The decrease in mean chlorophyll concentration between sites was largely attributed to algal sedimentation. Based on a range of assumptions, the sediment rate (*w*) of *Aulacoseira* was calculated from the fluorescence data using the following equation (Condie and Bormans, 1997):

$$w = \frac{\Delta C\, H}{C_0\, \Delta t} \qquad (2.1)$$

where ΔC is the observed change in abundance (chlorophyll-*a* decrease 10.9–9.4 mg m^{-3}), C_0 the initial algal concentration in the surface layer (10.9 mg m^{-3}), Δt the time interval (0.722 days) and H the mean water depth (5 m).

Using the above equation and values, Sherman *et al.* (1998) calculated a sedimentation value of 0.95 m d^{-1}, which (within the limits of their assumptions) meant that the entire water column would have been cleared of *Aulacoseira* in 5 days under stratified conditions.

Algal groups   Fluorimetry can detect and quantify the biomass of different algal groups in relation to their fluorescent emissions (from characteristic pigments) at specific wavelengths. This approach was used by Serra *et al.* (2007) to monitor the vertical distribution of four dominant groups of algae (green algae, cyanophytes, diatoms and cryptophytes) within the water of a stratified reservoir. The fluorimeter was operated at four different excitation wavelengths and was calibrated by the manufacturer using model algal species to convert fluorescence measurements to chlorophyll-*a* concentrations, from which biomass could be derived.

Potential limitations with this technique are:

- *Cross contamination.* Specificity of emission is not absolute. Chrysophytes, for example, emit fluorescence at a similar wavelength to cryptophytes, and (in the study of Serra *et al.*, 2007) would thus contribute to the signal obtained from the dominant cryptophyte community.

- *Photochemical quenching.* Under high irradiance values, excess light energy is dissipated as heat. This 'photochemical quenching' occurs in surface waters, down to a depth where light intensity is about 10% of the surface water value and leads to an underestimate of chlorophyll concentration – which needs to be corrected.

Colorimetric analysis   One of the simplest ways of estimating chlorophyll concentration is to make a direct colorimetric assessment of net biomass. Although this has been used particularly for marine phytoplankton (McQuatters-Gollop *et al.*, 2007), it also has potential for large freshwater bodies where an extensive area of lake is being analysed. The technique is as follows:

- Net-haul collection of samples using a recorder that is towed behind the ship. With marine systems, this is carried out over a speed of 28–37 km h^{-1}, at a mean depth of ~10 m, over a distance of 18 km. Phytoplankton biomass is collected on a fine silk mesh.

- Assessment of the intensity of the biomass green coloration by comparison to a standard colour chart. The phytoplankton colour index (PCI) can be directly related to chlorophyll-*a* concentrations obtained by conventional extraction and fluorimetric analysis procedures (see below).

PCI gives an immediate and *in situ* assessment of biomass. It has an advantage over algal count procedures, since small phytoplankton cells that cannot be counted under the light microscope contribute to the coloration of the filtering silk. PCI analyses are particularly useful for low nutrient (oligotrophic) waters where large areas are being analysed, and show good correlation with surface chlorophyll concentrations as determined by direct measurement and satellite methods. This approach has been used by various Water Authorities as part of lake management, providing a rapid and rough guide to phytoplankton levels. In this context, PCI does have a major disadvantage in not indicating which species are present and thus what potential problem algae are there.

Laboratory pigment analysis

Laboratory procedures for measuring pigments are accurate and straightforward, and can be carried out both *in vivo* (direct measurement from algae in container) and by extraction (from water samples).

Pigment analysis of laboratory samples can be carried out by fluorimetry (Table 2.1), spectrophotometry or high-performance liquid chromatography (HPLC). The latter techniques require filtration and pigment extraction prior to measurement of pigment concentrations.

Filtration   Water samples collected in the environment are normally brought back to the laboratory and filtered as rapidly as possible to remove all the algae – including the smallest organisms. Filtration is normally carried out using either a cellulose acetate filter membrane (0.45 μm) or a glass fibre filter (Whatman GF/C, 1.2 μm). Samples should not be exposed to high light or temperature prior to filtration (to minimise pigment destruction), and the filtration process should be carried out under light suction pressure (no greater than 3 atm) to avoid damage to delicate algal cells. Once filtration is complete, the filter should be immediately removed and pigment extraction ideally carried out straightway. If extraction is to be delayed, filters should be folded and stored at −20°C until required.

Pigment extraction   Chlorophylls and carotenoid pigments contained within algal cells are removed by physical break-up of the cells followed by extraction with organic solvents (acetone, hot or cold methanol). To carry out an acetone extraction:

1. Place the filters into the base of a Teflon/glass homogeniser, add 3–4 ml of 90% alkaline acetone plus a small amount of glass powder and grind the sample for a period of 45 s under a fume hood.

2. Decant the homogenate into a centrifuge tube, then wash out the remains of the homogeniser by adding a further 3 ml of the acetone solution and grind for a further 15 s. Add the rinse solution to the existing homogenate and maintain the stoppered centrifuge tube in darkness until centrifugation.

3. Centrifuge at 300–4000 rpm (about 1000 g) for 5 min, using a refrigerated centrifuge (5–10°C) if available. Remove and tap the tubes to dislodge any particulate matter that has stuck to the sides, then recentrifuge for a further 2 min.

4. Carefully remove the clear supernatant (without disturbing the sediment) and pipette the clear liquid into a clean (non-acidic) spectrophotometer cuvette. To avoid pigment dilution, the use of small (10 ml) cuvettes, with light paths of 5 or 10 cm, is recommended.

For a more detailed description, see Eaton *et al.* (2005).

Spectrophotometry   The concentrations of major algal pigments – chlorophylls (*a*, *b*, *c*) and carotenoids (carotenes and xanthophylls) – can be individually determined by measuring light absorption at their respective wavelengths of maximum absorption. In practice, chlorophyll-*a* is frequently used as the sole measure of algal biomass, since it is present in all taxonomic groups and is the most abundant of all the algal pigments. Chlorophyll-*a* concentration (*C*) can be determined (Jespersen and Christoffersen, 1987) by measuring UV absorption at 665 nm (chlorophyll-*a*) and 750 nm (turbidity) using the following equation:

$$C = AV/V_s f/L \qquad (2.2)$$

where *C* is the chlorophyll-*a* concentration per unit volume of lake water (mg l^{-1}), *V* the volume of solvent (ml), V_s the volume of sample (l), *L* the light path (cm), $f = (1/\text{specific extraction coefficient}) \times 1000$ and $A = \text{absorbance}^{665} - \text{absorbance}^{750}$.

Compensation for degradation products   Although chlorophyll concentrations are frequently estimated as a direct spectrophotometric reading, their degradation products should also be taken into account. Chlorophylls degrade to phaeophytins, which are structurally similar except that the magnesium atom is lost from the ring structure. Because absorption of light by these compounds is similar to chlorophylls (but less strong) they can interfere with chlorophyll analyses. They have to be determined separately and the estimate then deducted from the total 'chlorophyll' value. To do this the total amount of pigment (chlorophyll plus phaeophytin) is determined in alkaline acetone. The sample is then acidified, converting all chlorophyll to phaeophytin, the chlorophyll concentration determined by the change in absorbance (Wetzel and Likens, 1990).

Spectrophotometric analysis of chlorophylls-a, -b, and -c

Although chlorophyll-*a* is routinely used to monitor total algal biomass, analysis of other pigments can also be used for this and can also provide information on the taxonomic composition of phytoplankton communities. Spectrophotometric estimation of the range of chlorophylls can be carried out by determining extinction coefficients at 664, 647 and 630 nm, using the trichromatic method of Jeffrey and Humphrey

(1975). Total carotenoids can be estimated by measuring the extinction coefficient at 480 nm (Strickland and Parsons, 1972).

High-performance liquid chromatography

HPLC analysis of phytoplankton samples provides a useful supplement to the use of algal counts. The main limitation of HPLC is that it cannot determine individual algal species, but instead classifies the phytoplankton sample in terms of major taxonomic groups (phyla). The main advantage is that it can be used to analyse a large number of samples over an extended aquatic area, and has been employed particularly with marine (Goericke, 1998) and estuarine investigations. Paerl *et al.* (2005), for example, used this approach to monitor freshwater algae as bioindicators of eutrophication in various estuary systems of the Southern United States (see Section 3.5.2).

Taxonomic carotenoids   HPLC, coupled with photodiode array spectrophotometry (PDAS), can be used for chlorophyll and carotenoid analysis. The identity of pigments within the sample is established by comparison with elution times of pigments isolated from reference material. Taxonomic-specific carotenoids (alloxanthin, fucoxanthin, peridinin and zeaxanthin) are particularly useful as diagnostic biomarkers for determining the relative abundance of different algal groups (Fig. 2.11). Of these carotenoids, only peridinin and alloxanthin are specific to a single phylum (plus violaxanthin – used by some workers as a specific marker for green algae).

Quantitation   Absolute and relative abundances of major algal groups can be calculated using a matrix factorisation programme such as CHEMTAX® (Paerl *et al.*, 2005). Using this programme, input consists of a raw data matrix of phytopigment concentrations obtained from HPLC analyses, plus an initial pigment ratio file (standard ratios of secondary pigments to chlorophyll-*a* in different algal groups). The data matrix is subjected to a factor minimisation algorithm that calculated a best-fit pigment ratio matrix and a final phytoplankton group composition matrix. Relatively large errors in the initial estimates of pigment ratios have little influence on the final determination of algal group abundances. The relative chlorophyll-*a*

contribution of each algal group is particularly useful since it partitions the known total chlorophyll-*a* (overall phytoplankton biomass) into the separate phytoplankton groups (individual biomass determinations).

Although there are some quantitative limitations with this method – including variation in carotenoid ratios within groups, lack of absolute phylum-specificity and co-elution of some pigments (e.g. lutein and zeaxanthin) during HPLC separation – HPLC has been widely used in phytoplankton determinations. Various studies, using corroborative standard algal count procedures, have demonstrated the validity of using CHEMTAX® in assessing major taxonomic groups with mixed phytoplankton samples (Paerl *et al.*, 2005).

Numerous studies have also shown that examining phytoplankton community dynamics and successional changes at the phylum (rather than species) level often provides excellent insight into environmental changes. HPLC studies on the effects of estuarine eutrophication, for example, have shown clear alterations in taxonomic composition during the major phase of aquatic productivity and also in changes to seasonal succession (see Section 3.5.2). In a eutrophication-related study on marine phytoplankton, Goericke (1998) demonstrated that algal biomass in the Sargasso Sea varied over the annual cycle – probably in relation to nutrient availability. These overall biomass changes did not relate to any alteration in the composition of the eukaryote algal community or to the relative biomass of blue-green algae, as determined by HPLC analysis.

2.4 Flow cytometry: automated analysis of phytoplankton populations

Analysis of total algal biomass or biomass of major taxonomic groups gives little detailed information about the composition of algal populations in relation to taxonomic composition (species analysis), size range or numbers of individual algal units (cells or colonies).

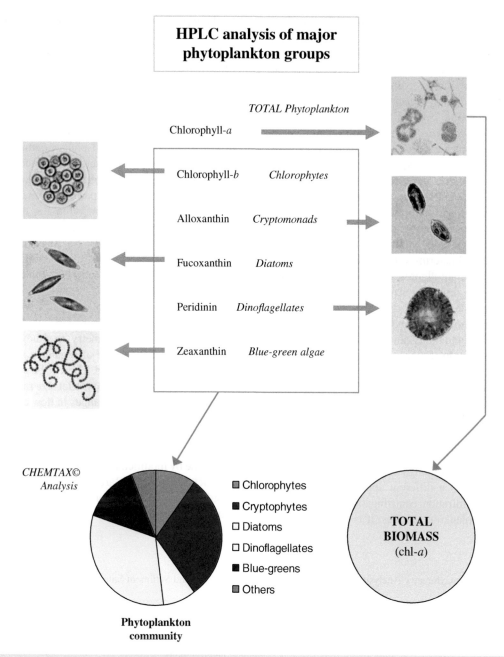

Figure 2.11 Phytoplankton community analysis based on HPLC separation and quantification of diagnostic pigments using the CHEMTAX© matrix factorisation programme. The phytoplankton pie chart shows a hypothetical population dominated by dinoflagellates and cryptophytes. Figure redrawn and adapted from Paerl *et al.*, 2005.

Flow cytometry provides a completely automated approach to analysis of particulate suspensions (0.5–40 μm diameter) such as phytoplankton populations. The phytoplankton sample is injected into the flow cytometer and is hydrodynamically focussed into a narrow stream that passes through a narrow laser beam. As the cells and colonies intercept the light source they scatter light and fluorochromes are excited to a higher state – releasing photons of light with spectral qualities unique to particular pigments. Flow cytometry thus measures fluorescence per cell or particle, in contrast to spectrophotometry – in which emission and absorption of particular wavelengths are measured for a bulk volume of sample.

Phytoplankton units within the mixed sample can be characterised in terms of forward-angle light scatter (FALS) and autofluorescence, which are related respectively to organism size and pigment concentration (Cavender-Bares et al., 1998). In addition to obtaining information about the phytoplankton sample, algal subpopulations can also be separated out as discrete fractions by the incorporation of a high-speed sorter.

Flow cytometry has been used by research workers to analyse phytoplankton in various ways (Table 2.2).

- *Collection of algal fractions for microscopy and bulk analysis.* Although the use of flow cytometry represents a different approach to classical techniques (filtration, centrifugation) for obtaining bulk phytoplankton samples, it is often preceded by these procedures for initial concentration of the sample.

Flow cytometry was used by Verspagen *et al.* (2005), for example, to obtain *Microcystis* fractions in their studies on benthic-pelagic coupling of this alga in a eutrophic lake. Bulk *Microcystis* populations were obtained from phytoplankton and sediment trap samples on the cytometric basis of size (0.5–2000 μm) and phycocyanin fluorescence. Sediment samples were fully mixed and initially centrifuged in a Percoll© mixture to separate algal cells from sediment particles. Final separation was carried out on a flow cytometer, calibrated to detect only particles containing chlorophyll-*a*. The biomass of *Microcystis* fractions from phytoplankton, sediment traps and sediment samples was then determined by chlorophyll-*a* analysis.

- *Determination of size range.* One of the problems in using flow cytometry to study natural samples is the wide range of organism sizes and concentrations encountered. This can be partly surmounted by prefiltration to restrict size range, otherwise time-consuming changes in flow cytometer configuration (lengthy changeover times) may be required to characterise the full phytoplankton size range. This instrumental problem can be largely overcome by modification of the conventional flow cytometer (Cavendar-Bares *et al.*, 1998), allowing simultaneous (real-time) analysis of the full spectrum of phytoplankton sizes.

Table 2.2   Flow Cytometry Analysis: Examples of Mixed Phytoplankton and Sediment Samples.

Target Organism	Sorting Criteria	Sample Analysis	Reference
Prochlorococcus	Fluorescence and light scattering	Molecular analysis of genetic heterogeneity	Urbach and Chisholm (1998)
Mixed phytoplankton	Forward angle light scatter	Real-time analysis of size spectra: modification of cytometer	Cavender-Bares *et al.* (1998)
Microcystis in water column, sediment traps and sediment	Unit size (0.5–2000 μm). Phycocyanin fluorescence	Blue-green algal biomass (chlorophyll-*a*)	Verspagen *et al.* (2005)

- *Species resolution of unicellular populations.* Flow cytometry has been particularly useful for collecting and analysing pico-phytoplankton fractions, where algae can be differentiated from heterotrophic bacteria in relation to size (Zubkov *et al.* 2003). It is also useful where mixed populations of different algal species are of a similar size but differ physiologically. Urbach and Chisholm (1998), for example, used flow cytometry to isolate populations of *Prochlorococcus* based on their red fluorescence (660–700 nm), absence of orange fluorescence (540–630 nm) and their characteristic forward light scatter. Cells of *Synechococcus*, which exhibit phycoerythrin autofluorescence (540–630 nm), were rejected during sorting. Molecular analysis of the purified samples of *Prochlorococcus* led these authors to demonstrate a high degree of genetic heterogeneity within local populations of this organism.

Although flow cytometry has considerable potential for freshwater studies, it has been used mainly in relation to marine phytoplankton, where automated collection of multiple samples over large areas of ocean has been particularly useful. The domination of many oceanic planktonic communities by unicellular blue-greens makes the technique specially relevant to marine biology.

2.5 Biodiversity of mixed-species populations: microscope counts and biovolumes

Microscopical analysis of species populations within the collected lake water sample involves algal identification (see Chapter 4) followed by counts of individual cells and colonies. In practice, it is often necessary to chemically preserve the algae and to concentrate the sample to make counting statistically valid.

Counts of algal species are normally carried out from volume (Section 2.2.2) or collecting tube (Section 2.2.3) samples, but net samples provide useful material to obtain a quick overview of major algal species present and to see algae in the living state.

2.5.1 Sample preservation and processing

Living phytoplankton

Visual examination of the fresh (live) phytoplankton sample is always a useful preliminary to analysis of fixed samples, particularly in relation to delicate flagellated algae – which may be distorted by the fixation process. Although smaller species are liable to be missed, the net sample is particularly useful for this – since the sample is concentrated during collection, and (if transported in the dark at 10°C) will remain fresh for a few hours.

Sample bottles should not be filled right to the top as there is no possibility for gas exchange once the bottle has been closed. If the bottle is only about 75% full then some exchange with the overlying air is possible for at least a short period. This is especially true for net hauls which, by their nature, are quite concentrated and have a high respiratory activity. These samples should be diluted to reduce oxygen demand and prevent the sample going anaerobic. If these precautions are not carried out there may be severe degradation in the sample by the time it is observed under a microscope. With live samples, zooplankton can be filtered off to avoid excessive grazing during transit.

Chemical fixation

Samples to be kept for longer storage require preservation, which normally involves treatment with either aldehyde or iodine fixative. Chemical fixation kills cells and prevents algal disintegration (caused by autolysis and bacterial degradation). It also prevents algal removal by any zooplankton present in the sample and increases algal cell density (which aids sample concentration by sedimentation). The major disadvantages of fixation are that algal shape and colour may be altered.

Aldehyde fixation   Although aldehyde fixatives tend to cause rupture, deformation and shrinkage of delicate algae, most species are preserved with minimal change to biovolume, shape and original coloration.

Fixation can be carried out using glutaraldehyde (final concentration 3% v/v, neutralised to pH 7.0 with Sorensen's phosphate buffer) or buffered 4% formaldehyde. The latter is prepared by adding concentrated buffered formaldehyde (40% v/v) to the phytoplankton sample at 1:9 v/v. Fixed samples should be kept in the refrigerator (at 4°C) in the dark, and may either be used for species counts (volume sampler) or for examination and photography (trawl-net sample).

Lugol's iodine   The preferred fixative for cell count samples is Lugol's iodine, which preserves cell shape/structure and also stains cells (yellowish to dark brown; Fig. 2.16) – so they can readily be seen in the counting chamber. Uptake of iodine (atomic weight 126.9) is also particularly effective in increasing cell density and aiding the process of sedimentation for concentration of the phytoplankton sample.

Lugol's iodine is prepared by dissolving 150 g potassium iodide plus 50 g iodine in 980 ml distilled water (DW), then adding 20 ml glacial acetic acid. Lugol's iodine is normally added to the lake water sample immediately after collection, using 2–3 ml of iodine per 100 ml of sample. A typical lake water volume collected for algal counts would be 200 ml, and this would normally be concentrated by sedimentation soon after collection and stored in the concentrated state.

Sample concentration

For phytoplankton counts, Lugol's iodine samples may need to be concentrated prior to enumeration in a counting chamber. This can be carried out by membrane filtration, centrifugation (1000 g for 20 min) or sedimentation.

Algal sedimentation   This is the preferred technique for sample concentration since it is non-selective in terms of size (unlike filtration) and non-destructive (unlike filtration and centrifugation).

Sedimentation is normally carried out by pouring the 100-ml sample into a measuring cylinder and allowing the cylinder to stand on a vibration-free surface over a 24–72 h period to allow all of the algae to

sediment. Sedimentation should be carried out in the dark and away from any sources of heat. The top 80 ml of algal-free liquid can then be carefully siphoned of using a pipette with a 'U' tip, to cause minimal disturbance. The remaining 20 ml (5× concentration from environmental sample) can be re-suspended in the bottom of the cylinder by agitation and then stored in an opaque glass or polyethylene container in the dark at constant temperature (4°C) until needed for analysis.

Although iodine increases cell density, some algal species still may not sediment. This problem applies especially to colonial blue-green algae that contain gas vacuoles, where their buoyancy can be great enough to overcome the additional weight of the iodine. It is very important to look carefully at the surface layer of the sample in the cylinder to make sure no buoyant species are present before decanting off the supernatant (90 ml). If buoyant blue-green species are present, one way to overcome their buoyancy is to transfer the fresh sample to a strong but flexible polythene bottle. The bottle should be filled right to the top with the phytoplankton sample, with no trapped air and the top screwed down tightly. The bottle should then be dropped onto a hard surface from a height of about 1.5 m. The sudden increase in pressure as the bottle hits the floor is usually enough to collapse the gas vacuoles in the cells so that they will then sediment without trouble.

An alternative procedure to the two-stage process of sedimentation followed by separate enumeration in a counting chamber is to make counts direct from the sedimentation chamber using an inverted microscope (Fig. 2.12).

2.5.2   Chemical cleaning of diatoms

Identification of diatoms to species level in environmental samples (phytoplankton, benthic communities and sediment samples) typically involves detailed light or scanning electron microscope examination of frustule ornamentation (Barber and Haworth, 1981). This requires the outer layer of organic material and associated debris to be removed, which can be carried out in various ways.

Cleaning is normally carried out chemically by treatment with acids and/or oxidising agents. Trial and error will show which method is best for particular samples. Mild treatments, which leave most frustules and some colonies intact, include treating samples with dilute hydrochloric acid (to remove calcareous matter) or soaking the material in 30% hydrogen peroxide at room temperature for 24 h.

More extreme cleaning procedures involve exposure to hot peroxide (see below) or treatment with strong acids (Barber and Howarth, 1981). These methods result in the loss of all non-siliceous algae, the fragmentation of colonial diatoms into separate cells and the break-up of diatom frustules. *These treatments result in the emission of toxic and corrosive fumes, and must be carried out in a fume cupboard.*

Hot hydrogen peroxide treatment

A very effective procedure, this is carried out as follows:

1. Approximately 2 g of environmental sample is placed in a 100-ml beaker, 20 ml of concentrated H_2O_2 solution added, and the vessel left for 20 min at room temperature. The beaker is then heated on a hotplate at 90°C) for 2 h, keeping topped-up with H_2O_2.

2. The beaker is removed from the hotplate and a few drops of HCl (50%) added, with release of chlorine (the HCl removes surplus H_2O_2 and carbonates.)

3. Samples are allowed to cool to room temperature, transferred to centrifuge tubes and spun for 4 min at 1200 rpm. The supernatant is removed and the sample resuspended in distilled water. The process is repeated five times, with the addition of a few drops of weak ammonia before the final wash to help clay removal, and final suspension in DW.

4. 0.5 ml of the suspension is placed in a centrifuge tube and diluted with DW to produce a final suspension that is neither totally clear nor milky, but has many fine particles when held up to the light.

5. 0.5 ml of the dilute suspension is placed on a coverslip and left overnight under sterile laminar flow to evaporate. A microscope slide is then placed on a hot plate (90°C), a drop of Naphrax© embedding medium placed on it and the coverslip bearing the dry diatom sample inverted onto it. The slide is subsequently removed from the hotplate and allowed to cool – forming a permanent preparation.

Numerous micrographs of hot peroxide-cleaned diatoms are shown in Chapter 4, with combined images of living and cleaned cells illustrated for *Aulacoseira* (Fig. 4.8), *Fragilaria* (Fig. 4.10), *Tabellaria* (Fig. 4.11), *Diatoma* (Fig. 4.12), *Asterionella* (Fig. 4.42), *Stephanodiscus* (Fig. 4.58), *Nitzschia* (Fig. 4.70a) and *Navicula* (Fig. 4.73a).

2.5.3 Species counts

Initial identification of major algal species within the fixed sample is a useful preliminary to the enumeration process, and may involve visual comparison to a fresh phytoplankton sample (Section 2.5.1). Species counts are usually made in relation to the 'living unit', which may be a single cell (unicellular alga), filament of cells or globular colony. There are no hard and fast rules as to whether colonies and filaments should be treated as single units or constituent cells, but clearly consistency is important within and between samples. The range of techniques available for making algal counts has been widely reported (Wetzel and Likens, 1991; Eaton *et al.*, 2005), including the use of statistical procedures to estimate precision, required number and randomness of algal enumeration (Lund *et al.*, 1958).

Microscopical counts of algal species (cells, filaments or colonies) can be carried out from

• Liquid samples contained in a sedimentation chamber

• Concentrated samples contained in a counting chamber

• Filtered samples present on a filter membrane.

Sedimentation chamber – use of inverted microscope

The combined use of a sedimentation chamber and inverted microscope allows a small volume of iodine-preserved sample to sediment and be analysed in a single operation. Cells that have collected at the bottom of the chamber are immediately viewed and counted from beneath. The essential differences between an inverted light microscope and a normal one is that the objective lenses face upwards, the condenser has a greater working distance and the slide is placed on the stage with the objectives beneath (Fig. 2.12a)

Originally designed by Utermöhl, sedimentation chambers have been widely used in phytoplankton studies (Rott,1981) and have been subsequently modified in various ways for both inverted (Evans, 1972) and upright (Hamilton *et al.*, 2001) microscopes. Sedimentation chambers typically consist of two main parts – a tower section (for holding the sedimenting plankton) and a settling chamber (contained within a base plate) to retain the sedimented algae for making counts (Fig. 2.12b). The tower section is positioned on top of the base plate with the central column directly above the settling chamber. A known volume (range 5–100 ml) of iodine-preserved phytoplankton suspension is poured into the column and allowed to settle for 24 h, leading to the deposition of algae onto a coverslip at the base of the settling chamber. The tower section can then be slid across the base plate, voiding the surplus liquid, and a coverslip finally placed on top of the settling chamber to limit evaporation. Sedimented phytoplankton can then be observed, identified and counted.

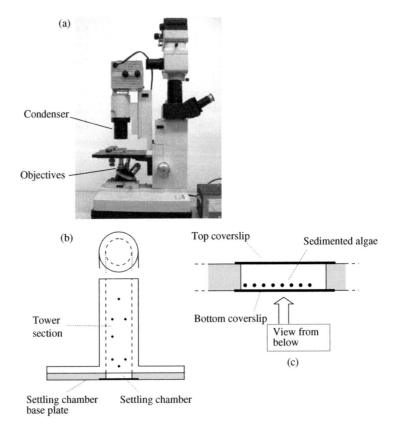

Figure 2.12 Combined use of inverted microscope and phytoplankton sedimentation chamber. (a) Inverted microscope: Substage objectives allow close optical access to algae that have sedimented to the bottom of the settling chamber. (b) General view of sedimentation chamber, composed of tower section and settling chamber base plate. (c) Detailed view of settling chamber after algal sedimentation and removal of the tower section. Sedimented algae lie on the lower coverslip (fixed in place by a retention ring) and can be observed from below using the inverted microscope.

Volume counting chambers: the
Sedgwick-Rafter slide and Lund
nanoplankton counting chamber

Sedgwick-Rafter slide   This contains a shallow
central chamber that has a capacity of 1 ml, an
engraved grid of 1000 squares (Fig. 2.13a) and is covered by a thick coverslip. The volume of the chamber
should be checked either by taking accurate measurements of its dimensions or by accurately weighing the
slide (including cover slip) dry and then reweighing
it filled with distilled water.

Species counts can be made as follows:

1. After removal from the storage container (e.g.
   refrigerator), allow the iodine sample to equili-
   brate to room temperature, and make sure that the
   sample is fully mixed by gentle inversion and agitation.

2. Remove a sample of fully mixed algal suspension using a wide-aperture pipette. With the coverslip placed across the Sedgwick-Rafter slide
   (Fig. 2.13b), fill the chamber with sample. Rotate
   the coverslip back to completely cover the chamber, taking care to excluding air bubbles.

3. Allow the algal cells to settle to the bottom of the
   chamber and examine with the 20× objective of
   an inverted-objective or conventional microscope.

4. Before making any counts, scan the whole chamber to identify major species present in the sample plus any other species of interest. If any

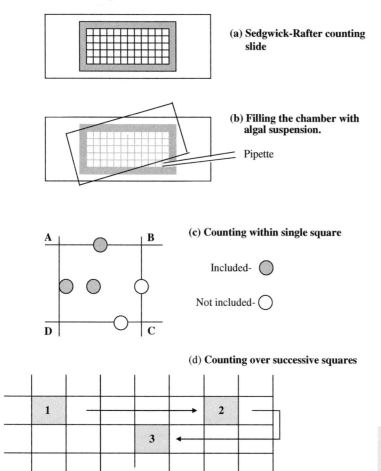

(a) Sedgwick-Rafter counting
slide

(b) Filling the chamber with
algal suspension.

Pipette

(c) Counting within single square

Included-

Not included-

(d) Counting over successive squares

Figure 2.13   The Sedgwick-Rafter
slide: design of counting chamber and
protocol for making counts.

problems arise, it can be useful at this stage to compare the iodine sample with the fresh sample (examined previously) or with the formaldehyde-fixed, trawl-net sample. Check that algae are dispersed randomly across the counting area and not localised to one particular region.

5. To make species counts, select one square at random and count all single cells and globular colonies within the square (see later for filamentous algae). Include all those algae that touch or overlap sides A–B and A–D of the square (Fig. 2.13c) but not those that are in contact with sides B-C and C-D. Further squares for counting should be selected on an objective basis, without any bias towards contents. To do this, move five squares from square 1 (the initial square) and count from the new square (square 2). Repeat the process as indicated in Fig. 2.13d. As an alternative, rather than a fixed interval of five squares, random numbers can be used to determine the spacing between counted squares. McAlice (1971) has shown that a count of 30 squares can be expected to reveal 90–95% of the species present. The number of squares counted should also be sufficient to give a statistically valid population estimate of major species or of minor species that are being studied. This will depend on which species are being investigated, and variability within the lake – but may exceed the count of 30 squares noted earlier.

6. Dead algal cells, including algae with no contents (Fig. 2.18) and the remains of diatom frustules, are particularly prominent at certain times of year (see Section 2.6.2a). These should be either ignored or recorded in the algal counts as a separate category.

7. To calculate the environmental populations (T) of individual species from the Sedgwick-Rafter counts:

If C is the number of organisms counted in N squares and there is a 10× concentration from the original aquatic sample,

$$T = \frac{1000\,C}{10\,N} \qquad (2.3)$$

where T is expressed as the number of organisms (single cell or colonies) per ml of original sample.

If statistical evaluation of phytoplankton species diversity is required, the entire counting chamber may be screened to record as many species as possible within the sample. If possible, at least 400 units (cells or colonies) of each species should be counted to keep the counting error at <10% (Lund *et al.* 1958).

Counts of single cells and small colonies (with defined numbers of cells) are relatively straightforward, but problems may be encountered with filamentous and large globular (particularly blue-green) algae, where unit size and cell number are highly variable.

Filamentous algae　Although filamentous algae can simply be recorded as numbers of units (individual filaments), a more accurate approach is to measure filament length. This is because filaments of these algae may vary considerably in length (within and between samples), so counting as individuals becomes meaningless. Filaments may also extend across more than one side of the counting square, making enumeration difficult.

To measure filament length, it is necessary to calibrate an eyepiece micrometer, which is a small scale engraved onto a circle of glass. Once this has been inserted into the eyepiece of the microscope, its scale becomes visible at the same time as the specimen. Align the micrometer scale against the filament by rotating the eyepiece and measure the length of the filament in eyepiece units (Fig. 2.14a). The eyepiece units can then be converted to absolute units (μm) by removing the specimen slide, inserting a stage micrometer slide, aligning the two scales in the field of view, then reading off the number of absolute units equivalent to a block of eyepiece units (Fig. 2.14b).

In practice, rather than converting each filament to absolute units, the total length of filaments within each square, and across 30 squares can be summed – and the final cumulative value (eye piece units) then converted using the determined conversion factor.

Large globular algae　Large globular algae such as *Microcystis* and *Gomphosphaeria* are frequently encountered in lake samples, but are difficult to enumerate because of their irregular form and variable

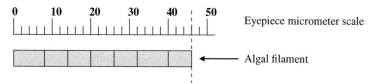

(a) Measurement of filament length in eyepiece scale units

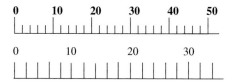

(b) Calibration of the eyepiece scale

Figure 2.14   Measurement of algal filament length: use of eyepiece and stage micrometers. In this illustration, the algal filament (composed of 7 cells) has a length of 46 eyepiece units (a); 50 eyepiece units are equivalent to 33µm (b), so the length of the filament in absolute units is 30.4µm.

size. Simple colony counts are difficult to interpret, and more useful approaches for enumeration are either to measure colony sizes and determine biovolumes (see Section 2.5.4) or to estimate the number of cells contained within the colonies. Estimating cell numbers is not normally carried out on a routine basis, but can be carried out either by breaking up colonies into their constituent cells or by estimating cell numbers from colony size (Reynolds and Jaworski, 1978).

- *Colony fragmentation.* Colonies of algae such as *Microcystis* can be broken up by chemical (alkaline hydrolysis) or physical (sonication) procedures. Reynolds and Jaworski (1978) carried out alkaline hydrolysis by adding a single pellet of sodium hydroxide to 100 ml of phytoplankton suspension, then heating the alkaline suspension at constant temperature (90°) for 30 min. This normally causes complete disruption of the mucilaginous colonies, allowing individual cells of *Microcystis* to be counted. The use of sonication can also be very effective in disrupting colonies. Reynolds and Jaworski (1978) found that 1 min ultrasound treatment (~12 µm, 20 kHz) of phytoplankton suspension resulted in homogeneous cell suspensions, but the required duration will clearly vary with the specific details of phytoplankton sample and sonication apparatus.

- *Cell estimates from colony size.* According to the observations of Reynolds and Jaworski (1978) the

number of cells (y) in a healthy quasi-spherical colony of *Microcystis* can be approximated from its diameter (x) by the regression equation:

$$Log_{10}y = 2.99log_{10}x - 2.80 \qquad (2.4)$$

Although all of the above methods give comparable estimates of single cell counts of *Microcystis*, the authors consider that the regression method is least useful because of variation both in colony shape (often not approximating to a sphere) and also of cell packing (cell number per unit volume) within the mucilaginous matrix.

The Lund nanoplankton counting chamber

Although the Sedgwick-Rafter slide is ideal for medium- to large-sized algal species, it is difficult to visualise and count smaller species – which may thus be entirely missed. This problem arises because high-power lenses, with their small depth of focus, cannot be used due to the depth of the counting chamber and the thickness of the coverslip (non-inverted microscope) or the thickness of the slide itself (inverted microscope).

The optical problems encountered with the Sedgwick-Rafter slide have been largely overcome with the introduction of the Lund nanoplankton counting chamber, which is ideal for counting small phytoplankton cells (Fig. 2.15). This has reduced the

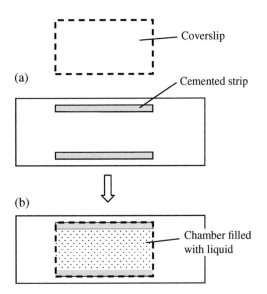

Figure 2.15 Lund nanoplankton counting chamber. (a) Preparation of chamber. Cemented strips on glass slide with separate coverslip. (b) Coverslip in place. Chamber filled with water (for weighing and volume determination) or phytoplankton sample (cell/colony counts).

depth of liquid combined with a thinner coverslip, allowing the use of high-power objectives and was originally developed by Lund (1951).

The chamber can be constructed as follows. Take a large no.1 cover slip. Using a diamond pencil cut two strips about 3.0 mm wide. These are cemented to a normal microscope slide about 1.6 cm apart and in parallel with each other. Weigh the slide with its side walls together with the complete cover slip that will be used. A drop of DW is now introduced between the cover slip ridges and the main complete cover slip placed in top. The water should fill the entire chamber. Any excess outside the chamber can be removed with a piece of absorbent filter paper. Weigh the whole. The difference between the dry weight and the wet weight with a full chamber gives the volume of the cell. Dry the cell and now introduce a well-mixed aliquot of the sample to be collected. Count all cells by carefully traversing the chamber backwards and forwards, using the eye-piece cross hairs as a guide. It is advisable to pre-filter the sample through a phytoplankton net mesh first to remove large colonial algae and long filaments as these may clog the counting chamber.

Algal counts from filter membranes

This method involves passing a measured volume of water sample through a membrane filter, and making algal counts direct from the membrane. The procedure does not require pre-concentration of the water sample as the required population density needed for counting can be obtained by adjusting the volume of sample filtered. This technique has been used particularly in relation to marine phytoplankton, with various modifications for staining and enumeration (McNabb, 1960).

Filtration A suitable volume of sample (either preserved or not) is filtered through a membrane filter paper having a pore size of 0.2 or 0.45 μm. The recommended sample volume (Vollenweider, 1969) is between 1 and 200 ml cm^{-2} of filter membrane surface although experience will tell the investigator what volume is best for them. Cellulosic, polycarbonate or any other suitable filter material may be used. To avoid uneven distribution of algae on the filter membrane surface, the sample should be carefully added to the filter holder and allowed to stand for 2–3 min before applying the vacuum.

Gentle suction is now applied, avoiding the use of excessive force since this can cause collapse of cells onto the filter surface. Turn off the suction when there is about 5 mm depth of liquid left above the filter and allow the last sample to pass through very gently. If an iodine-preserved sample has been used it may be advantageous to wash away any excess iodine by passing a small amount of distilled water through the membrane, taking care not to disturb the distribution of cells on the membrane during the washing process. When filtration is complete, dry air should be gently passed through the filter for a further minute to remove excess moisture.

Preparation for counting Three different approaches may now be used, either singly or in combination.

1. For the first procedure it is essential to have used membrane filters that are unaffected by ethanol. The algal cells on the filter are dehydrated by

passing small amounts (10–15 ml) of increasing concentrations of ethanol through the filter. When dehydration is complete the algal cells are stained with alcoholic fast green (0.1% in 95% ethanol) by allowing the filter to stand with the stain overlying it for 20 min before sucking it through. The filter is finally washed under gentle vacuum with a small amount of ethanol (Vollenweider, 1969).

2. Take a drop of either microscope immersion oil or good quality cedar wood oil and place it on a microscope slide. Take an unused dry membrane filter paper and place it on the oil droplet. After a few minutes the oil penetrates the filter which then goes completely clear. If this does not happen gentle warmth (about 30–35°C) can be tried to hasten the clearing process. If the filter still does not go transparent then this method cannot be used as the filter is unsuitable. If the filter used for the sample is the correct type then the process is repeated but this time the filter paper plus filtered sample is placed, face up, on the oil droplet. The slide + filter are now placed in a dust-free atmosphere with or without gentle warming until the filter is transparent. A cover slip is now placed over the clear filter avoiding trapping any air bubbles. To help avoid air bubbles a small drop of oil can be placed on the underside of the cover slip which is then slowly and carefully lowered onto the filter. Care must be taken to avoid disturbing the cells on the filter surface. Cedar wood oil is less expensive than immersion oil but is darker in colour so that subsequent microscopic observations are not quite as clear.

3. This approach is basically the same as in option (ii) but helps preserve the randomness of cell distribution on the filter a little better. In this case, a drop of immersion oil is placed on a large coverslip (large enough to take the whole filter pad) and allowed to spread. The filter is then placed face down on the coverslip, transferred to a dust-free environment and, with or without warming, allowed to go transparent. The coverslip with filter are then carefully inverted onto a microscope slide.

Algal counts The algae present on the filter membrane can now be observed and counted using a normal light microscope or by epifluorescence microscopy (see end of section). Two counting methods are commonly used.

1. Most Probable Number (MPN) estimate

This method can be used for species that are particularly abundant in the sample. Using the MPN estimate runs the risk that uncommon or infrequent species might be overlooked, and these should be estimated using an alternative technique. The MPN technique is also used in bacteriology and records not only the presence but also the absence of organisms.

The procedure is as follows.

- First view the slide under low power to confirm that the cells present are randomly distributed. Now, using whatever magnification is appropriate, identify the more common species present. It is always helpful to have previously looked at a fresh unfiltered sample to help identification as less robust species can be damaged by the filtration process making identification a little more difficult. Robust species such as diatoms are not affected by the filtration process. It can be easier to spot and identify cells if they have been stained (e.g. with Lugol's iodine).

- Select an appropriate magnification objective so that only the most common species are present in the field of view for about 80% of the time. It may be necessary to either concentrate the sample, that is filter a larger volume through the membrane, or dilute it to achieve the required concentration. If the membrane is marked with a grid, then each square can be numbered and a random number table used to select particular squares for algal recording. If gridded membranes are not available the slide is moved about and observed in a random manner. Fifty randomly selected fields should be observed. Note the number of fields of view in which the different species occur. If there are less than 80% occurrences of the key species, then select a lower magnification objective, thus giving a larger field of view. If there is greater than 90% occurrence, then a higher magnification objective giving a smaller field of view can be used. If selecting a different objective does not work, then either a greater or lesser volume of sample should be filtered. Cell counts are not needed, only presence or absence.

- Calculate the frequency (F) for each species using the following formula:

$$F = \frac{\text{Total number of occurrences of species}}{\text{Total number of quadrats/fields observed}} \times 100$$

$$(2.5)$$

The percentage frequency for a species can now be converted to a theoretical number of individuals per field observed, assuming they were randomly distributed. This is called the theoretical density (d) and can be obtained using the conversion factors given by Fracker and Brischle (1944).

Once the value for d has been estimated, the population count (N) can be calculated using the following formula:

$$N = \frac{d A_1}{A_2(V_1+V_2)},$$

$$(2.6)$$

where d is the theoretical density (see above), A_1 the area of filter (note this is the area through which filtration takes place, not the total size of the filter), A_2 the area of the field of view, N the number of cells per litre, V_1 the volume of sample filtered in litres and V_2 the volume of preservative added in litres (if used, but this is very small in relation to the total volume). Conversion of F to d is only valid if the cells are randomly distributed.

2. Single-field counts

The filter area and the field area need to be known as in method (1). Organisms also need to be randomly distributed across the membrane. For this technique, a gridded filter membrane should be used and all organisms within a single grid are counted. The total number of organisms (N) per unit volume of sample can be determined as follows:

$$N = \frac{YAD}{av}$$

$$(2.7)$$

where Y is the mean count per graticule area; A is the filtration area; v is the sample volume filtered and d the dilution factor for the sample.

A full description of this technique together with modifications for staining, fluorescence and statistical validation is given by Jones (1979).

Epifluorescence microscopy   This technique uses a specific fluorescent stain known as a fluorophore that is taken up into living cells (Paddock, 2007). One of the most commonly used stains, acridine orange, is excited at wavelengths of 502 nm and has an emission maximum of 526 nm (green).

Epifluorescence is particularly useful for counting small size plankton (diameter <5 μm), where organisms may be difficult to resolve taxonomically (into major groups) and also difficult to distinguish from non-living particulate material. Tittel *et al.* (1998), for example, used epifluorescence microscopy to count and measure picoplankton in their studies on community structure and plankton size distribution in north German lakes. Water samples were filtered through black Nuclepore membranes (0.2 μm), and stained to visualise bacteria (acridine orange), unicellular algae and heterotrophic flagellates (stained with proflavin, examined with blue excitation) and picocyanobacteria (counted unstained with green excitation, then distinguished from eukaryotic picoalgae by changing to blue excitation). Multicellular algae and larger unicellular algae (>5 μm diameter) were counted and measured by bright field microscopy, using an inverted microscope.

2.5.4 Conversion of species counts to biovolumes

Counts of separate organisms – single cells or colonies, can provide useful information on the populations of individual algal species. These counts can be used, for example, to monitor the seasonal changes in a particular species or assess vertical distribution of a species within the water column. Species counts do not, however, give comparative information on the relative contribution of different species to overall phytoplankton biomass, since unit size (μm) and therefore overall unit sizes (biovolume – μm^3), vary considerably. Unit size can be transformed (using appropriate conversion factors) to cell carbon (gC) or cell biomass (g). Such calculations are potentially very useful, but should be treated with caution, since they depend on a sequence of approximations – including mean algal dimensions, conversion of linear to three dimensions (Fig. 2.17) and variation in conversion factors from biovolume to biomass and cell carbon. In practice, algal counts are routinely converted to biovolumes, which serve as an approximate index of algal biomass.

Species count and biovolume within phytoplankton populations

Algal size varies considerably from one species to another and even within the same species at different stages in its growth cycle. A species of *Chlorella*, for example, whose shape is spherical and has a mean

cell diameter of 4 μm would have a mean cell volume of 30 μm^3. In contrast to this, one cell of *Stephanodiscus* sp., a centric diatom shaped like a short drum, would have a volume of nearly 16,000 μm^3. In these examples, the mean unit volume of one species is 530 times larger than the other so their contribution to phytoplankton biomass (per cell) would be significantly different.

The ecological significance of algal size in relation to biomass is seen in Fig. 2.16, which shows a lake phytoplankton sample with algae ranging in size (greatest axial linear dimension) from 100 μm (*Ceratium*) to 3 μm (*Synechococcus*). Species counts from the same sample (Table 2.3) show that *Stephanodiscus* has the highest overall population (78% total phytoplankton count), but determination of biovolumes shows that it occupies only 13% of the total phytoplankton level.

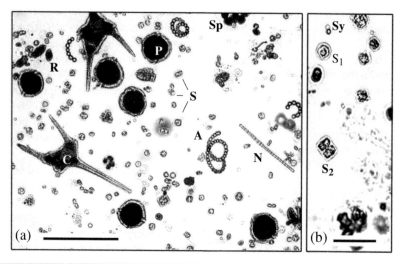

Figure 2.16   Taxonomic and size diversity in a mixed phytoplankton sample. (a) Low-power view, showing the presence of algae belonging to the blue-green algae (*Anabaena*, A; *Aphanizomenon*, N), green algae (*Sphaerocystis*, Sp), dinoflagellates (*Ceratium*, C; *Peridinium*, P), cryptomonads (*Cryptomonas*, R) and diatoms (*Stephanodiscus*, S). Scale bar 100 μm. (b) High-power view showing details of some of the smaller phytoplankton – including *Stephanodiscus* in valve (S1) and girdle view (S2, pair of cells), and the unicellular blue-green alga *Synechococcus* (Sy, pair of cells). Scale bar 25 μm. Considerable variation in size and shape occurs within the sample. The cells have been fixed in iodine in preparation for cell/colony counts (see Table 2.3). In these fields of view, cell size (greatest axial dimension) ranges from ∼180 μm (*Ceratium*) to ∼2 μm (*Synechococcus*). Cell/colony shape ranges from spherical (*Peridinium*) to oval (*Cryptomonas*) to the extended forms of *Ceratium* (unicell) and *Aphanizomenon* (colony).

Table 2.3   Species Counts and Biovolumes from the Lake Water Sample Shown in Fig. 2.16.

Species	Count (Cells or colonies ml^{-1})	% Total Phytoplankton Count	Unit Species Biovolume (μm^3)	Total Species Biovolume $\times 10^5$ (μm^3 ml^{-1})	% Total Phytoplankton Biovolume
Stephanodiscus minutula	12 000	78	380	46	13
Cryptomonas ovata	775	5	1050	8.1	2
Rhodomonas minuta	680	4	145	1.0	<1
Anabaena flos-aquae*	120	1	2165	2.6	<1
Synechococcus aeruginosa	30	<1	20	0.005	<1
Aphanizomenon flos-aquae*	480	3	1520	7.3	2
Ceratium hirundinella	30	<1	40 000	12	4
Peridinium cinctum	578	4	48 000	280	77

Unit species biovolume: mean volume of a single cell or colony (*), estimated from linear measurements; total species biovolume: volume of the entire population of a single species, determined as the product of unit biovolume and cell count.
Percentage contributions: These are shown for each species in relation to total phytoplankton count and total phytoplankton biovolume (shaded columns) and differ markedly in relation to unit (cell or colony) size.

Calculation of biovolumes

The biovolume of particular algal species within mixed lake phytoplankton samples can be considered in relation to

- mean unit biovolume – the average volume of individual organisms (cells or colonies), expressed as μm^3

- species population biovolume – the volume occupied by single-species populations per unit volume of lake water, expressed as μm^3 l^{-1}, and

- total phytoplankton biovolume – the cumulative volume occupied by all the species within the phytoplankton sample, expressed as μm^3 l^{-1}

Mean unit biovolume   Mean unit biovolume depends on size and shape, which vary greatly from species to species (Fig. 2.16). In some cases, algal shape approximates to a simple three-dimensional figure such as a sphere (e.g. *Chlorella*) or rod (e.g. *Melosira*) whilst others are more complicated (*Ceratium, Campylodiscus*). Calculation of mean unit biovolume for colonial algae such as *Microcystis* may present particular problems, since they are typically highly irregular in shape, vary considerably in size

but often make a major contribution to phytoplankton biomass.

Calculations of mean unit biovolumes for individual species can be carried out from measured dimensions of the organisms, treating them as simple geometric shapes or as combinations of geometric shapes. Some examples of standard cell shapes, key dimensions, geometric formulae and representative species are given in Fig. 2.17, and a list of typical biovolumes for some common planktonic algae in Table 2.4. A comprehensive reference set of geometric shapes and mathematical equations for calculating biovolumes of >850 pelagic and benthic microalgae from microscopic observations has been published by Hillebrand *et al.* (1999).

Although typical unit biovolumes for particular species based on geometric shapes can be obtained from the literature (Bellinger, 1974; Reynolds and Bellinger, 1992; Stephen, 1997), it is always preferable to make calculations afresh for each phytoplankton sample, preferably using live material. It should be noted that variations in mean unit biovolume for a particular species can vary spatially within a freshwater environment (e.g. lake water column), temporally with season (Bellinger, 1977) and with conditions of water quality – including heavy metal pollution (Fig. 3.7) and acidity (see Section 3.2.3).

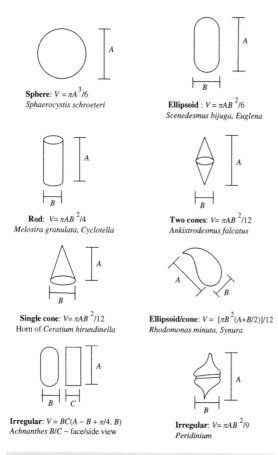

Sphere: $V = \pi A^3/6$
Sphaerocystis schroeteri

Ellipsoid : $V = \pi AB^2/6$
Scenedesmus bijuga, Euglena

Rod: $V= \pi AB^2/4$
Melosira granulata, Cyclotella

Two cones: $V= \pi AB^2/12$
Ankistrodesmus falcatus

Single cone: $V= \pi AB^2/12$
Horn of *Ceratium hirundinella*

Ellipsoid/cone: $V = [\pi B^2(A+B/2)]/12$
Rhodomonas minuta, Synura

Irregular: $V = BC(A - B + \pi/4 . B)$
Achnanthes B/C – face/side view

Irregular: $V= \pi AB^2/9$
Peridinium

Figure 2.17 Standard shapes for calculating algal biovolumes: some common examples.

In calculating species mean unit biovolume, linear measurements should be made from a range of cells within the species population, and images should be obtained of cells with different orientations to decide the best combination of geometric shapes or measurements that fit the cells in that population. In cases where cell shape cannot be easily represented by means of a geometric formula, for example *Cymbella* and *Amphora*, best estimates have to be made. It may also be useful to make plastic models to determine unit volume (Bellinger, 1974).

Species population biovolume   Calculation of total species population biovolume (S_b) in a particu-lar environment can be determined from the species count (C) and the mean unit biovolume (B'):

$$S_b = CB' \qquad (2.8)$$

where S_b is the total species volume per unit volume of water ($\mu m^3\ ml^{-1}$), C the number of organisms per unit volume of water (number ml^{-1}) and B' the mean unit biovolume (average volume of an individual organism, μm^3).

Some of the limitations involved in the determination of B' for individual species (variation in size and shape, approximation to three-dimensional model), have already been noted and will clearly influence the accuracy of species population biovolume estimation. The need for accurate species counts (C) is also of paramount importance.

The use of biovolume as an index of species biomass (dry weight, calculation of particulate organic carbon) is problematic in algae that have cellular compartments with a high water content. In this respect, cell volume may be overestimated in larger cells with large central vacuoles (Smayda, 1978) and also in mucilaginous algae, which have a high proportion of extracellular watery matrix. In these cases, alternative biomass estimates can be carried out from cell dimensions on the basis of cellular surface area or 'plasma volume'. The latter is the volume of dense peripheral protoplasm and cell wall material and can be calculated for algae with prominent vacuoles by subtraction of the central vacuolar volume from the total cell volume (Hillebrand *et al.*, 1999).

Total phytoplankton biovolume   For a particular mixed phytoplankton sample, the total phytoplankton biovolume (P_b) can simply be derived as the sum of all the species biovolumes (S_b^1, S_b^2, etc.):

$$P_b = \sum S_b^1, S_b^2, S_b^3 \cdots S_b^x \qquad (2.9)$$

Biovolumes (P_b – expressed as $\mu m^3\ ml^{-1}$) can be converted to fresh weight or dry weight ($\mu g^3\ ml^{-1}$) using appropriate conversion factors if required.

Calculation of total phytoplankton biovolume (P_b) provides an indirect assessment of phytoplankton biomass since it is based on separate determinations

Table 2.4   Typical Unit Biovolumes for Some Common Planktonic Algae.

Class	Species	Colony/Cell/Filament Biovolume (μm^3)	Reference
Bacillariophyta	*Asterionella formosa*	5040*	1
	Aulacoseira granulata var. angustissima	8500*	1
	Cyclotella sp. (large)	1000	1
	Cyclotella sp. (small)	160	1
	Melosira sp.	16,000*	1
	Nitzschia spp.	300	1
	Synedra sp.	600	1
	Stephanodiscus sp.	380	1
	Stephanodiscus rotula	25,000	1
	Tabellaria fenestrata	950	1
	Tabellaria fenestrata var. asterionelloides	7125*	1
Chlorophyta	*Actinastrum* sp.	1050*	1
	Ankyra sp.	40	1
	Chlamydomonas sp.	100	2
	Chlorella sp.	30	3
	Coelastrum sp.	6500	3
	Dictyosphaerium sp.	1500*	1
	Elakatothrix gelatinosa	170	1
	Monoraphidium sp.	45	1
	Micractinium sp.	1440*	3
	Eudorina sp.	5600*	3
	Pediastrum duplex	16,000*	3
	Scenedesmus obliquus	160*	1
	Scenedesmus quadricauda	160*	1
	Sphaerocystis sp.	160*	1
	Staurastrum sp.	3100 (2 semicells)	2
Cryptophyta	*Cryptomonas* sp.	1050	1
	Rhodomonas minuta	140	1
Cyanophyta	*Anabaena* sp.	2165*	2
	Aphanizomenon flos-aquae	1520*	1
	Aphanocapsa sp.	6000*	4
	Gloeocapsa sp.	500	1
	Gomphosphaeria	55,000*	1
	Microcystis sp.	77,120*	2
	Oscillatoria sp.	800	1
	Synechococcus sp.	20	1
Dinophyta	*Ceratium hirundinella*	41,400	2
	Peridinium cinctum	48,000	1

Biovolumes are for single cells unless indicated (*), when they are for a 'typical' colony or filament. Values are from the following sources:
[1] Dean (2004); The *Gomphosphaeria* biovolume was obtained by multiplying cell volume by the number of cells in an average colony.
[2] Reynolds and Bellinger (1992).
[3] Bellinger (1974).
[4] Stephen (1997).

of unit biovolume and species count. The use of phytoplankton biovolume as an index of overall phytoplankton biomass complements the different, more direct bulk analysis techniques noted earlier (see Section 2.3), and avoids some of the pitfalls associated with each of these methods. A further advantage of biovolume calculation is that total changes in P_b can be related to individual algal species and genera, thus providing insight into phytoplankton dynamics within the lake ecosystem.

If biovolume determinations are valid, then changes in P_b (e.g. with seasonal progression) should parallel bulk determinations of phytoplankton biomass. In many situations (as shown in Fig. 2.8) there does appear to be a good correlation, with total biovolumes showing corresponding increases to chlorophyll-a concentration during spring and summer algal blooms, and an inverse relationship to Secchi depth.

2.5.5 Indices of biodiversity

Determinations of species counts (see Section 2.5.3) and biovolumes (see Section 2.5.4) can be used to estimate biodiversity within mixed-species algal populations (Sigee, 2004). Most estimates of algal diversity are based on species counts, which can be used to generate bioindices of three main types – species richness, species evenness/dominance and a combined index of species diversity.

Species richness

This index relates to the total number of species present in the population sample – the greater the number of species the greater the measure of biodiversity. Species richness may be assessed in two main ways:

Total number of species (S)   Species richness is often determined simply in relation to the total number of species. The major problem with this index is that the value S may depend on sample size – the bigger the sample the more species there are likely to be.

Margalef index (d)   This index (Margalef, 1958) avoids the complication of sample size by incorporating the number of individuals (N) in the sample. The index thus provides a measure of the number of species relative to the overall sample size, where

$$d = (S - 1)/\log_e N \qquad (2.10)$$

Species evenness/dominance

The evenness of species occurrence is also an important measure of biodiversity. Considering two hypothetical samples (I and II), each composed of four species (A, B, C and D):

Sample I. Total 100 individuals; species A (25), B (25), C(25) and D (25).

Sample II. Total 100 individuals; species A (94), B (2), C (2) and D (2).

Sample I would be regarded as more diverse than sample II, though species richness is the same. Sample I has a greater evenness of species occurrence, but a lower dominance of any species. Sample II is the converse.

Species evenness may be determined as Pielou's evenness index (J'):

$$J' = H'(\text{observed})/H'_{\text{max}} \qquad (2.11)$$

where H'_{max} is the maximum possible diversity that would occur if all species were equally abundant.

Dominance is the converse of evenness and can be determined using the Simpson index (D), which relates the number of individuals in each species (n_i) to the total number of individuals in the sample (N) :

$$D = \sum (n_i/N^2) \qquad (2.12)$$

Comparison of indices for species evenness and dominance over a number of population samples shows a clear inverse relationship.

Combined index of biodiversity

Clearly, species richness and evenness (or dominance) both contribute towards mixed population diversity, and a most useful approach would be to bring these together as a single value. The Shannon–Wiener diversity index is the most common combined measure of diversity, taking into account richness, evenness and abundance of the community structure and assumes that individuals are randomly sampled from an infinitely large population (Mason, 2002). The Shannon–Wiener index may be expressed as

$$H' = \sum P_i(\log_n P_i) \qquad (2.13)$$

where H' is diversity and $P_i = n_i/N$, with n_i and N as in Equation (2.12).

In practice, plankton diversity is typically linked to productivity, as indicated by total algal biomass. A recent study by Skácelová and Lepš (2014) on standing waters in the Czech Republic, for example, has shown that diversity was normally low in conditions of low and high productivity – where algal growth was respectively restricted by low nutrients and competition for light. Maximum algal diversity was found in situations of intermediate productivity, but even here there were examples of low diversity indices – indicating that low phytoplankton diversity can be caused by a wide range of factors. Further aspects of algal diversity in lakes are discussed in Section 3.2.3.

2.6 Biodiversity within single-species populations

Although classical limnology has concentrated particularly on the species composition of phytoplankton assemblages, new analytical techniques can also provide information on variation within species and the environmental presence of intra-specific subpopulations. Studies on diversity at the subspecies level are complicated by the fact that natural phytoplankton populations are typically a complex mixture of different species – so bulk analysis procedures cannot be used. Two major approaches have the resolution to study intra-specific variation within this complex situation – molecular studies and microscopic analysis.

2.6.1 Molecular analysis

Molecular analysis of environmental algal populations involves techniques such as oligopeptide analysis, gene sequencing and use of molecular probes.

Oligopeptide analysis

Different algal strains within a particular species population can be analysed in relation to the diversity and pattern of their oligopeptides. In the case of the colonial blue-green alga *Microcystis*, for example, Welker *et al.* (2007) have demonstrated the occurrence of distinct oligopeptide groupings or chemotypes.

These *Microcystis* chemotypes reflect variants of particular gene clusters and are major factors influencing the toxin (microcystin) content of blooms. They can be considered as evolutionary units (with interaction and competition within species between phenotypes) and show significant differences between planktonic and sediment populations. Chemotypes undergo seasonal changes which reflect their ecophysiological characteristics. Individual *Microcystis* strains differ considerably in their functional responses to nutrient and light availability, particular growth factors and susceptibility to grazing by herbivores.

Gene sequencing

The recent introduction of comparative sequence analysis of ribosomal RNA genes (phylogenetic marker genes) has indicated high molecular variation within individual species based on morphological criteria (morphotypes). Recent studies by Pfandl *et al.* (2009) on small subunit ribosomal RNA (SSU rRNA) sequences in samples of the chrysophyte *Spumella*, have demonstrated considerable local heterogeneity – with individual populations composed of different ecotypes and genotypes. Similarly, genetic analysis of the chlorophyte *Desmodesmus* by Vanormelingen *et al.* (2009) has also shown heterogeneity, with different clones varying in the number of cells per colony. This may represent a genetic response to local environmental evolutionary pressure (zooplankton grazing), since an increase in the number of cells per colony results in larger colony size and increases resistance to ingestion.

2.6.2 Analytical microscopical techniques

Light and electron microscopes have the resolution to study sub-populations of cells and colonies within single-species populations. These will be considered from two main aspects – detection of algal senescence in algal populations and use of specialised microanalytical techniques.

Detection of algal senescence

Senescence can be defined as the complement of ageing processes that lead to cell death.

In mixed phytoplankton samples, individual species populations invariably contain a proportion of cells that are undergoing senescence and will eventually die. These senescent cells may occur as single cells within unicellular populations or as groups of cells within colonial algae. In either case, they constitute an important sub-population that is in dynamic equilibrium with actively growing cells, the balance between the two determining whether the overall population is in net growth or decline.

Senescence of algal cells can be monitored in various ways, including testing cell viability, detecting programmed cell death (PCD) and recording dead cells within the algal population.

Cell viability   Cells undergoing senescence are non-viable and are unable to exclude dyes such as Evans blue from entering the protoplast. Treatment of a mixed population of viable and non-viable cells with Evans Blue therefore results in some cells (non-viable) becoming stained, while others (viable) are not stained.

The presence of non-viable algal cells is seen in Fig. 2.18 where late-season (November) colonies of *Microcystis* and *Pediastrum* contain cells showing clear senescent changes. This preparation was carried out in the laboratory by incubating a freshly collected trawl-net sample in Evans blue dye for 4 h. Fresh dye solution (2 mg Evans blue in 10 ml phosphate-buffered saline [PBS] solution) needs to be made up prior to each batch of analyses.

Programmed cell death   Death of lake microalgae, as with other organisms, falls into two main categories – necrosis and programmed cell death (PCD). Necrosis occurs as an injury response to harmful external conditions (e.g. high light levels, chemical perturbation, ingestion by zooplankton) and does not depend on cellular activation (Hochman, 1997). In contrast to this, PCD is triggered by gene activation as a rapid physiological response to external factors – resulting in the elimination of 'unwanted' cells (Penell and Lamb, 1997). In the case of blue-green algae, for example, PCD may occur at the end of a protracted growth phase (summer bloom) as a preparation for over-wintering – for which the eliminated cells are not required.

There is no clear test for necrosis, but PCD can be detected as specific cleavage of the nuclear genome, yielding DNA fragments of roughly 200 bp in size (Kressel and Groscurth, 1994). These DNA fragments can be visualised either in bulk samples as a DNA ladder by gel electrophoresis or at the microscopical level by Hoechst staining (condensed regions of nucleoid) or Tdt-mediated dUTP-biotin nick-end labelling (TUNEL) analysis. The latter involves Tdt-mediated dUTP-biotin nick end (TUNEL) labelling of the fragmented DNA with *in situ* detection using fluorescein as part of the TUNEL kit.

Relatively few studies have been carried out on the occurrence of senescence in natural phytoplankton populations. Sigee *et al.* (2007) demonstrated the presence of non-viable cells (Evans blue reaction) in freshly isolated colonies of *Microcystis*, and showed that this was caused by PCD (rather than necrosis) by Hoechst staining and a positive TUNEL reaction.

Dead cells   Dead phytoplankton cells (necrosis or PCD) are often observed under the light microscope as cells with a cell wall but no contents. Dead cells are particularly noticeable at the end of a major bloom of large algae such diatoms and dinoflagellates

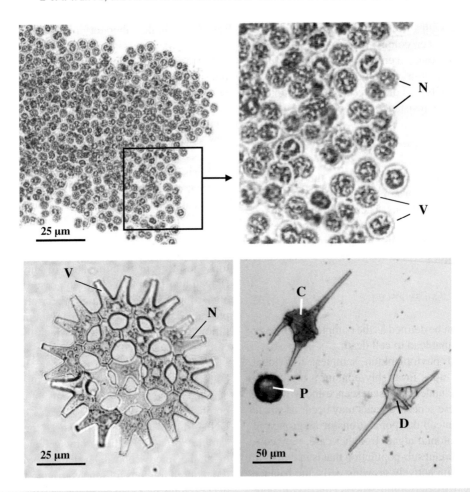

Figure 2.18 Observation of viable and non-viable algal cells in Evans Blue–stained preparations of freshly isolated phytoplankton. Top: Low-power and detailed views of *Microcystis* colony showing a mosaic of stained (non-viable: N) and unstained (viable: V) cells. Bottom left: Colony of *Pediastrum* with distinct non-viable cells, some with the remains of cell contents. Bottom right: Live cells (with dense contents) of *Ceratium* (C) and *Peridinium* (P), plus a dead (completely empty) cell of *Ceratium* (D).

(Fig. 2.18). In other cases, the dead cells may have totally disintegrated, leaving a decomposing mass of organic debris (Fig. 2.7). Another direct indication of dead cells within the lake is obtained by *in situ* fluorimeter traces of chlorophyll-*a* and particulate matter (Fig. 2.10), where there is a clear lag in the particulate trace compared with chlorophyll. In this situation, dead cells have lost their internal biomass (including chlorophyll) but remain in suspension as they sediment out of the surface waters of the water column.

The phase of greatest phytoplankton population decline and cell death in temperate eutrophic lakes occurs towards the end of the summer bloom, when non-viable cells, PCD and dead cells are particularly evident. At this time the lake is still stratified, and senescent changes are triggered primarily by adverse environmental alterations within the epilimnion.

In addition to its ecological importance, the occurrence of senescence and cell death also has practical implications for the determination of biomass. The partial or complete loss of cell contents means that

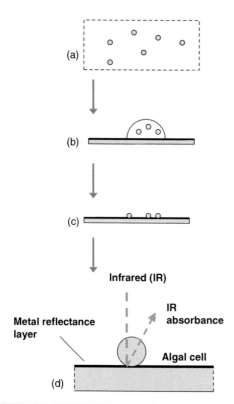

Figure 2.19 Preparation of phytoplankton sample for FTIR analysis. (a) Fresh phytoplankton suspension collected from lake. (b) Suspension droplet deposited on slide. (c) Rapid air-drying under sterile laminar flow. (d) Collection of infra-red absorbance spectrum – view of single cell on reflectance slide (enlarged).

simple use of cell/colony counts and unit biovolume to monitor biomass may be misleading, and determination of particulate counts (Section 2.3.1) may also lead to errors in estimating biomass.

Specialised microanalytical techniques

Combined microscopical and spectroscopic analysis has become a powerful tool in the study of intra-specific diversity and algal sub-populations. This can operate at the level of either the light (e.g. FTIR microspectroscopy) or electron (e.g. X-ray micro-analysis) microscope.

FTIR microspectroscopy Fourier transform infrared (FTIR) microspectroscopy has particular potential for analysing and resolving phytoplankton populations, since the technique quantifies single cells or colonies over a broad range of macromolecular constituents.

Freshly isolated phytoplankton samples are deposited and air-dried onto a reflectance slide (Fig. 2.19), and individual cells or colonies (identified under a light microscope) are then analysed in relation to their infrared absorption spectrum. This can be obtained using either a laboratory infrared system or a high intensity synchrotron source – which has the advantage of higher probe resolution and better signal-to-noise ratios. A typical synchrotron-generated FTIR spectrum from phytoplankton (Fig. 2.20) shows a sequence of absorption bands derived from various excitation states of lipids, proteins, nucleic acid and carbohydrates.

Multiple FTIR analyses of different cells or colonies ($n = 20$–30) from single species within a particular phytoplankton sample provide information on localised micropopulations. FTIR analysis can be used to demonstrate:

- Considerable molecular heterogeneity within individual species. Studies on colonies of *Pediastrum*, for example, showed wide variation in lipids, starch and nucleic acids (Sigee *et al.*, 2002). Concentrations of constituents within individual bands can be measured as either band area or band height, and must be normalised to protein (amide I band) to eliminate variation due to differences in specimen thickness.

- Resolution of species populations. Although FTIR absorption spectra from different species of algae look broadly similar, with the same range of bands shown in Fig. 2.20, mixtures of spectra from different species can be resolved into separate groups using multivariate analysis. This is shown in Fig. 2.20, where spectra from co-dominant algae in a lake (*Ceratium* and *Microcystis*) were subjected to principal component analysis of a 'fingerprint region' within the spectra (1750–900 cm^{-1}). Further analysis of these spectra showed that differences in the principal components related to

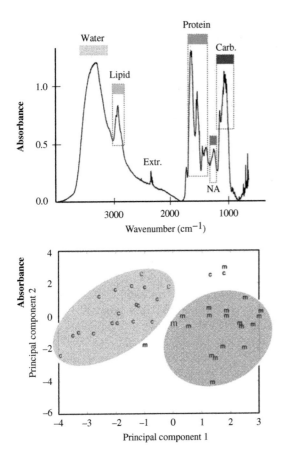

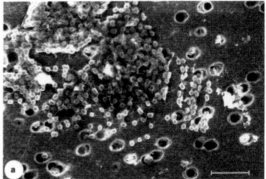

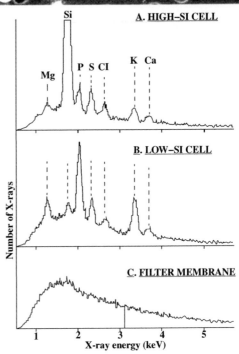

Figure 2.20 FTIR resolution of phytoplankton populations. Top: FTIR spectrum from single colony of *Microcystis*. The absorption bands are derived from different radicals or molecular groups in water, lipid, protein, nucleic acid (NA) and carbohydrate (Carb) within algal biomass. One band (Extr.) is derived from extraneous (atmospheric) CO_2. Bottom: Principal Component Analysis (PCA) loading plot, separating spectra of *Microcystis* (m) and *Ceratium* (c) into two separate clusters. PCA was carried out on all spectra ($n = 40$) for the spectral region 1750–900 cm^{-1}. Adapted from Dean *et al.*, 2007.

Figure 2.21 High-Si and low-Si subpopulations in a colony of *Microcystis*. Top: Scanning electron micrograph of carbon-coated, freeze-dried colony of *Microcystis*. Bottom: X-ray emission spectra from single cells of *Microcystis* with high (A) and low (B) above sample levels of Si. The control spectrum (C) from a clear area of filter membrane next to cell A does not have any elemental peaks. Sigee and Levado, 2000.

differences between the two algae in relation to protein and carbohydrate levels (Dean *et al.*, 2007).

• Resolution of species sub-populations. Multivariate analysis can also be applied to single-species samples and has the potential to demonstrate diversity within the same standing population, at different depths in the water column and within a time sequence during seasonal progression. Studies by Dean *et al.* (2006), for example, on epilimnion and hypolimnion populations of *Aphanizomenon* and *Anabaena* within a water column demonstrated distinctive subpopulations in the former case – with surface algae (depth 0–5 m) having significantly higher levels of carbohydrate compared to those sampled at 10–15 m.

X-ray microanalysis X-ray microanalysis (XRMA) provides information on the elemental composition of microsamples (such as algal preparations) and has considerable potential for studying phytoplankton populations. The technique combines the high spatial resolution of electron microscopy (ability to locate and analyse single algal cells) with the use of a fine-electron probe and an X-ray detector to collect generated X-rays (Sigee *et al.*, 1993). These can then be analysed (multi-channel analyser) to produce an X-ray emission spectrum. Analysis of phytoplankton samples is more readily carried out using a scanning electron microscope (SEM) rather than in transmission mode. SEM preparation is carried out by depositing freshly isolated, washed phytoplankton samples onto a cellulose membrane, freeze-drying the sample, then coating with a fine layer of carbon. Chemical fixation must be avoided (to retain the natural elemental composition) and gold coating (normally used for conventional SEM) should not be carried out.

A typical X-ray emission spectrum (Fig. 2.21) shows clear peaks of major cationic (Mg, K, Ca)

and anionic (P, S, Cl) elements, with other elements – such as Si, also detectable in some preparations. As with FTIR microspectroscopy, XRMA can be used to characterise single-species micropopulations (20–30 cells) within the mixed phytoplankton sample, and to make comparisons both within and between species. Micropopulations can be characterised simply in terms of elemental frequency distributions and mean elemental compositions, or by multivariant analysis – carrying out correlation plots, principal component analyses and hierarchical cluster analysis.

The potential use of XRMA to resolve intraspecific phytoplankton populations is illustrated in Fig. 2.21, where cells within a colony of *Microcystis* have either a high-Si or low-Si content. The distinct sub-populations are demonstrated by a clear bimodal frequency distribution of Si concentration, which occurred in all colonies analysed and at all depths sampled within the water column (Sigee and Levado, 2000). In high-Si cells, Si is associated with Al and appears to be mainly located at the cell surface, as indicated by the elemental correlation pattern and by a greater mean diameter compared to low-Si cells. Both high-Si and low-Si populations include dividing and non-dividing cells, and the biological significance of this bimodality is not known. It is interesting to note that similar data have also been obtained for another blue-green alga (*Anabaena flos-aquae*) in the same water column (Sigee *et al.*, 1999), but analysis of other algae including *Ceratium* and *Staurastrum* did not show Si-bimodality.

B.   NON-PLANKTONIC ALGAE

Non-planktonic algae comprise a diverse range of organisms, differing from phytoplankton in terms of their substrate-association – which may involve attachment (fixed location) or free motility (movement over the substrate surface). The distinction between planktonic and non-planktonic algae is not absolute, and benthic forms may simply occur as a

seasonal phase or be in dynamic equilibrium with planktonic populations (Fig. 2.1).

Although many of the non-planktonic algae occur in the low-light benthic zones of lakes, rivers and estuaries, they can also be found in environments of high light exposure. These include the edge of lakes (littoral zone), shallow streams (Fig. 2.23), exposed surface of sand and mud (estuaries – Fig. 3.8) and the surface waters of lakes and wetlands, attached to macroalgae (Fig. 2.28) and higher plants (Fig. 2.29).

The techniques used for collecting non-planktonic algae depend upon the depth of the water, the nature of the substrate and the type of algal community. Once these algae have been collected, the principles of biomass and species quantitation are essentially the same as those previously described for phytoplankton. The application of these quantitative techniques to periphyton, including cell counts, use of vital stains, determination of biovolumes and assay of dry/ash-free weight is described by Eaton *et al.* (2005). In this section, the techniques involved in algal collection and analysis are considered from two main viewpoints – deep water benthic algae and shallow water communities.

2.7 Deep-water benthic algae

Deep-water benthic algae may be defined as those algae at the bottom of lakes, deep rivers and estuaries that are relatively inaccessible and have to be sampled or monitored by remote procedures. These include the

use of sediment traps, bulk samplers and remote sensors. The use of such techniques is illustrated in relation to studies (Table 2.5) on benthic algae in lakes (benthic-pelagic coupling) and estuaries (sediment stability, invertebrate grazing). The algae being sampled may occur below the zone of light penetration (photic zone) and include attached biofilm organisms as well as unattached algae that have sedimented to the bottom of the water body or been carried there by circulation within the water column.

2.7.1 Benthic–pelagic coupling

In temperate lakes, many algae have distinct benthic and pelagic phases, overwintering on sediments as resting stages before migrating into the water column (Spring/early Summer) under more favourable conditions to form an actively growing planktonic population. This has been studied particularly with bloom-forming colonial blue-green algae, which are able to dominate the surface waters of both rivers (Baker, 1999) and lakes (Verspagen *et al.*, 2005) in summer, but occur as resting benthic populations during the rest of the year. The transition from planktonic

Table 2.5   Some Recent Studies on Deep Water Benthic Algae.

Site	Methods	Reference
Benthic-pelagic coupling		
River benthos	Measurement of planktonic and benthic (akinete) populations of *Anabaena*	Baker (1999)
Eutrophic lake	Measurement of benthic populations of *Microcystis* and recruitment levels into water column	Verspagen *et al.* (2005)
Eutrophic lake	Measurement of recruitment of *Gloeotrichia* into the water column. Laboratory experiments on akinete colonies removed from sediment.	Karlsson (2003)
Eutrophic lake: shallow and deep waters	Measurement of benthic populations of *Anabaena* and *Aphanizomenon* and recruitment levels into water column	Karlsson-Elfgren and Brunberg (2004)
Sediment stability		
Estuarine sediments	Correlation of algal biomass (chlorophyll-*a*) and biofilm matrix (carbohydrate level) with sediment stability	Sutherland *et al.* (1998)
Benthic grazing		
Estuarine sediments	Measurement of algal removal by filter-feeding bivalves. Collection of near-sediment samples and remote fluorimeter monitoring.	Jones *et al.* (2009)

(actively photosynthetic, vegetative cells) to benthic form (resistant spores) begins in the upper part of the water column, and algal spores can be collected in sediment traps (Fig. 2.7) as they begin their descent to the lower part of the water column. These overwintering benthic populations are present both on shallow and deep (>10 m) sediments, the latter well below the photic zone, and can be sampled by direct collection of bulk sediment samples or via recruitment traps.

Bulk samples

Bulk samples of sediment algae can be obtained using either a grab sampler (removing a parcel of sediment – e.g. Ekman or Peterson grabs) or a core sampler (removing a cylinder of sediment). Sediments are often very fine, so samples need to be taken as carefully as possible to make sure that the substratum is not disturbed and that organisms are retained *in situ*.

In their studies on *Microcystis*, Verspagen *et al.* (2005) collected core samples of lake sediment, using a Perspex corer. The depth of sample was adjusted to collect a core of sediment that did not extend beyond the surface algal layer – ranging from 2 cm (shallow parts of the lake) to 8 cm (deep regions). Subsequent laboratory analysis of the sediment involved homogenisation, suspension in mineral liquid medium followed by centrifugation – leaving cells and colonies of *Microcystis* in the supernatant. A purified suspension of *Microcystis* was then obtained via flow cytometry, with cytometric selection based on particle size (0.5–2000 μm) and fluorescence (phycocyanin). The biomass of lake sediment populations were compared with those from surface waters and sedimentation traps in relation to dry weight and chlorophyll-*a*.

Microscopic examination of sediments, using similar procedures to those with phytoplankton, is also important for counting benthic populations. Populations of sediment surface akinetes of *Anabaena* (Karlsson-Elfgren and Brunberg, 2004; Baker, 1999) and *Gloeotrichia* (Karlsson, 2003) have been estimated in this way. In a related laboratory study, akinetes of *Gloeotrichia* were isolated from sediments by Karlsson (2003) using a Pasteur pipette to determine the sediment growth period.

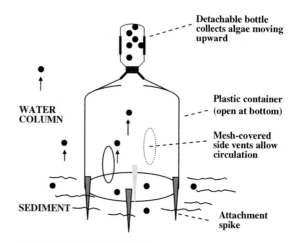

Figure 2.22 Phytoplankton recruitment trap – used to collect benthic algae that are rising from the lake sediment. Figure redrawn and adapted from Karlsson-Elfgren and Brunberg, 2004.

Recruitment traps

The vertical release of benthic algae into the water column (recruitment) has typically been studied using samplers modified from sedimentation traps (turned upside down). The traps used by Karlsson-Elfgren and Brunberg (2004), for example, were large 20 l vessels – open at the bottom and anchored to the sediment by long spikes (Fig. 2.22). Deep traps were also attached to an anchor, to which a surface buoy was connected. To allow lateral exchange of water, but not the studied algal species, two openings were cut into the side of the vessel and covered by 40-μm mesh. On top of each trap, a 500 ml plastic bottle filled with filtered (40 μm mesh) lake water was attached to collect algal filaments moving upward. Sample bottles were collected by divers on a weekly basis, and replaced with new ones containing filtered lake water.

The suspension of algae collected from recruitment traps can be analysed in terms of biomass and species composition as with phytoplankton samples.

2.7.2 Benthic algae and sediment stability

Analysis of benthic algae is an important aspect of investigations on sediment erosion. The studies of

Sutherland *et al.* (1998) on estuarine benthic algae, for example, involved the use of a Van Veen Grab sampler to collect intact sediment samples at various points along a line transect. Individual syringe cores were taken from each grab sample, and algal biofilm biomass analysed at the core surface in terms of chlorophyll concentration (direct measure of algal biomass) and colloidal carbohydrate (biofilm matrix). Significant relationships occurred between erosional thresholds (tested using a Sea Carousel) and chlorophyll-*a*/carbohydrate levels, suggesting that both algal biomass and mucilage biofilm matrix serve as good indicators of sediment stability.

In these estuarine waters, diatoms concentrate in the upper 2 mm of sediment, forming a mucilage biofilm that alters the geophysical and biochemical properties of the sediment, acting as a stabilising influence. With this type of sample it is essential to maintain the structure of the sediment core and to analyse biomass within the top 2 mm of the sediment surface to avoid dilution of the near surface signal by diatom-free underlying sediment.

2.7.3   Invertebrate grazing of benthic algae

In estuaries, removal of suspended algae by bivalve filter-feeders at the bottom of the water column may have a major effect on phytoplankton populations. Losses of estuarine planktonic algae at the sediment interface were estimated by Jones *et al.* (2009) via bulk sampling or by *in situ* monitoring.

Bulk sampling involved the construction of water sampling frames to collect algal samples at discrete heights (0.1, 0.15, 0.2, 0.25 m) close to the sediment bed. Water samples were collected via intake tubes, and algal biomass estimated as fluorometrically determined chlorophyll-*a* concentration. In addition to discrete water sample determinations, a continuous series of chlorophyll-*a* measurements was remotely recorded from a fluorimeter (see Section 2.3.3) placed in fixed position close to the sediment bed. The study demonstrated a much higher removal of algae by benthic grazers than had been anticipated from laboratory filtration studies and shows how dynamic activities of benthic algae can be directly monitored under field conditions.

2.8   Shallow-water communities

In shallow waters such as streams (Fig. 2.23) where algal material can be collected directly by hand, the sampling procedure depends to a large extent on the nature of the substrate and the type of community that develops.

2.8.1   Substrate

This varies considerably (Table 2.6) and algae can be classified as epilithic (living on stones or rocks, epidendric (on woody material), epipelic (loosely associated with bottom deposits), epipsammic (on or amongst sand grains), epiphytic (on plant surfaces) and epizooic (on animal surfaces). Collection of algae at a particular site may involve sampling from all of these substrates (general ecology) or may be selective. King *et al.* (2006), for example, recommended lake diatom sampling from stones and macrophytes (littoral zone) rather than from mud and silt in the assessment of water quality (see Section 3.2.1). In all cases, algae may simply be collected in specimen tubes, and retained in the living (fresh) state or fixed in aldehyde or iodine preservatives (see Section 2.5.1).

Rocks and stones

The structure of the epilithic algal community (typically periphyton and biofilms) on stable substrata such as submerged stones or rocks is often quite complex, so care must be taken in detaching the organisms for microscopic examination if the original community organisation is to be retained. This is shown in Fig. 2.23, where the epilithic algae in a shallow, fast-flowing stream form a complex mat of blue-green alga (*Phormidium*) with clusters of attached diatoms also present.

Epilithic algae may be sampled by collecting the rocks and stones from the river bottom and placing them in a plastic bottle or a polythene bag for subsequent laboratory analysis. In situations where the stones are too large, or for other reasons, the sample should be collected *in situ*. This can be carried out by careful scraping with a sharp knife, which

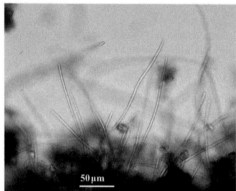

Figure 2.23 River periphyton. Top: general view of fast-flowing stream (left) with detail of rocky substratum (right). Bottom left: Detached periphyton sample obtained by scraping the stone surfaces, showing a filamentous mat of *Phormidium* (blue-green) attached to dense organic material. Bottom right: Stellate cluster of diatom (possibly *Synedra*) cells radiating out from a single point of attachment.

Table 2.6 Shallow-Water Communities: Variations in Algal Collecting Method Depending on Substrate Type.

	Substrate Type	Collecting Method
Epilithic.	Stones and rocks	Remove stone and scrape surface with knife or brush with stiff brush. If stone too large to remove hold a phytoplankton net downstream of the rock and then scrape or brush so that the material removed is swept into the net.
Epidendric	Woody material	As above
Epipelic	Sediment surface	Gently suck up sediment surface using a pipette with a rubber teat or gently scoop up the top layer of sediment into a suitable container.
Epiphytic	Plant surfaces	Scrape algae off the macrophyte surface, place in a tube with water and/or preservative and observe later under a microscope.
Epipsammic	Sand	Collect a small quantity of sand from the surface region. Place in a tube with a small quantity of water and shake vigorously. Allow to settle and pipette off some of the settled material and place on a slide for observation. The selection of sand grains should also be observed in case there are species that have not been detached by shaking.
Epizooic	Hard-shelled animals	Carefully collect the host animal. Use the same technique as for the epilithon.

typically results in the detachment of fragments of intact periphyton community attached to a base of organic debris that covered the stone (Fig. 2.23). Attached algae can also be sampled by taking an open-ended tube (2 or 4 cm in diameter, about 10 cm long), attaching a small ring of sponge rubber to one end of the tube and pressing the tube, sponge end down, firmly onto the surface of the stone or rock being sampled. Insert a stiff brush down the tube and brush the surface of the stone to dislodge any algae living there. Remove the brush and now suck up the dislodged algae through a pipette and transfer to a sample bottle. As long as all of the dislodged material is sucked up then, knowing the area of the stone that has been brushed, a quantitative estimate of the populations present can be made. These epilithic communities are typically stratified, so different degrees of rubbing or scraping can remove different algal layers.

Standard protocol for epilithic algae Where algae are being sampled as bioindicators for environmental monitoring, a standard protocol needs to be implemented to ensure comparability between different collection sites. This is required, for example, where benthic diatoms are being sampled and bioindices calculated to assess river quality (Section 3.4.2). The system used by Kelly *et al.* (1995), adapted from Round (1993), is as follows:

- Five different boulders were chosen at any one river sampling site, taken from five different positions within a 10-m reach. As far as possible, boulders (>256 mm) were selected that were free of filamentous algae. In lowland streams, where this was typically not possible, diatom material was removed from parts of the boulder free of such contaminants.

- Selected boulders were washed briefly in stream water at the site to remove lightly attached stream organisms and then diatoms removed to provide a composite sample.

- Diatoms were removed by scraping the upper surface of the boulder with a stiff toothbrush, then collecting the epilithon, suspended in stream water, in a 250-ml sample bottle.

- In the laboratory, samples were allowed to settle for 24 h. The supernatant was then removed and sedimented algae were initially examined live.

- Algae were then chemically processed (see also Section 2.5.2) to reveal frustule surface ornamentation. Samples were oxidised in a mixture of concentrated sulphuric acid, potassium permanganate and oxalic acid, then mounted in Naphrax©. Diatoms (cleaned frustules) were identified under oil immersion at ×1000 magnification. To ensure a statistically valid count for calculating bioindices, between 200 and 250 frustules were identified for each sample.

The above protocol is based on taking diatoms from a single habitat (boulders) within a reach of the river. This ensures comparability between analyses and is an efficient means of summarising key aspects of the benthic algal community for use with European Water Framework Directive (WFD; European Union, 2000) legislation. The above protocol has led to the development of European standards for the collection and analysis of benthic material from rivers (CEN, 2003, 2004). A similar restricted habitat (stones and macrophytes) protocol has been proposed for shallow-water sampling in lakes (King *et al.*, 2006).

Loose sediments Algae present on sand (epipsammic) and mud (epipelic) surfaces can usually be collected by taking a portion of sediment with a scoop and transferring this to a wide-necked sample bottle.

Algae growing on loose sediments may be disturbed and mixed with the sediment whilst being collected, so it is important to make sure that only the thin surface layer of sediment is scooped up when collecting the sample. A sub-sample of the sediment and algae can either be directly observed under a microscope or algae within the sample can be concentrated. This can be achieved by placing the sample in a Petri dish and allowing it to settle for 12–24 h in the light. During this period, many (but not all) benthic algae are able to migrate through the sediment towards the light, becoming concentrated in the surface layer. These surface algae can be collected by placing a small piece of microscope lens tissue on the mud surface and left for a few hours in the light.

The motile algae will migrate through the tissue to the light, and the tissue can then be carefully removed and placed on a microscope slide for observation. Alternatively, the surface layer can be carefully removed for examination using a fine pipette.

2.8.2 Algal communities

The collection and quantitation of algae from different microenvironments also depends on the type of community that is being sampled. Algae associated with underwater solid surfaces may be either freely motile (moving by various mechanisms) or attached.

The development of attached algal communities follows a general sequence from pioneer colonies to biofilms and periphyton growths (Fig. 2.24). The initial colonisation of freshly exposed surfaces involves the settlement of planktonic algae and other microorganisms onto a fine layer of organic material (conditioning layer) that has adsorbed to the solid surface. The importance of associated organic material continues with the development of a mucilaginous matrix during the formation of some

biofilms, and the accumulation of organic debris as a mature periphyton community develops.

Diatoms play a key role in the establishment and development of attached communities, with a succession of different species involved at different phases of community transition (Table 2.7). Early colonising (phase I) diatoms attach rapidly to exposed surfaces and are able to grow rapidly under conditions of high irradiance. Mid successional (phase II) diatoms are tolerant of a wider range of irradiances, while late successional (phase III) species are able to maintain a high growth rate at very low irradiances (shaded conditions). Many of the phase III diatoms found in mature algal communities are unattached, occurring as tangled cells within the mature filamentous algal growths. Filamentous blue-green algae are a major component of these mature communities, tending to dominate the periphyton community to form dense algal mats under high light conditions.

A wide range of physicochemical factors control attached algal community development – including substrate stability, water flow rate, degree of turbulence, light availability and inorganic nutrient supply. Biological factors (e.g. constituent organisms, extent

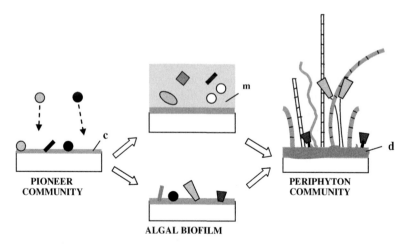

PIONEER COMMUNITY

ALGAL BIOFILM

PERIPHYTON COMMUNITY

Figure 2.24 Succession of attached algal communities in streams and shallow parts of lakes. The initial pioneer community typically develops into a biofilm, which can be a simple monolayer of algae (lower figure) or a mixture of organisms within a mucilaginous matrix (top figure). A dense mat of filamentous growths (periphyton) can develop from either of these. Each community has its own region of associated organic material: conditioning layer (c), mucilage matrix (m) and organic debris (d). Diagrammatic representation (not to scale) with algae shown as solid objects or elongate filaments.

Table 2.7   Diatom Succession in Algal Biofilms.

Species	Growth Form	Growth Dynamics	Light Adaptation
Early colonisers (phase I)	Attached diatoms		High irradiance species High G_{max} and I_s values
Gomphonema angustatum	Short chains of cells	Settlement from high population level in water column	
Meridion circulare	Attached rosettes	Fast immigration due to rapid reproduction after attachment	
Surirella ovata	Single cells or pairs	Fast immigration due to rapid attachment	
Mid-successional species (phase II)	Attached diatoms		Low irradiance species Low G_{max} and I_s values
Cymbella sp. *Gomphonema olivaceum*	Vertical growth via long mucilaginous stalks	High growth rate maintained in dense algal conditions	
Late-successional species (phase III)	Unattached, tangled diatoms		Low irradiance species Moderate G_{max} and very low I_s values
Fragilaria vaucheriae	Unicellular	High growth rate maintained in dense algal conditions	
Aulacoseira varians	Chain forming		
Cocconeis placentula	Prostrate		

The transition from early colonising diatoms to mid- and late successional species is related to growth form, growth dynamics and light adaptation (see text). Light adaptation parameters: G_{max} – estimated maximum growth rate; I_s – estimated light intensity at half G_{max}.

of grazing by invertebrates) are also a key determinant. The sequence shown in Fig. 2.24 gives an indication of general trend (from pioneer community to mature periphyton) and may terminate as a mature community at any point in the succession, depending on conditions. The intermediate biofilm community varies considerably, ranging from a mucilaginous biofilm (typically rich in bacteria, but containing algae) to a non-mucilaginous community composed largely of attached algae – particularly diatoms. In conditions which promote active algal growth, the sequence may thus progress from pioneer to periphyton community, with little development of a mucilaginous biofilm.

Factors affecting community development can be studied under controlled laboratory conditions. Gainswain *et al.* (2006), for example, looked at the influence of substrate particle size – fine material (<2 mm), gravel (2–20 mm) and stones (>20 mm) – on biofilm development and phosphorus release in a laboratory fluvarium. Progression to periphyton (substantial filamentous mass) only occurred with gravel and stone substrate – the fine material was not sufficiently stable to support attached filamentous algae.

2.9   Algal biofilms

Biofilms are monolayer communities of microorganisms occurring at a physical interface, which in freshwater systems includes water/air boundaries (neuston biofilm at the water surface) and substratum/water boundaries – such as the surfaces of fine sediments, rocks and plants. Microalgae are important constituents of biofilms, both in the developing biofilm and in the mature community. This may remain at an early stage of development, such as the diatom biofilm on estuarine sediments (Sutherland *et al.*,

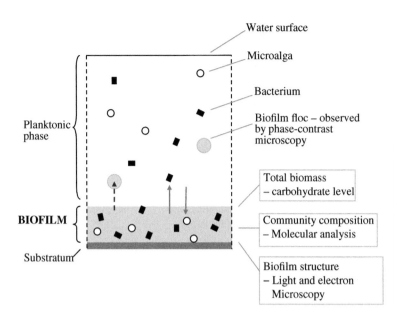

Figure 2.25 Experimental approaches to biofilm species enumeration and biomass determination.
Hypothetical biofilm composed of
bacteria and algae embedded in extracellular polymeric matrix (shaded).
The biofilm is in equilibrium with
a planktonic phase, which includes
bacteria, microalgae and globules of
detached matrix (flocs). Details of
biofilm structure not shown.

1998), or may develop to produce dense growths of
attached filamentous algae (periphyton). Organisms
present in biofilms, such as algae and bacteria, are
also in dynamic equilibrium with surrounding planktonic populations (Fig. 2.1) and detached floccules of
matrix (Fig. 2.25) so any study of biofilm biomass and
community change should also take these planktonic
constituents into account.

Submerged biofilms vary considerably in terms of
the development of carbohydrate matrix (Fig. 2.24),
with mucilaginous communities requiring specialised techniques of sampling, identification and
enumeration.

2.9.1 Mucilaginous biofilms

Quantitative assessment of mucilaginous biofilms
can be considered in relation to biomass, community composition and matrix structure (Fig. 2.25,
Table 2.8).

2.9.2 Biomass

Biofilm biomass includes organic matrix, inorganic
constituents plus the whole range of microorganisms
contained within the matrix – including algae.

Algal biomass

The combined biomass of eukaryotic and prokaryotic algae within biofilms is normally assessed as
chlorophyll-*a* content per unit area or per unit carbon, using laboratory extraction and pigment analysis techniques outlined in Section 2.3.3. This can be
carried out directly on environmental samples or on
experimental biofilms.

Sutherland *et al.* (1998) used chlorophyll-*a* analysis to monitor algal biomass in estuarine diatom
biofilms, showing a significant positive correlation
with biofilm biomass measured as colloidal carbon
(biofilm matrix). Lyon and Ziegler (2009) analysed
algal biomass in epilithic biofilm communities of four
headwater streams (Arkansas, USA). Mean values for
total biofilm chlorophyll content ranged from 2.8 to
19.9 µg mg^{-1} (chlorophyll/carbon), directly relating
to stream trophic status and biofilm nitrogen content
(0.7–2.9% total mass).

With experimental biofilms, Jarvi *et al.* (2002)
studied an algal biofilm using artificial plastic substrates placed on the river bed and within the water
column. Rates of algal biofilm production were determined as chlorophyll-*a* increase cm^{-2} day^{-2}. These
studies demonstrated a river algal biofilm bloom in
early spring, prior to peak suspended chlorophyll-*a*

Table 2.8   Mucilaginous Biofilms: Examples of Matrix and Community Analysis.

Environment/Experiment	Analysis	Reference
(a) Total biomass		
Experimental biofilm	Total carbohydrate (matrix), PLFA (organisms) and chlorophyll-*a* (algae) analysis	Droppo *et al.* (2007)
Estuarine sediments	Colloidal and bulk carbohydrate	Sutherland *et al.* (1998)
Epilithic biofilms in headwater streams	Total algal biomass (chlorophylls-*a*, -*b*,-*c*) and eukaryotic algal biomass (PLFA analysis) expressed per unit biofilm carbon	Lyon and Ziegler (2009)
(b) Cell counts		
River biofilm food web	Biofilm disrupted by sonication or shaking. Bacteria, protozoa and algae counted by light microscopy.	Augspurger *et al.* (2008)
(c) Community molecular analysis		
Experimental; biofilm	PLFA analysis and nucleotide sequencing.	Droppo *et al.* (2007)
River biofilm	Seasonal changes in enzyme activity	Romani *et al.* (2013)

PLFA, phospholipid fatty acid.

concentrations (phytoplankton) that developed in the water column.

In addition to chlorophyll determinations, the biomass of eukaryotic algae can be estimated by phospholipid fatty acid (PLFA) analysis (Fig. 2.26).

Matrix biomass: carbohydrate analysis

Biofilm matrix (often taken as an index of total biofilm biomass) is typically measured as total carbohydrate (units of monosaccharide) per unit area. This can be determined as colloidal and bulk carbohydrate – expressed as µg glucose cm^{-2}.

Sutherland *et al.* (1998) used a modified phenol-sulphuric acid procedure with spectrophotometry to monitor biofilm carbohydrate levels in estuarine sediment samples as follows:

- Half slices of sediment core were placed in test tubes and washed in 10% HCl followed by rinses in purified water.

- The test tubes were shaken thoroughly, left for 1 h (to extract colloidal carbohydrate) and then centrifuged for 10 min at 2000 rpm.

- The supernatant (colloidal carbohydrate fraction) was transferred to a clean set of test tubes.

- 1 ml of 5 ml concentrated sulphuric acid and 5% phenol were added to each sediment pellet (bulk carbohydrate fraction) and each supernatant.

- Samples were shaken, allowed to stand for 1 h and then spectrophotometrically analysed at 485 nm. Readings were expressed in reference to a glucose calibration curve.

Assessment of biofilm biomass can provide useful information on the overall growth of biofilm and the external factors that affect this. Sutherland *et al.* (1998) established that stabilisation of estuarine sediments was directly related to levels of surface biofilm. More recently, Droppo *et al.* (2007) measured biomass as carbohydrate per unit area (µg glucose cm^{-2}) to assess the impact of wave action on biofilms under laboratory experimental conditions. Exponential increase in biomass occurred over a 15-day period under relatively undisturbed conditions (Fig. 2.26), whereas erosion and biomass loss occurred when shear forces were imposed by increased wave action.

2.9.3   Taxonomic composition

Community composition may be defined as the total numbers and balance of individual microbial

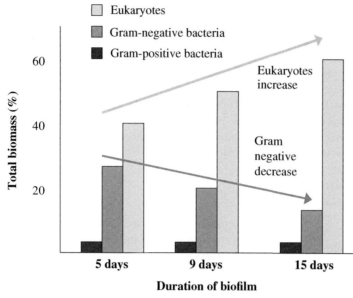

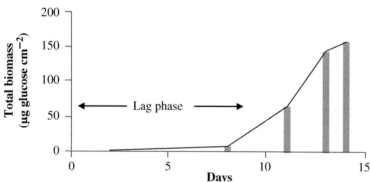

Figure 2.26 Mixed experimental biofilm: Changes in constituent organisms and total biomass. Top graph: Proportions of eukaryotic and prokaryotic organisms after three periods of biofilm growth (PLFA analysis). The 15-day biofilm is dominated by eukaryotes – mainly fungi and algae. Bottom graph: Typical biofilm growth curve. Biomass increase determined from estimation of carbohydrate accumulation. Adapted and redrawn from Droppo et al., 2007.

populations (including a variety of algae) within the biofilm. This can be assessed in two main ways – by direct cell counts and by use of molecular techniques.

Biofilm cell counts

Cell counts can only be made after the biofilm matrix has been broken up and biofilm cells released into suspension. This approach was used by Augspurger et al. (2008) for biofilms grown on tiles. These biofilms were scraped off into filter-sterilised tap water, fixed in formaldehyde or Bouin's fluid to stabilise the cells and then biofilm cells released from the matrix either by sonication (small cells – bacteria and flagellates)

or via shaking by hand (larger cells – ciliates, diatoms, other non-filamentous algae and filamentous algae). Suspensions of the latter were transferred to sedimentation chambers and counted after 24 h based on their morphological characteristics.

Molecular techniques

Information on the community composition of biofilms can be readily obtained using molecular or biochemical techniques such as phospholipid, nucleotide and chlorophyll analyses. These provide information ranging from broad taxonomic categories down to genus level.

Phospholipids PLFA analysis (gas chromatography) can be used to estimate total biofilm biomass as well as biomass levels of different taxonomic groups.

In the biofilm studies of Droppo *et al.* (2007), for example, the pattern of PLFA accumulation (expressed as pmol PLFA mm^{-2} surface area) was similar to that of polysaccharides, showing that the increase in biomass of biofilm organisms was closely similar to the entire biofilm (including matrix).

PLFA analysis can also be used to assess changes in the biomass of major groups of biofilm organisms, including eukaryotes (determination of polyenoics), gram-negative biota (monosaturated fatty acids) and gram-positive bacteria (tertiary and branched fatty acids). PLFA analysis was used to demonstrate a major switch from prokaryote to eukaryote organisms (Fig. 2.26) in the experimental biofilm studies of Droppo *et al.* (2007), paralleling similar changes in the plankton community.

The biomass of eukaryotic algae in epilithic biofilms was monitored by Lyon and Ziegler (2009) by determination of polyunsaturated PLFAs (16:2ω4, 18:2ω6, 18:3ω3, 20:4ω6, 20:5ω3). Other eukaryotes may also contain these fatty acids, so it is important to remove biofilm invertebrates prior to analysis and to check that other eukaryotic microorganisms are absent or present only at low levels.

Nucleotide analysis Sequence analysis of 16s RNA genes, using denaturing gradient gel electrophoresis can lead to the identification of a wide range of microbial species within biofilms. These prokaryote nucleotide sequences provide data on genera of bacteria, blue-green algae and also eukaryote algae via their chloroplast genomes (Table 2.9).

In some cases, the biofilm shows a major temporal shift in dominant organisms. Within the cohort of organisms identified by this method, Droppo *et al.* (2007) demonstrated a clear transition from bacteria (5 day biofilm) to algae, with *Scenedesmus* (green alga) dominating at 9 days and *Phormidium* (blue-green alga) at 15 days.

2.9.4 Matrix structure

The organic matrix of biofilms is secreted by different organisms within the community and is composed largely of carbohydrate, with DNA, proteins and uronic acids also present (Bura *et al.*, 1998). It is colloidal in nature, with the 10–20 nm colloidal particles being aggregated into a fibrillar network.

Various types of microscopy have been used to visualise the structure of the biofilm matrix and the microorganisms (including algae) in it. One major problem with studying the biofilm matrix is that it has a high water content, and the specimen dehydration associated with conventional light and electron microscopy leads to complete loss of integrity. Alternative procedures for examination therefore need to be used:

- Detached globules of biofilm matrix (flocs) floating in the planktonic phase can be sedimented and examined by phase-contrast microscopy.

Table 2.9 Identification of Major Biofilm Genera by 16s RNA Gene Sequence Analysis.

Main Taxonomic Group	DNA Analysed	Genera
Blue-green algae	Main genomic DNA	*Phormidium, Chroococcidiopsis, Anabaena, Synechococcus*
Eukaryote algae	Chloroplast genome	*Scenedesmus, Chlorella*
Bacteria		
α *Proteobacteria*	Main genomic DNA	*Caulobacter*
Cytophaga/Flavobacterium/Bacteroides Group	Main genomic DNA	*Flavobacterium*
Proteobacterium	Main genomic DNA	*Sphingobacterium*

Source: Droppo *et al.*, 2007. Limnology and Oceanography.
Genera identified in experimental (laboratory) biofilms, with a database (GenBank) sequence match of at least 90%.

- Surface structure of biofilms can be viewed using the environmental scanning electron microscope (ESEM). This allows direct visualisation of the biofilm in the natural wet state.

- Matrix variation with depth in the biofilm can be observed using the scanning confocal laser microscope (SCLM). Using a range of lectins to label different polysaccharide components, Droppo *et al.* (2007) used this approach to demonstrate a highly stratified matrix structure, with a change in carbohydrate composition through the matrix as different organisms contributed to the matrix along the temporal succession.

- The extracellular polysaccharide fibrils that constitute the bulk of the biofilm matrix can be viewed by cryoscanning electron microscope (Defarge *et al.*, 1996).

2.10   Periphyton – algal mats

The development of periphyton (extended filamentous growths) represents an end point in attached community development (Fig. 2.24) and is typical of shallow waters in lakes and streams. The substratum can be either inorganic (e.g. stones) or organic (e.g. macrophyte surfaces) – Table 2.10.

2.10.1   Inorganic substratum

The composition of mature biofilm and periphyton populations can be studied in terms of existing natural communities or by inserting fresh surfaces into environmental or laboratory conditions.

Natural communities

Mature natural algal communities on rocks and stones (Fig. 2.23) are often dominated by mats of filamentous blue green algae (such as *Phormidium*), with large attached diatoms that radiate out from a basal attachment point (e.g. *Synedra*) or project out on stalks (e.g. *Gomphonema*). As with mature terrestrial communities, competition for light results in extended growths and dense area coverage. Various studies have been carried out on diatom populations within periphyton and mature biofilms for the determination of bioindices and assessment of water quality (see Section 3.4.5). These require consistent protocols for sampling and analysis at different sites in the study area (see Section 2.8.1).

Experimental conditions

Development and growth of periphyton can be studied experimentally by placing fresh surfaces into the stream or other shallow water environment. Glass slides are particularly useful for this, since they can be exposed for set periods of time, then brought back to the laboratory and directly examined by light microscopy. Etched rather than smooth slides are often used, as some algae appear not to stick to the smooth surface for attachment.

The periphyton sampler shown in Fig. 2.27 (containing glass slides) is designed to be placed midstream, and has floating baffles and a current deflector to reduce the full force of the current. Other materials can also be used for algal colonisation. Hill *et al.* (2008) studied the growth of periphyton by placing ceramic tiles in both indoor (experimental) and outdoor (natural) streams. Periphyton was collected by carefully removing the tiles, then placing them in Petri dishes which were frozen and stored at $-85°C$. Periphyton was later harvested by brushing the thawed tiles, collecting the slurry on preweighed filter membranes and then drying at $60°C$ for 24 h to determine dry mass.

2.10.2   Plant surfaces

Epiphytic algae are common in many shallow river and standing waters attached to plant surfaces – including macroalgae, mosses and higher plants (Table 2.10). As with other attached communities (Fig. 2.1), these algae interact with planktonic populations in terms of recruitment and loss. This section commences with some examples of epiphytic communities, followed by an account of experimental procedures.

Table 2.10   Recent Studies: Biofilm and Periphyton Development on Exposed Surfaces.

Experiment	Methods	Reference
Inorganic substrata		
Stream communities[a]	Collection of epilithic diatoms from periphyton and biofilm communities for water quality assessment	Kelly *et al.* (1995)
	Algae grown on tiles inserted into experimental and natural streams. Biomass assessed as dry weight.	Hill *et al.* (2008)
	Algal biofilms grown on artificial plastic substrates. Algal biomass assessed as chlorophyll-*a*	Jarvi *et al.* (2002)
Laboratory fluvarium	Effect of substrate particle size on periphyton development and phosphorus release	Gainswin *et al.* (2006)
Plant surfaces		
Shallow eutrophic lake	Measurement of biomass and productivity of algae attached to *Myriophyllum*	Jones (1984)
Shallow river	Measurement of composition and abundance of epiphytic algae on river mosses	Sahin and Ozdemir (2008)
Shallow lake	Seasonal variation and vertical zonation of algae on *Phragmites* stems	Muller (1999)
Commercial algal cultivation	Identification of epiphytic filamentous red algae on the cultivated red alga *Kappaphycus*	Vairappan (2006)
Oligotrophic lake – littoral macrophytes	Determination of epiphytic algal biomass and productivity on leaves of *Potamogeton*	Sheldon and Boylen (1975)
Urban waterways	Study of epiphytic algal diversity on four ecological groups of vascular plants	Sosnovskaya *et al.* (2008)
Marine seagrass meadows	Ecological role of epiphytic algae – high productivity and preferential ingestion by invertebrates	Kitting *et al.* (1984)
Laboratory study	Effect of epiphytic algae on the growth and production of Potamogeton	Asaeda *et al.* (2004)

[a]Stream communities. See also references to the collection and evaluation of diatoms from biofilm/ periphyton communities for water quality assessment in Sections 3.4.2–3.4.6.

Epiphytic communities

Attachment to other algae, such as *Cladophora*, is particularly important in parts of lakes and rivers where macrophytes are absent. The example shown in Fig. 2.28 was taken from an extensive growth of *Cladophora* that was attached to a buoy in the middle of a lake. The presence of *Cladophora* at this site presented an opportunity for establishment and extensive growth of attached diatoms, including chains of *Tabellaria* and sessile *Cocconeis*. Although these attached algae will show some detachment to planktonic forms, with subsequent recolonisation of fresh *Cladophora* filaments, they were not detected in routine plankton samples collected from the adjacent epilimnion.

The limited surface area of *Cladophora* did not support dense periphyton communities, but submerged surfaces of higher plants (macrophytes) can. This is seen in Fig. 2.29, where the submerged stems of *Phragmites* have an extensive community of attached blue-green algae (*Phormidium*), stalked (*Gomphonema*) and sessile (*Rhoicosphenia*) diatoms. Unattached diatoms such as *Nitzschia* are also caught up in the tangled filaments of the blue-green mat. The large stand of *Phragmites* occurring around much of the periphery of the lake supports an extensive community of these epiphytic algae and scrapings from

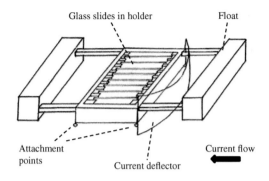

Glass slides in holder Float

Attachment points Current flow

Current deflector

Figure 2.27 Floating periphyton sampler – with upstream deflecting baffle and slide rack holding eight microscope slides. The rack can also be directly anchored to suitable substrate (without floats) and slides can be mounted both vertically and horizontally.

stems invariably showed periphyton development. Differences clearly exist between macrophytes in relation to their attached algae, since scrapings from submerged stems of *Iris pseudacorus* (another major littoral macrophyte) revealed very little in terms of attached algae.

Experimental procedures

Studies on epiphytic algae typically involve collection of host material, removal of algae and then experimental or taxonomic analysis.

Collection of host material Where algae attached to macrophytes are being sampled, the situation is often complicated by the diversity of higher plants in the ecosystem. Studies by Sheldon and Boylen (1975) on the contribution of epiphytic algae to the primary productivity of an oligotrophic freshwater lake, for example, encountered over 40 species of rooted macrophytes. These authors simplified the investigation by carrying out experiments on algal assemblages taken from a single macrophyte species (*Potamogeton amplifolius*), which was

20 μm

200 μm

Figure 2.28 Chained and sessile diatom epiphytes on a macroalga: Low-power view of *Cladophora*, with numerous attached diatom chains. Left inset: Zigzag chain of epiphytic *Tabellaria*, attached to by a mucilaginous plug (arrow). Right inset: Two cells of the sessile diatom *Cocconeis* attached to a dead algal filament. The *Cladophora* was attached to a buoy in the centre of a eutrophic lake. Formaldehyde-fixed preparation.

Figure 2.29 Epiphytic algae on lake macrophytes. Top left: Dense reed beds (*Phragmites*) at edge of a eutrophic lake (Rostherne Mere, UK). Samples of epiphytic algae obtained by scraping the surfaces of submerged stems are shown. Top right: Sessile diatom *Rhoicosphenia* (R) with filament of the blue-green alga *Phormidium* (P). Bottom left: Mixed periphyton community, including the stalked (arrow) diatom *Gomphonema* (G) and *Phormidium* (P). Bottom right: Dense tuft of *Phormidium*, with unattached diatoms, possibly *Navicula* (N).

common throughout the littoral zone of the lake. Leaves of *Potamogeton* (fourth to sixth position on the plant stem) were collected by divers at depths of 3 m and then carefully stored underwater in inverted Erlenmeyer flasks to minimise loss of epiphytes from the leaf surfaces.

Studies by Jones (1984) on biomass and productivity of algae in a shallow eutrophic lake also targeted a single macrophyte species (*Myriophyllum*), while a broad study on epiphytic algal diversity by Sosnovskaya *et al.* (2008) was carried out on four different ecological groups of higher plants.

Removal of epiphytic algae Epiphytic algae that occur attached to other algae (Fig. 2.28) can normally be examined and identified *in situ*. In contrast, the study of algae attached to higher plants requires their removal from the host surface by mechanical or manual procedures.

In the study of Sheldon and Boylen (1975), algal epiphytes were removed from the *Potamogeton* leaves by swirling in lake water, followed by rinsing the leaf surface with a water spray. Microscopic examination showed that >95% of attached algae were removed by this procedure. Simple rinsing of macrophyte leaves with tap water was not effective in epiphyte removal from *Myriophyllum* (Jones, 1984), and a laborious shaking technique was adopted. Other researchers have adopted a manual approach to removing algae. Asaeda *et al.* (2004), for example, scraped epiphytic algae off *Potamogeton* leaves using a soft toothbrush, prior to analysing leaf growth (in the absence of algae) and epiphyte biomass per unit leaf area. Algal removal without damage to leaf tissues was confirmed by microscopic observation. In a study on moss epiphytes, Sahin and Ozdemir (2008) scraped algae from the surface of small pieces of host material with a needle under a binocular microscope.

Experimental and taxonomic studies Once epiphytic algae have been removed from their host plant, they can then be identified and enumerated in a similar way to planktonic organisms. As with phytoplankton, taxonomic studies are best carried out on fresh unfixed samples (to avoid colour change), though diatom identification may require the use of cleaned material.

Algal biomass can be determined as dry weight or, chlorophyll-*a*, normalised to leaf surface area. Chlorophyll-*a* is a better index of algal standing crop since it does not include other microbes and detritus also associated with leaf surfaces (Jones, 1984).

Physiological studies on epiphytic algae should be carried out *in situ* rather than on separated suspensions, to avoid artefacts caused by the separation procedure. The radioactive tracer experiments of Sheldon and Boylen (1975), for example, were carried out on intact *Potamogeton*/epiphyte preparations. This was followed by algal removal and separate analysis of host/algal incorporated carbon (net photosynthetic rates).

The ecological impact of epiphytic algae has been examined particularly in the seagrass meadows of brackish and marine environments. Studies by Kitting *et al.* (1984), for example, have shown that epiphytic algae are the primary basis of the food web in the Gulf of Mexico, with algae having a higher productivity than the seagrass and being preferentially ingested by invertebrate herbivores.

3

Algae as Bioindicators

Biological indicators (bioindicators) may be defined as particular species or communities, which, by their presence, provide information on the surrounding physical and/or chemical environment at a particular site. In addition to the diagnostic presence of indicator species, cell size may also be used to monitor environmental conditions in lakes (Fig. 3.7) and wetlands (Section 3.3).

In this book, bioindicator algae are considered particularly in relation to aquatic chemistry or 'water quality'. The basis of individual species as bioindicators lies in their preference for (or tolerance of) particular habitats, plus their ability to grow and outcompete other algae under particular conditions of water quality. Ecological preferences and bioindicator potential of particular algal phyla are discussed in Chapter 1. This chapter considers water quality monitoring and algal bioindicators from an environmental perspective, dealing initially with general aspects of algae as bioindicators and then specifically with algae in the four main freshwater systems – lakes, wetlands, rivers and estuaries.

3.1 Bioindicators and water quality

Freshwater algae provide two main types of information about water quality:

- **Long-term information, the *status quo*.** In the case of a temperate lake, for example, detection of an intense summer bloom of the colonial blue-green alga *Microcystis* is indicative of pre-existing high-nutrient (eutrophic) status.

- **Short-term information, environmental change.** In a separate lake situation, detection of a change in subsequent years from low to high blue-green dominance (with increased algal biomass) may indicate a change to eutrophic status. This may be an adverse transition (possibly caused by human activity) that requires changes in management practice and lake restoration.

In the context of change, bioindicators can thus serve as early-warning markers that reflect the 'health' status of an aquatic system.

3.1.1 Biomarkers and bioindicators

In the above example, environmental change (to a eutrophic state) is caused by an environmental stress factor – in this case the influx of inorganic nutrients into a previously low-nutrient system. The resulting loss or dominance of particular bioindicator species is preceeded by biochemical and physiological changes in the algal community referred to as 'biomarkers'. These may be defined (Adams, 2005) as short-term indicators of exposure to environmental stress, usually expressed at suborganismal levels – including biomolecular, biochemical and physiological responses. Examples of algal biomarkers include DNA damage (caused by high UV

Freshwater Algae: Identification, Enumeration and Use as Bioindicators, Second Edition. Edward G. Bellinger and David C. Sigee.
© 2015 John Wiley & Sons, Ltd. Published 2015 by John Wiley & Sons, Ltd.

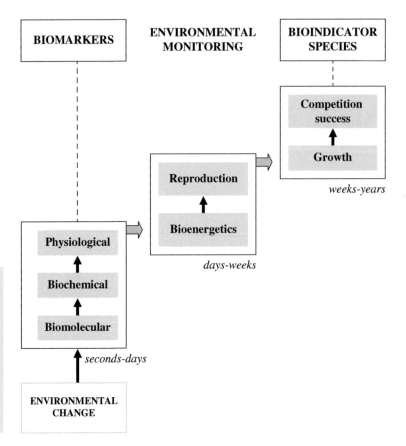

Figure 3.1   Hierarchical responses of algae to environmental change, such as alterations in water quality. The time-response changes relate to sub-organismal (left), individual (middle) and population (right) aspects of the algal community. Environmental monitoring of the algal response can be carried out at the biomarker or bioindicator species level. Adapted from Adams, 2005.

irradiation, exposure to heavy metals), osmotic shock (increased salinity), stimulation of nitrate and nitrite reductase (increased aquatic nitrate concentration) and stimulation of phosphate transporters/reduction in alkaline phosphatase secretion (increased aquatic inorganic phosphate concentration).

The timescale of perturbations in the algal community that results from environmental change (stress) can be expressed as a flow diagram (Fig. 3.1), with monitoring of algal response being carried out either at the biomarker or at the bioindicator species level. Although the rapid response of biomarkers potentially provides an early warning system for monitoring environmental change (e.g. in water quality), the use of bioindicators has a number of advantages (Table 3.1) including high ecological relevance and the ability to analyse environmental samples (chemically fixed) at any time after collection.

3.1.2   Characteristics of bioindicators

The potential for freshwater organisms to reflect changes in environmental conditions was first noted by Kolenati (1848) and Cohn (1853), who observed that biota in polluted waters were different from those in non-polluted situations (quoted in Liebmann, 1962).

Since that time much detailed information has accumulated about the restrictions of different organisms (e.g. benthic macroinvertebrates, planktonic algae, fishes, macrophytes) to particular types of aquatic environment, and their potential to act as environmental monitors or bioindicators. Knowledge of freshwater algae that respond rapidly and predictably to environmental change has been particularly useful, with the identification of particular indicator species or combinations of species being widely used in assessing water quality.

Table 3.1 Main Features of Biomarkers and Bioindicators in the Assessment of Environmental Change.

Major Features	Biomarkers	Bioindicators
Types of response	Subcellular, cellular	Individual community
Primary indicator of	Exposure	Effects
Sensitivity to stressors	High	Low
Relationship to cause	High	Low
Response variability	High	Low
Specificity to stressors	Moderate-high	Low-moderate
Timescale of response	Short	Long
Ecological relevance	Low	High
Analysis requirement	Immediate, on site	Any time after collection (fixed sample)

Source: Adapted from Adams, 2005.

Single species

In general, a good indicator species should have the following characteristics:

- A narrow ecological range

- Rapid response to environmental change

- Well-defined taxonomy

- Reliable identification, using routine laboratory equipment

- Wide geographic distribution

Combinations of species

In almost all ecological situations it is the combination of different indicator species or groups that is used to characterise water quality. Analysis of all or part of the algal community is the basis for multivariate analysis (Section 3.4.3), application of bioindices (Sections 3.2.2 and 3.4.4) and use of phytopigments as diagnostic markers (Section 3.5.2)

3.1.3 Biological monitoring versus chemical measurements

In terms of chemistry, water quality includes inorganic nutrients (particularly phosphates and nitrates),

organic pollutants (e.g. pesticides), inorganic pollutants (e.g. heavy metals), acidity and salinity. In an ideal world, these would be measured routinely in all water bodies being monitored, but constraints of cost and time have led to the widespread application of biological monitoring.

The advantages of biological monitoring over separate physicochemical measurements to assess water quality are that it:

- Reflects overall water quality, integrating the effects of different stress factors over time; physicochemical measurements provide information on one point in time.

- Gives a direct measure of the ecological impact of environmental parameters on the aquatic organisms.

- Provides a rapid, reliable and relatively inexpensive way to record environmental conditions across a number of sites.

Biological monitoring has been particularly useful, for example, in implementing the European Union Directive relating to surface water quality (94C 222/06, 10 August 1994), where Member States were obliged to establish freshwater monitoring networks by the end of 1998.

Table 3.2   Trophic Classification of Temperate Freshwater Lakes, Based on a Fixed Boundary System.

	Trophic Category				
	Ultra-oligotrophic	Oligotrophic	Mesotrophic	Eutrophic	Hypertrophic
Nutrient concentration ($\mu g\ l^{-1}$)					
Total phosphorus (mean annual value)	<4	4–10	10–35	35–100	>100
Ortho-phosphate[a]		<2	2–5	5–100	>100
DIN[a]		<10	10–30	30–100	>100
Chlorophyll-*a* concentration ($\mu g\ l^{-1}$)					
Mean concentration in surface waters	<1	1–2.5	2.5–8	8–25	>25
Maximum concentration in surface waters	<2.5	2.5–8	8–25	25–75	>75
Total volume of planktonic algae[b]	0.12	0.4	0.6–1.5	2.5–5	>5
Secchi depth (m)					
Mean annual value	>12	12–6	6–3	3–1.5	<1.5
Minimum annual value	>6	>3.0	3–1.5	1.5–0.7	<0.7

Lakes are classified according to mean nutrient concentrations and phytoplankton productivity (shaded area). Boundary values are mainly from the OECD classification system (OECD, 1982), with the exception of orthophosphate and dissolved inorganic nitrogen (DIN), which are from Technical Standard Publication (1982).
[a]Orthophosphate and DIN are measured as the mean surface water concentrations during the summer stagnation period.
[b]Total volumes (% water) of planktonic algae are for Norwegian lakes, growth season mean values (Brettum, 1989).

3.1.4   Monitoring water quality: objectives

Environmental monitoring of aquatic systems, particularly in relation to water quality, provides information on:

- *Environmental impacts – changes in hydrology.* Changes in phytoplankton population can be used to monitor major alterations in lake hydrology. Inundation of floodplain lakes during periods of high rainfall, for example, is characterised by diatoms tolerant to water column mixing (e.g. *Cyclotella, Asterionella*) with the presence also of coccoid green algae (e.g. *Schroederia, Kirchneriella*) associated with turbid and mixed conditions (Stević *et al.*, 2013). Regression of flood conditions leads to water column stability, with increased phytoplankton biomass dominated by colonial blue-green algae.

- *Seasonal dynamics.* In temperate lakes these include hydrological measurements (water flows, residence time), thermal and chemical stratification and changes in nutrient availability at the lake

surface. Epilimnion concentrations of nitrates and phosphates can fall to very low levels towards the end of stratification (late summer), and the lake could then be dominated by algae such as dinoflagellates (e.g. *Ceratium*) and colonial blue-greens (e.g. *Microcystis*) which are able to carry out diurnal migrations into the nutrient-rich hypolimnion.

- *Classification of ecosystems in relation to water quality, productivity and constituent organisms.* The most widely used classification (for both lotic and lentic systems) is based on inorganic nutrient concentrations, with division into oligotrophic, mesotrophic and eutrophic systems (Table 3.2). Detection and analysis of indicator algae (Table 3.3) provides a quick indication of trophic status and possible human contamination of freshwater bodies.

- *Dynamics of nutrient and pollutant entry into the aquatic system via point or diffuse loading.* Localised or diffuse entry of contaminants can be studied by analysis of benthic algal communities. The potential use of these littoral algae in lakes,

Table 3.3 Lake Trophic Status: Phytoplankton Succession and Algal Bioindicators.

Lake Type	Spring	Summer	Autumn	Mid-summer Algal Bioindicators	Example
Oligotrophic	DIATOMS Cyclotella	→	DINO Ceratium → BG Gomphosphaeria	**DIA** *Cyclotella comensis Rhizosolenia* spp. **G** *Staurodesmus* spp.	Carinthian Lakes[1] Wastwater[2] Ennerdale
Mesotrophic	DIATOMS Asterionella	CHRYSO Mallomonas CRYPT Cryptomonas	DINO Ceratium BG Gomphosphaeria GREEN Sphaerocystis → DIATOMS Asterionella Gomphosphaeria Sphaerocystis	**DIA** *Tabellaria flocculosa* **CHR** *Dinobryon divergens, Mallomonas caudata* **G** *Sphaerocystis schroeteri, Dictyosphaerium elegans, Cosmarium* spp, *Staurastrum* spp **DINO** *Ceratium hirundinella* **BG** *Gomphosphaeria* spp.	Lunzer Untersee[1] Bodensee[3] Erken[4] Windermere[2] Grasmere
Eutrophic	DIATOMS Asterionella	GREEN Eudorina CRYPT Cryptomonas	BG Anabaena Aphan. → DINO Ceratium → BG Microcystis → DIATOMS Steph.	**DIA** *Aulacoseira* spp., *Stephanodiscus rotula* **G** *Eudorina* spp., *Pandorina morum, Volvox* spp. **BG** *Anabaena* spp., *Aphanizomenon flosaquae, Microcystis aeruginosa*	Prairie Lakes[5] Norfolk Broad[2] Rostherne Mere[2]
Hypertrophic	SMALL DIATOMS Steph.	GREEN Scenedesmus	GREEN Pediastrum → BG Aphanocapsa	**DIA** *Stephanodiscus hantzschii* **G** *Scenedesmus* spp., *Ankistrodesmus* spp., *Pediastrum* spp. **BG** *Aphanocapsa* spp., *Aphanothece* spp., *Synechococcus* spp.	Fertilised waters e.g. Třeboň fishponds[6]

Main phytoplankton groups: BG, blue-green algae; CHRYSO, chrysophytes; DIA, diatoms; CRYPT, cryptomonads; DINO, dinoflagellates; G, green algae; Genera: Steph., *Stephanodiscus*; Aphan., *Aphanizomenon*.

Location of lakes: [1]Austria, [2]United Kingdom, [3]Germany, [4]Sweden, [5]United States, [6]Czech Republic.

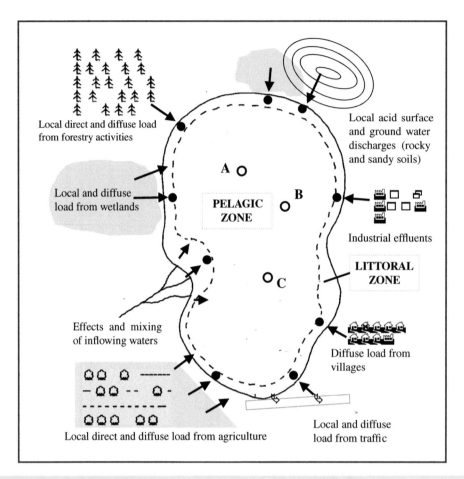

Local direct and diffuse load
from forestry activities

Local and diffuse
load from wetlands

**PELAGIC
ZONE**

A ○

B
○

○ C

**LITTORAL
ZONE**

Local acid surface
and ground water
discharges (rocky
and sandy soils)

Industrial effluents

Effects and mixing
of inflowing waters

Diffuse load from
villages

Local direct and diffuse load from agriculture

Local and diffuse
load from traffic

Figure 3.2 Lake water quality: phytoplankton and periphyton as bioindicators. General lake water quality: phytoplankton in pelagic zone (○ sites A, B, C). Local water quality at edge of lake (• periphyton in littoral zone), with inputs (→) from a range of surrounding terrestrial sources. Figure adapted and redrawn from Eloranta, 2000.

where water quality at the lake edge directly relates to different types of inflow from the catchment area, is shown in Fig. 3.2.

- *Human impacts.* Long-term monitoring of anthropogenic effects within the ecosystem – including changes (Section 3.2.3) such as eutrophication, increase in organic pollutants, acidification and heavy metal contamination. Analysis of indicator diatoms within sediment cores has been particularly useful in monitoring long-term changes in water quality and general ecology (Section 3.2.2).

- *Suitability of water for human use.* This includes compliance of water quality with regulations for human consumption and recreation. Build-up of colonial blue-green algal populations, with increased concentrations of algal toxins, can lead to closure of lakes for production of drinking water and recreation. The use of a reactive monitoring programme for blue-green algal development over the summer months is now an essential part of water management for in many aquatic systems (e.g. Hollingworth Lake, UK – Sigee, 2004).

- *Conservation assessment.* Analysis of freshwater algae has become an important part of the survey and data collection programme used in the evaluation of lakes for their nature conservation value (Duker and Palmer, 2009). Evaluation of water quality is also important for existing conservation sites. In the United Kingdom, a large number of freshwater Sites of Special Scientific Interest (SSSIs) are believed to be affected by eutrophication (Nature Conservancy Council, 1991; Carvalho and Moss, 1995). In case of lakes, the role of algal bioindicators in this assessment is based on both contemporary organisms (Section 3.2.1) and fossil diatoms within sediments (Section 3.2.2).

3.2 Lakes

The use of algae as bioindicators of water quality is influenced by the long-term retention of water in lake systems and also by the age of the lake. Retention of water can be quantified as 'water retention time' (WRT), which is the average time that would be taken to refill a lake basin with water if it were emptied. WRT for most lakes and reservoirs is about 1–10 years, but some of the world's largest lakes have values far in excess of this – including extreme values of 6000 and 1225 years for lakes Tanganyika (Africa) and Malawi (Africa), respectively.

Characteristics of lake hydrology and age result in:

- *Phytoplankton dominance.* In moderate- to high-nutrient deep lakes, phytoplankton populations are able to grow and are retained by the system (not flushed out). Their dominance over non-planktonic algae and macrophytes means they are routinely used as bioindicators.

- *Long-term exposure.* In many lakes, planktonic and benthic algae have relatively long-term exposure to particular conditions of water quality and relate to specific chemical and physical characteristics over extended periods of time (weeks to years).

- *Endemic species.* Some of the world's largest lakes have existed over a long period of time – including lakes Tanganyika (Africa: 2–3 Million years),

Malawi (Africa: 4–9 My) and Baikal (Russia: 25–30 My). Long-term evolution within these independent and enclosed aquatic systems has led to the generation of high proportions of unique species (endemism), with over 50% endemic fauna and flora in lakes Tanganyika and Baikal (Martens, 1997). The presence of substantial levels of endemism in these large water bodies, together with the fact that even non-endemic species may have unique adaptations, means that conventional bioindices will need to be adjusted to suit particular situations.

3.2.1 Contemporary planktonic and attached algae as bioindicators

Both planktonic algae (present in the main water body of the lake – pelagic zone) and attached algae (occurring around the edge of the lake – littoral zone) have been used to monitor water quality (Fig. 3.2).

Phytoplankton: general water quality

Most studies on lakes (see Section 3.2.3) have used the phytoplankton (rather than benthic) community for contemporary environmental assessment, since it is the main algal biomass, is readily sampled at sites across the lake and many planktonic species have defined ecological preferences. Analysis of the phytoplankton community from a number of sites across the lake also provides information on aquatic conditions in general and is the basis of broad categorisation of lakes in relation to water quality, particularly trophic state (Table 3.3).

Attached algae: ecological status and localised water quality

Although there have been relatively few studies using attached (benthic and epiphytic) algae to assess water quality, analysis of non-planktonic (mainly littoral) algae can provide useful information on general ecology and local water quality.

General ecological status As well as planktonic algae, attached algae can also be important in providing a measure of the general ecology of the lake. This is recognised in the European Union Water Framework Directive (WFD; European Union, 2000), which requires Member States to monitor the ecology of water bodies to achieve 'good ecological status'. Macrophytes and attached algae together form one 'biological element' that needs to be assessed under this environmental programme (see also 'Multi-proxy approach' – Section 3.2.2).

Local water quality Various authors have analysed benthic or epiphytic algal populations in relation to water quality, including the extensive periphyton growths that occur in the littoral region of many lakes. These algae are particularly useful in relation to local water conditions (e.g. localised accumulations of metal toxins, point source and diffuse loading at the edge of the lake), since their permanent location at particular sites gives a high degree of spatial resolution within the water body.

Localised metal accumulations. Cattaneo *et al.* (1995) studied periphyton growing epiphytically in macrophyte beds of a fluvial lake in the St. Lawrence River (Canada), to see if they could resolve periphyton communities in relation to water quality (toxic and non-toxic levels of mercury) under differing ecological conditions (e.g. fine vs. coarse sediment). The periphyton, composed of green algae (40%), blue-greens (25%), diatoms (25%) and other phyla, was collected from various sites and analysed in terms of taxonomic composition and size profile. Multivariate (cluster and biotic index) analysis of periphyton communities gave greatest separation in relation to physical ecological (particularly substrate) conditions rather than water quality. The authors recommended that the use of benthic algae as aquatic bioindicators should involve sampling from similar substrate sites to eliminate ecological variation other than water quality.

Point source and diffuse loading at the edge of the lake. Water quality in the littoral zone may differ considerably from that in the main part of the lake. This is partly due to the proximity of the terrestrial ecosystem (with inflow from the surrounding catchment area) and partly due to the distinctive zone of littoral macrophyte vegetation, making an important buffer zone between the shore and open water (Eloranta, 2000). Analysis of littoral algae, either by multivariate analysis or by determination of bioindices, has the potential to provide information on water quality at particular sites along the edge of the lake in relation to point discharges (stream inflows, industrial and sewage discharges) and diffuse loadings. The latter include input from surrounding agricultural land, discharges from domestic areas, traffic pollutants and loading from local ecosystems such as forests and peat bogs (Eloranta, 2000; Fig. 3.2).

Sampling and analysis of littoral algae Although attached algal communities (as with phytoplankton) can theoretically be related to water quality in terms of total biomass, this does not correlate well with nutrient loading (King *et al.*, 2006) – chiefly due to grazing and (in eutrophic waters) competition with phytoplankton. Also, nutrients in the water can be supplemented by nutrients arising from the substratum.

Species counts, on the other hand, can provide a useful measure of water quality. Recent recommendations for littoral sampling (King *et al.*, 2006; see also Section 2.10) concentrate particularly on diatoms – collecting specimens from stones and macrophytes (Fig. 2.29), since these substrata are particularly common at the edge of lakes. Epipelic diatoms (present on mud and silt) are probably less useful as bioindicators since they are particularly liable to respond to substrate 'pore water' chemistry rather than general water quality. The epipelic diatom community of many lowland lakes also tends to be dominated by *Fragilaria* species, which take advantage of favourable light conditions in the shallow waters, but are poor indicators of water quality – having wide tolerance to nutrient concentrations. Having obtained samples and carried out species counts of diatoms from habitats within the defined littoral sampling area, weighted-average indices can be calculated as with river diatoms (Sections 3.4.5 and 3.4.6) and related to water quality.

3.2.2 Fossil algae as bioindicators: lake sediment analysis

Recent water legislation, including the US Clean Water Act (Barbour *et al.*, 2000) and the European

Union Water Framework Directive (WFD; European Union, 2000) have required the need to assess current water status in relation to some baseline state in the past. This baseline state (referred to as 'reference conditions') defines an earlier situation when there was no significant anthropogenic influence on the water body. In the United Kingdom, this reference baseline is generally set at about 1850, prior to the modern era of industrialisation and agricultural intensification. Having defined the reference conditions, contemporary analyses can then be used to make a comparative assessment of human influences on lake biology, hydromorphology and water chemistry. For a particular water body, the absence of long-term contemporary data means that reference conditions have to be assessed on a retrospective basis, including the use of palaeolimnology – the study of the lake sediment record. The use of lake sediments to generate a historical record only gives reliable results under optimal conditions of optimal algal preservation (see below) and if the sediments are undisturbed by wind, bottom-feeding fish and invertebrates.

Lake sediments – algal accumulation and preservation

Continuous sedimentation of phytoplankton from the surface waters (photic zone) of lakes leads to the build-up of sediment, with the accumulation of both planktonic and benthic algal remains at the bottom of the water column. In a highly productive lake such as Rostherne Mere, UK (Fig. 3.5), the wet sedimentation rate in the deepest parts of the water body has been estimated at 20 mm year^{-1} (Prartano and Wolff, 1998), with subsequent compression as further sedimentation and decomposition occur. Decomposition of algal remains leads to the loss of most organic biomass, and algal identification is largely based on the relatively resistant inorganic (siliceous) components of diatoms and chrysophytes (Section 1.9). Optimal preservation of this cell wall material requires anaerobic conditions, and sediment samples are best taken from central deep parts of the lake rather than from shallow regions such as the littoral zone (Livingstone, 1984).

Diatom bioindicators within sediments

Within lake sediments, diatoms have been particularly useful as bioindicators (Section 1.10) of past lake acidification (Battarbee et al., 1999), point sources of eutrophication (Anderson et al., 1990) and total phosphorus concentration (Hall and Smol, 1999). The widespread use of lake sediment diatoms for reconstruction of past water quality is supported by the European Diatom Database Initiative (EDDI). This is a web-based information system designed to enhance the application of diatom analysis to problems of surface water acidification, eutrophication and climate change. The EDDI has been produced by combining and harmonising data from a series of smaller datasets from around Europe and includes a diatom slide archive, electronic images of diatoms, new training sets for environmental reconstruction (see below) and applications software for manipulating data. In addition to the EDDI, other databases are also available – including a large-scale database for benthic diatoms (Gosselain et al., 2005).

Diatoms within sediments are chemically cleaned to reveal frustule structure (Section 2.5.2), identified and species counts expressed as percentage total. Numerous examples of cleaned diatom images from lake sediments are shown in Chapter 4. As with fossil chrysophytes (Section 1.9), subsequent analysis of diatom species counts to provide information on water quality can involve the use of transfer functions, species assemblages and may be part of a multi-proxy approach. The data obtained, coupled with radiometric dating of sediment cores, provide information on times and rates of change and help in setting targets for specific restoration procedures to be carried out.

Transfer functions These are mathematical models that allow contemporary data to be applied to fossil diatom assemblages for the quantitative reconstruction of (otherwise unknown) past water quality. Various studies (Bennion et al., 2004; Taylor et al., 2006; Bennion and Battarbee, 2007) have used this approach, which involves

- Generation of a predictive equation (transfer function) from a large number of lakes, in each case relating the dataset of modern surface-sediment diatom samples to lake water quality data

(Bennion *et al.*, 1996). The 'training set' of lakes should match the lake under study in terms of geographic region and lake morphology and should have a range of water quality characteristics extending beyond the investigation site. In the study by Bennion *et al.* (2004), for example, a north-west European training set of 152 relatively small shallow lakes was used for the smaller productive Scottish lochs, and a training set of 56 relatively large, deep lakes was used for the larger, deeper, less-productive test sites.

- Application of the training set transfer functions to fossil diatom assemblages in sediment cores to derive past water chemistry. Reconstruction of past environmental conditions involves weighted averaging (WA) regression and calibration, with WA partial least squares (WA-PLS) analysis (Bennion *et al.*, 2004).

Species assemblages Changes in water quality from historic times to the present day can be inferred by comparison of species associations at the top (current time) and bottom (reference state) of the core. In the study by Bennion *et al.* (2004) on Scottish freshwater lochs, diatom species counts were expressed as percentage relative abundances, and a detrended correspondence analysis (DCA) carried out of the top and bottom of the cores to assess the direction and magnitude of the floristic (algal) change. DCA results reveal two clear axes of variation (accounting for 62% and 49% of sample variance) in the species data. The DCA biplot (Fig. 3.3) shows clear separation of species into distinctive groups (diatom assemblages) and clear floristic changes for some sites (sites A, B and C in Fig. 3.3) that indicated clear alterations in water quality. In a number of deep lochs (C), limited eutrophication has occurred, with transition from a *Cyclotella/Achnanthes* assemblage to a species combination (*Asterionella/Aulacoseira*) typical of mesotrophic waters. Some shallow lochs (B) also showed nutrient increase, indicated by transition from a non-planktonic (largely benthic) to a plankton-dominated diatom population. In other cases, deep oligotrophic (E) and shallow (D) lochs showed little change in diatom assemblage, indicating minimal alteration in water quality.

Multi-proxy approach In a multi-proxy approach, diatoms are just one of a number of groups of organisms that are counted and analysed within the lake sediments (Bennion and Battarbee, 2007). For European limnologists, the stimulus for a multi-proxy approach has come with the most recent Water Framework Directive (WFD; European Union, 2000). This focuses on ecological integrity rather than simply chemical water quality, for which the use of hydro-chemical transfer functions and diatom species assemblage analysis are not sufficient.

Multi-proxy analysis uses as broad a range of organisms within the food web (e.g. pelagic food web) as possible, commensurate with those biota with remains that persist in the sediment in an identifiable form. In addition to micro-algae (diatoms, chrysophytes), fossil indicators also include macro-algae (Characeae), protozoa (thecamoebae), higher plants (pollen and macro-remains), invertebrates (chironomids, ostracods, cladocerans) and vertebrates (fish scales).

This approach is illustrated by the study of Davidson *et al.* (2005) on Groby Pool, UK, a shallow lake that has undergone nutrient enrichment in the past 200 years. Comparison of 20-year slices from the sediment surface (recent: 1980–2000) and base (reference: 1700–1720) indicate major changes in lake ecology (Fig. 3.4), driven primarily by alterations in water quality (Bennion and Battarbee, 2007). The ecological reference state is one of dominance by benthic diatoms, colonisation by low nutrient-adapted macro-algae and higher plants, with detectable invertebrates restricted to plant-associated Chydoridae. In contrast to this, the current-day ecosystem is much more productive – dominated by planktonic algae, high nutrient-adapted macrophytes and abundant zooplankton populations.

Other examples of a multi-proxy approach to lake sediment analysis include the studies of Cattaneo *et al.* (1998) on heavy metal pollution in Italian Lake d'Orta (Fig. 3.7) and studies on eutrophication in six Irish lakes (Taylor *et al.*, 2006). The latter study involved sediment analysis of cladocera, diatoms and pollen from mesotrophic–hypertrophic lakes, to reconstruct past variations in water quality and catchment conditions over the past 200 years. The results showed that five of the lakes were in a

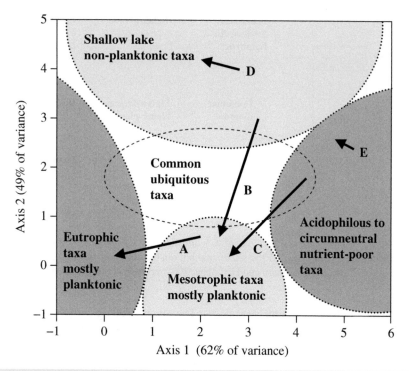

Figure 3.3 Diatom assemblages in sediment cores taken from 24 fresh water Scottish lakes. Diagrammatic view of a detrended correspondence analysis (DCA) biplot showing: (a) Individual sites (arrows). Each arrow denotes the transition from the bottom of the sediment core (reference sample) to the top (current state). Five sites are shown: A. mesotrophic to eutrophic plankton. B. mainly benthic to planktonic diatoms. C. oligotrophic to mesotrophic state. D. shallow lake, little change. E. deep oligotrophic lake, little change. (b) Diatom assemblages, in different regions of the plot. Shallow lake, non-planktonic (*Achnanthes, Cocconeis, Cymbella, Fragilaria, Navicula*); acidophilous, nutrient-poor (*Achnanthes minutissima, Brachysira vitrea, Cyclotella comensis, Cyclotella kuetzingiana, Tabellaria flocculosa*); mesotrophic, planktonic (*Asterionella formosa, Aulacoseira subarctica, Fragilaria crotonensis*); eutrophic planktonic (*Aulacoseira granulata, Cyclostephanos, Stephanodiscus*); common ubiquitous taxa. Figure adapted and redrawn from Bennion *et al.*, 2004.

far more productive state compared to the beginning of the sediment record, with accelerated enrichment since 1980.

3.2.3 Water quality parameters: inorganic and organic nutrients, acidity and heavy metals

A wide range of chemical parameters can be considered in relation to general lake water quality – including total inorganic ion content (conductivity), inorganic nutrients (nitrogen and phosphorus), soluble organic nutrients, acidity, heavy metal contamination and presence of coloured matter

(caused particularly by humic materials). The diversity and inter-relationships of different lakes in relation to these characteristics were emphasised by the study of Rosen (1981), evaluating lake types and related planktonic algae from August sampling of 1250 Swedish standing waters. A summary from this study of algae characteristic of Swedish acidified lakes, oligotrophic forest lakes (varying in phosphorus content and conductivity), humic lakes, mesotrophic and eutrophic lakes is given by Willen (2000).

In this section, the role of algae as bioindicators is considered in reference to four main aspects of lake

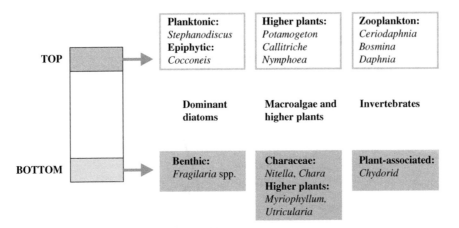

Figure 3.4 Multi-proxy palaeoecological analysis of a sediment core: Groby Pool, United Kingdom. Analysis of 20-year slices from the top (recent ecology, ~1980–2000) and bottom (reference ecology, ~ 1700–1720) samples of the core. Figure adapted and redrawn from Bennion and Battarbee, 2007, original data from Davidson *et al.*, 2005.

water quality – inorganic nutrients, soluble organic nutrients, acidity and heavy metal contamination.

Inorganic nutrient status: oligotrophic to eutrophic lakes

The trophic classification of lakes, based primarily on inorganic nutrient status, has become the major descriptor of different lake types. Its importance reflects:

- the key role that nutrients have on the productivity, diversity and identity of algae and other lake organisms

- a major distinction between deep mountain lakes (typically oligotrophic) and shallow lowland lakes (typically eutrophic)

- the major impact that humans have on changing the ecology of lakes, typically from oligotrophic to eutrophic

- ecological problems that may arise as lakes change from eutrophic to hypertrophic, leading to degenerative changes that can only be reversed by human intervention

The diverse ecological effects that increasing nutrient concentrations have on lake ecology have been widely reported (e.g. Kalff, 2002; Sigee, 2004), including the ecological destabilisation that occurs at very high nutrient concentrations.

Definition of terms Lake classification, from oligotrophic (low nutrient) to mesotrophic and eutrophic (high nutrient), is based on the twin criteria of productivity and inorganic nutrient concentration – particularly nitrates and phosphates. Various schemes have been proposed to define these terms, including one developed by the Organisation for Economic Cooperation and Development (OECD, 1982). This scheme (Table 3.2) uses fixed boundary values for nutrient concentration (mean annual concentration of total phosphorus) and productivity (chlorophyll-*a* concentration, Secchi depth). In this scheme, for example, the mean annual concentration of total phosphorus ranges from 4 to 10 μg l^{-1} for oligotrophic lakes, and 35 to 100 μg l^{-1} for eutrophic waters. In addition to total phosphorus, boundaries for the main soluble inorganic nutrients – orthophosphate and dissolved inorganic nitrogen (nitrate, nitrite, ammonia) – have also been designated (Technical Standard Publication, 1982).

Phytoplankton productivity is based on either chlorophyll-*a* concentration (mean/maximum annual concentration in surface waters) or Secchi depth (mean/maximum annual value). Examples of total volumes of planktonic algae at different trophic levels (Norwegian lakes) are also given in Table 3.2, together with characteristic bioindicator algae.

Algae as bioindicators of inorganic trophic status

Planktonic algae within lake surface (epilimnion) samples can be used to define lake trophic status in terms of their overall productivity (Table 3.2) and species composition (Table 3.3). Species composition can be related to trophic status in four main ways – seasonal succession, biodiversity, bioindicator species and determination of bioindices.

1. **Seasonal succession**. In temperate lakes, the development of algal biomass and the sequence of phytoplankton populations (seasonal succession) directly relate to nutrient availability. In all cases, the season commences with a diatom bloom, but subsequent progression (Reynolds, 1990) can be separated into four main categories (Table 3.3).

- Oligotrophic lakes. In low-nutrient lakes, the spring diatom bloom is prolonged, and diatoms may dominate for the whole growth period. Chrysophytes (*Rhizosolenia*) and desmids (*Staurastrum*) may also be present, and in some lakes, *Ceratium* and *Gomphosphaeria* may be able to grow in the nutrient-depleted waters by migrating down the water column higher nutrient conditions.

- Mesotrophic lakes. These have a shorter diatom bloom (dominated by *Asterionella*), often followed by a chrysophyte phase then mid-summer dinoflagellate, blue-green and green algal blooms.

- Eutrophic lakes. In high-nutrient lakes, the spring diatom bloom is further limited, leading to a clear water phase (dominated by unicellular algae), followed by a mid-summer bloom in which large unicellular (*Ceratium*), colonial filamentous (*Anabaena*) and globular (*Microcystis*) algae predominate.

- Hypertrophic lakes. These include artificially fertilised fish ponds (Pechar *et al.*, 2002) and lakes with sewage discharges, and are dominated throughout the season by small unicellular algae with short life cycles. The algae form a succession of dense populations, out-competing larger colonial organisms which are unable to establish themselves.

2. **Species diversity**. The use of algal counts to characterise species richness (Margalef index), species evenness/dominance (Pielou index, Simpson index) and a combination of richness and dominance (Shannon–Wiener index) is discussed in Section 2.5.5.

Analysis of species richness is particularly useful, combining data on the total number of species identified and total number of individuals in the population (Equation 2.10). During the summer growth phase, species diversity is typically low in oligotrophic lakes, rising progressively in mesotrophic and eutrophic lakes, but falling again in some eutrophic hypertrophic lakes where small numbers of species may out-compete other algae. The effects of increasing nutrient levels on algal diversity (d) are illustrated by Reynolds (1990), with summer-growth values of 3–6 for the nutrient-deficient North Basin of Lake Windermere (UK) – falling to levels of 2–4 in a nutrient- rich lake (Crose Mere, UK) and 0.2–2 for a hypertrophic water body (fertilised enclosure, Blelham Tarn, UK).

3. **Bioindicator species**. Some algal species and taxonomic groups show clear preferences for particular lake conditions and this can act as potential bioindicators. In a broad comparison of oligotrophic versus eutrophic waters, desmids (green algae) tend to occur mainly in low-nutrient waters while colonial blue-green algae are more typical of eutrophic waters. Such generalisations are not absolute, however, since some desmids (e.g. *Cosmarium meneghinii*, *Staurastrum* spp.) are typical of meso- and eutrophic lakes, while colonial blue-green algae such as *Gomphosphaeria* are also found in oligotrophic waters.

Although it is not possible to pin-point individual algal species in relation to particular trophic states, it

Figure 3.5 Eutrophic lake (Rostherne Mere, UK). The high nutrient status of the lake is indicated by water analyses (mean annual total phosphorus >50 μg l⁻¹), high productivity (maximum chl-*a* concentration typically >60 μg l⁻¹) and characteristic bioindicator algae. These include planktonic blooms of *Anabaena*, *Aphanizomenon*, *Microcystis* (colonial blue-greens) plus other eutrophic algae (see text). Attached macroalgae (*Cladophora*, Fig. 2.28) and periphyton communities (present on the fringing reed beds, Fig. 2.29) are also well-developed.

is possible to list organisms that are typical of summer growths in different standing waters (Table 3.3). Identification of such indicator species, particularly at high population levels, gives a good qualitative indication of nutrient state. As an example of this, the high-nutrient lake illustrated in Fig. 3.5 is characteristic of temperate eutrophic water bodies, with high productivity, characteristic seasonal progression (Fig. 2.8) and with the eutrophic bioindicator algae listed in Table 3.3. In addition to phytoplankton bioindicators, the trophic status of the lake is also reflected in extensive growths of attached algae such as *Cladophora* (Fig. 2.28) and in the dense periphyton communities (Fig. 2.29) that occur in the littoral reed beds. Analysis of lake sediments (Capstick, unpublished observations) indicates increased eutrophication in recent historical times, with higher proportions of the diatoms *Asterionella formosa* plus *Aulacoseira granulata* var. *angustissima* and marked decreases in *Cyclotella ocellata* and *Tabellaria flocculosa* (more typical of low-nutrient waters) over the last 50 years.

Although individual algal species can be rated primarily in terms of trophic preferences, they are also frequently adapted to other related ecological factors.

- *Acidity*. Oligotrophic waters are frequently slightly acid with low Ca concentrations, and vice versa for eutrophic conditions.

- *Nutrient balance*. Mesotrophic waters may be nitrogen-limiting (high P/N ratio), promoting the growth of nitrogen-fixing (e.g. *Anabaena*) but not non-fixing (e.g. *Oscillatoria*) colonial blue-green algae.

- *Long-term stability*. In hypertrophic waters, domination by particular algal groups may vary with the long-term stability of the water body. High-nutrient lakes, with established populations of blue-greens and dinoflagellates, often have these as dominant algae during the summer months. Small newly-formed ponds, however, are often dominated by rapidly growing chlorococcales (green algae) and euglenoids. The latter are particularly prominent at high levels of soluble organics (e.g. sewage ponds), using ammonium as a nitrogen source. Some of the most hypertrophic and ecologically unstable waters are represented by artificially fertilised fish ponds, such as those of the Třeboň wetlands, Czech Republic (Pokorny *et al.*, 2002a,b).

In addition to considering individual algal species, taxonomic grouping (assemblages) may also be useful environmental indicators. Reynolds (1980) considered species assemblages in relation to seasonal changes and trophic status, with some groupings (e.g. *Cyclotella comensis/ Rhizosolenia*) typical of oligotrophic waters and others were typical of eutrophic (e.g. *Anabaena/ Aphanizomeno/ Gloeotrichia*) and hypertrophic (*Pediastrum/ Coelastrum/ Oocystis*) states. Consideration of algae as groups rather than individual species leads on to quantitative analysis and determination of trophic indices.

4. **Phytoplankton trophic indices**. In mixed phytoplankton samples, algal counts can be quantitatively

Table 3.4 Lake Trophic Indices.

Index	Calculation	Result	Reference
(a) Major taxonomic groups: numbers of species			
Chlorophycean index	Chlorococcales spp./Desmidiales spp.	<1 = oligotrophy >1 = eutrophy	Thunmark (1945)
Myxophycean index	Cyanophyta spp./Desmidiales	<1 = oligotrophy >1 = eutrophy	Nygaard (1949)
Diatom index	Centrales spp./Pennales		
Euglenophycean index	Euglenophyta/ Cyanophyta + Chlorophyta		
A/C diatom index	Araphid pennate/Centric diatom spp.	<1 = oligotrophy >2 = eutrophy	Stockner (1972)
Compound index	Cyanophyta + Chlorococcales + Centrales + Euglenophyta spp./ Desmidiales spp.	<1 = oligotrophy 1–3 = mesotrophy	Nygaard (1949)
(b) Indicator algae: species counts or biovolumes			
Species counts	Eutrophic spp./Oligotrophic spp.	<8 = oligotrophy	Heinonen (1980)
Species biovolumes	Eutrophic spp./Oligotrophic spp.	<35 = oligotrophy	
(c) Indicator algae: species given weighted scores			
Trophic index	$I_L = \sum(fI_S)/\sum f$	10–100, high values indicate higher pollution	Hörnström (1981)
Trophic index—biovolumes	$I_T = \sum(vI_S)/\sum v$	Index values for each trophic level	Brettum (1989)

Hörnström's index: I_L, trophic index of lake; f, species frequency on a 5° scale; I_s, trophic index of species; **Brettum's index**: I_T, index of trophic level T; v, volume of species per litre; I_s, trophic index of species.

expressed as biotic indices to characterise lake trophic status (Willen, 2000). These indices occur at three levels of complexity (Table 3.4).

- *Indices based on major taxonomic groups.* Early phytoplankton indices (Table 3.4a) used major taxonomic groups that were considered typical of oligotrophic (particularly desmids) or eutrophic (chlorococcales, blue-greens, euglenoids) condition. The proportions of eutrophic/oligotrophic species generated a simple ratio which could be used to designate trophic status. Using the chlorophycean index of Thunmark (1945), for example, counts of chlorococcalean and desmid species can be expressed as a ratio, which indicates trophic status over the range oligotrophy (<1) to eutrophy (>1).

Although such indices provided useful information (see below), they tended to lack environmental resolution since many algal classes turn out to be heterogeneous – containing species typical of oligo- and eutrophic lakes. Problems were also encountered in some of the early studies with sampling procedures, where net collection of algae resulted in loss of small-sized (single cells or small colonies) species. Such algae are often dominating elements in the plankton community, and their loss from the sample meant that the index was not representative.

Example: The A/C diatom index of Stockner (1972). The ratio of araphid pennate/centric diatoms (A/C ratio) provides a good example of the successful use of a broad taxonomic index in a particular lake situation. Studies by Byron and

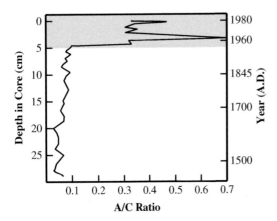

Eloranta (1984) on the sediments of Lake Tahoe (USA) showed a clear change in the diatom community during the late 1950s (Fig. 3.6), consistent with eutrophication. The increase in A/C ratio (<0.7 to 0.7) at this time was related to increases in the indicator species *Fragilaria crotonensis* and *Synedra* spp. (with decreases in *Cyclotella ocellata* and *Melosira* spp.), and occurred at a time of increasing human population around the lake. Inferences of eutrophication from the fossil record (diatom A/C ratio) are confirmed by contemporary lake measurements carried out from 1959 to 1990 (Carney et al., 1994) – showing an increase in primary productivity (annual rise of 5.6%), decrease in Secchi depth and an increase in available nitrogen (as NO_3).

- *Indices based on indicator species.* Improvements in sampling procedure, recording the full spectrum of species sizes, led to quantitative information being obtained on all planktonic algae and the development of indices based on separate indicator species rather than taxonomic groups.

  The most direct indices are based simply on the recorded presence of eutrophic and olig-

otrophic indicator species (Table 3.4b). Heinonen (1980), for example, studied the occurrence of algae in a range of Finnish lakes, listing about 100 species indicating eutrophy and 25 indicating oligotrophy. Species were classified as eutrophic indicators when their occurrence in eutrophic/oligotrophic waters was >2. The corresponding ratio for oligotrophic indicators was 0.7. The trophic state of lakes was calculated as the ratio of eutrophic/oligotrophic species counts or as biovolumes. Index values <8 indicated oligotrophic conditions in the case of the species count, and <35 for the biovolume ratio.

- *More complex indices.* These are based on quantitative rating of individual indicator species. Brettum (1989) developed a system in which 150 algal species were scored on their proportion of total biomass for various classes of environmental variable (pH, total P content, total N content, N/P ratio, etc.). A final trophic index could then be constructed for individual trophic levels from ultraoligotrophy to hypertrophy.

Organic pollution

According to Palmer (1969), organic pollution tends to influence the algal flora more than many other factors in the aquatic environment such as water hardness, trophic status, light intensity, pH, DO (dissolved oxygen), rate of flow, size of water body and other types of pollutants. Organic pollution resembles trophic status (previous section) in relating to nutrient availability, but differs in being soluble organic rather than inorganic nutrients. The terms oligo-, meso- and eutrophic are used specifically for inorganic nutrients and not for soluble organics (Table 3.2).

Palmer (1969) carried out an extensive literature survey to assess the tolerance of algal species to organic pollution, and to incorporate the data into an organic pollution index for rating water quality. Algal genera and species were listed separately in order of their pollution tolerance (Table 3.5), and included a wide range of taxa (euglenoids, blue-greens, green algae and diatoms) as well as planktonic and benthic forms. The assessment of genera was determined as

Table 3.5   Organic Pollution: Most Tolerant Algal Genera and Species.

No.	Taxon	Class	Pollution Index	Freshwater Habitat
	Genus			
1	*Euglena*	Eu	5	Planktonic
2	*Oscillatoria*	Cy	5	Planktonic or benthic
3	*Chlamydomonas*	Ch	4	Planktonic
4	*Scenedesmus*	Ch	4	Planktonic
5	*Chlorella*	Ch	3	Planktonic
6	*Nitzschia*	Ba	3	Benthic or planktonic
7	*Navicula*	Ba	3	Benthic
8	*Stigeoclonium*	Ch	2	Attached
9	*Synedra*	Ba	2	Planktonic and epiphytic species
10	*Ankistrodesmus*	Ch	2	Planktonic
	Species			
1	*Euglena viridis*	Eu	6	Ponds and shallow lakes
2	*Nitzschia palea*	Ba	5	Lakes and rivers
3	*Oscillatoria limosa*	Cy	4	Stagnant or standing waters
4	*Scenedesmus quadricauda*	Ch	4	Lake phytoplankton
5	*Oscillatoria tenuis*	Cy	4	Ponds and shallow pools
6	*Stigeoclonium tenue*	Ch	3	Epiphyte, shallow waters
7	*Synedra ulna*	Ba	3	Lake phytoplankton
8	*Ankistrodesmus falcatus*	Ch	3	Lake phytoplankton
9	*Pandorina morum*	Ch	3	Lake phytoplankton
10	*Oscillatoria chlorina*	Cy	2	Stagnant or standing waters

Source: Palmer, 1969. Reproduced with permission of John Wiley and Sons Ltd.
Ten most tolerant algal genera and species listed in order of decreasing tolerance. Algal phyla –
Cyanophyta (Cy), Chlorophyta (Ch), Euglenophyta (Eu) and Bacillariophyta (Ba). Pollution index –
see text.

the average of all recorded species within the genus, and is perhaps less useful than the species rating - where single, readily identifiable taxa can be directly related to pollution level.

The species organic pollution index developed by Palmer uses the top 20 algae in the species list (top 10 shown in Table 3.5). Algal species are rated on a scale 1 to 5 (intolerant to tolerant) and the index is simply calculated by summing up the scores of all relevant taxa present within the sample. In analysing the water sample, all of the 20 species are recorded, and an alga is considered to be 'present' if there are 50 or more individuals per litre.

Examples of environmental scores are given in Table 3.6, with values of >20 consistent with high organic pollution and <10 signifying lack of nutrient enrichment. Palmer's index was used by Cattaneo *et al.* (1995) in their studies on a fluvial lake of the St. Laurence River system (Canada), with values from 15 to 24 at different sampling sites indicating moderate to high levels of organic pollution. Care should be taken in applying this index, since many sites with high organic pollution (e.g. soluble sewage organics) also have high inorganic nutrients (phosphates, nitrates), and algae characteristic of such sites are typically tolerant to both.

In addition to the general application of Palmer's index using a wide range of algal groups, the index may also be more specifically applied to benthic

Table 3.6   Organic Pollution at Selected Sites: Application of Palmer's (1969) Indices.

Aquatic Site	Recorded Genera	Pollution Rating	High Organic Pollution
Sewage stabilisation pond	*Ankistrodesmus, Chlamydomonas, Chlorella, Cyclotella, Euglena, Micractinium, Nitzschia, Phacus, Scenedesmus*	25	Clear supporting evidence
Greenville Creek, Ohio	*Euglena, Nitzschia, Oscillatoria, Navicula, Synedra*	18	Probable
Grand Lake, Ohio	*Anacystis, Ankistrodesmus, Melosira, Navicula, Scenedesmus, Synedra*	13	No evidence
Lake Salinda, Indiana	*Chlamydomonas, Melosira, Synedra*	7	No organic enrichment

Source: Palmer, 1969. Reproduced with permission of John Wiley and Sons Ltd.
See text for calculation of pollution rating.

diatoms in the assessment of river water quality (Section 3.4.5).

Acidity

Acidity becomes an important aspect of lake water quality in two main situations – naturally occurring oligotrophic waters and in cases of industrial pollution. Algal bioindicators have been important for monitoring lake pH change both in terms of lake sediment analysis (fossil diatoms, Section 3.2.2) and contemporary epilimnion populations – see below.

Oligotrophic waters The tendency for oligotrophic lakes to be slightly acid has already been noted in relation to nutrient status (Section 3.2.3), and for this reason, many algae typical of low-nutrient waters – including various desmids and species of *Dinobryon* (Table 3.3) – are also tolerant of acidic conditions. Acidic, oligotrophic waters tend to be low in species diversity. In a revised classification of British lakes proposed by Duker and Palmer (2009), naturally acid lakes include highly acid bog/heathland pools (group A), acid moorland pools and small lakes (B) and acid/slightly acid upland lakes (group C).

Industrial acidification of lakes Industrial atmospheric pollution during the last century led to

acid deposition and acidification of lakes in various parts of Central and Northern Europe.

Central Europe. In Central Europe, regional emissions of S(SO_4) and soluble inorganic N(NO_3, NH_4) compounds reached up to ~280 mmol m^{-2} year^{-1} between 1940 and 1985, then declined by ~80% and ~35% respectively during the 1990s (Kopacek *et al.*, 2001). This atmospheric deposition led to acid contamination of catchment areas and the resulting acidification of various Central European mountain lakes, including a group of eight glacial lakes in the Bohemian Forest of the Czech Republic (Vrba *et al.*, 2003; Nedbalova *et al.*, 2006).

Studies by Nedbalova *et al.* (2006) on chronically acidified Bohemian Forest lakes have demonstrated some recovery from acid contamination. This is now beginning, about 20 years after the reversal in acid deposition that occurred in 1985, with some lakes still chronically acid – but others less acid and in recovery mode (Table 3.7). Chronically acid lakes have low primary productivity, with low levels of phytoplankton and zooplankton and domination by heterotrophic bacteria. Lakes in recovery mode have a higher plankton standing crop, which is dominated by phytoplankton and zooplankton rather than bacteria.

Phytoplankton species composition is characterised by acid-tolerant oligotrophic unicellular algae. Lakes in recovery mode are still acid and have a

Table 3.7   Bioindicator Algae of Acid Lakes: Comparison of Chronically Acidified and Recovery-Mode Oligotrophic Bohemian Forest Lakes.

	Chronically Acidified Lakes	Recovery-Mode Lakes
Lakes	Cerne jezero, Certova jezero, Rachelsee	Kleiner Arbersee, Prasilske jezero, Grosser Arbersee, Laka
pH and buffering of surface water	pH 4.7–5.1 Carbonate buffering system not operating	pH 5.8–6.2 Carbonate buffering system now operating
Total plankton biomass	~100 µg C l^{-1} Very low, dominated by bacteria	~200 µg C l^{-1} Higher, dominated by phytoplankton and crustacean zooplankton
Phytoplankton biomass (Chl-*a* concentration)	0.6–2.8 µg l^{-1}	4.2–17.9 µg l^{-1}
Dominant algae	No differences in species composition of phytoplankton: -Dinoflagellates: *Peridinium umbonatum, Gymnodinium uberrimum* -Chrysophyte: *Dinobryon* spp.	
Other algae typically present in all lakes	Blue-green: *Limnothrix* sp., *Pseudanabaena* sp. Dinoflagellates: *Katodinium bohemicum* Cryptophytes: *Cryptomonas erosa, Cryptomonas marssonii* Chrysophytes: *Bitrichia ollula, Ochromonas* sp. *Spiniferomonas* sp. *Synura echinulata,* Green algae: *Carteria* sp., *Chlamydomonas* sp.	
Phytoplankton biodiversity (total taxa)	No significant differences in biodiversity: 19–22 taxa in chronically acidified lakes, 15–27 in recovery-mode ones.	

Source: Nedbalova *et al.*, 2006. Reproduced with permission from Springer Verlag.
All data obtained during a September 2003 survey of the lakes.

phytoplankton composition closely similar to chronically acid standing waters. These are dominated by two dinoflagellates (*Peridinium umbonatum, Gymnodinium uberrimum*) and a chrysophyte (*Dinobryon* sp.), which serve as bioindicators. Other algae present in the Czech acid lakes (Table 3.7) included many small unicells (particularly chrysophytes and cryptophytes). Diatoms were not present, presumably due to the chemical instability of the silica frustule under highly acid conditions.

Northern Europe. Acidification of lakes in Southern Sweden follows a similar pattern in terms of algal species, with domination of many acid lakes by the same bioindicator algae seen in Central Europe – *P. umbonatum, G. uberrimum* and *Dinobryon* sp. (Hörnström, 1999). In an earlier study of acid Swedish lakes (typically total phosphorus <5 µg l^{-1}, pH <5), Rosen (1981) also identified *Mougeotia*

spp., *Oocystis submarina* and small naked chryso- and dinoflagellates as typical of such conditions.

In addition to atmospheric pollution, lake acidification has also been caused by industrial effluents – where it is frequently linked with heavy metal pollution.

Heavy metal pollution

Planktonic algae are considerably influenced by heavy metal pollution, which can arise in a variety of ways – including sewage discharge (Seidl *et al.*, 1998), resuspension of toxic sediments (Nayar *et al.*, 2004) and industrial effluent discharge.

Cattaneo *et al.* (1998) studied the response of lake diatoms to heavy metal contamination, analysing sediment cores in a northern Italian lake (Lake d'Orta) subject to industrial pollution:

Environmental changes Lake d'Orta had been polluted with copper, other metals (Zn, Ni, Cr) and acid (down to pH4) for a period of over 50 years – commencing in 1926 and reaching maximum pollutant levels (30–100 μg Cu l^{-1}) between 1950 and 1970. Lake sediment cores collected after 1990 were analysed for fossil remains of diatoms, thecamoebians (protozoa) and cladocerans (zooplankton), all of which showed a marked reduction in mean size during the period of industrial pollution.

Diatom response to pollution The initial impact of pollution, recorded by contemporary analyses, was to dramatically reduce populations of phytoplankton, zooplankton, fish and bacteria. Subsequent sediment core analyses of diatoms showed that heavy metal pollution:

- Caused a marked decrease in the mean size of individuals. The proportion of cells with a biovolume of <10^2 μm^3 increased from under 10% of the total population in 1920 to over 60% in 1950.

- The decrease in mean diatom size was caused primarily by a change in taxonomic composition from assemblages dominated by *Cyclotella comensis* (10^2–10^3 μm^3) and *Cyclotella bodanica* (10^3–10^4 μm^3) to populations dominated by *Achnanthes minutissima* (<10^2 μm^3).

- The change in mean size was also caused by a shift in the size within a single taxon – *A. minutissima*. The mean length of this diatom decreased significantly from about 14 μm before pollution to a minimum value of 9 μm during 1950–1970 (Fig. 3.7).

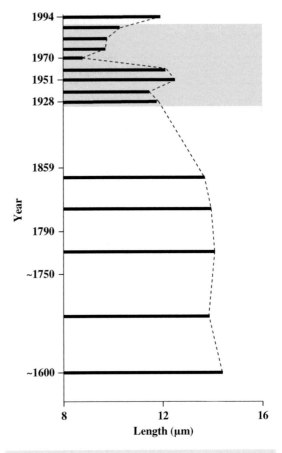

Figure 3.7 Changes in the average valve length of the diatom *Achnanthes minutissima* measured along a sediment core collected in Lake d'Orta (Italy). The size of this heavy metal-tolerant diatom showed a marked decrease during the period of industrial contamination (shaded area), falling to 9 μm by the end (1970) of the most acute period of toxicity. Figure adapted and redrawn from Cattaneo *et al.*, 1998.

Achnanthes minutissima **as a bioindicator** Dominance of Lake d'Orta under conditions of heavy metal pollution by *A. minutissima* is in agreement with the known ability of this diatom to withstand strong metal stress. Other studies (Takamura *et al.*, 1990) have shown it to be dominant in streams subject to heavy metal contamination. The diatom is cosmopolitan, however, often common in benthic assemblages of neutral waters, and its domination of a particular environment is not therefore directly indicative of heavy metal pollution. In spite of this, the presence of dominant populations (coupled with a decrease in mean cell size) is consistent with severe environmental stress – and would corroborate other environmental data indicating heavy metal or acid contamination.

3.3 Wetlands

Wetlands are habitats that contain shallow water and can be divided into two main groups – marshes and peatlands.

3.3.1 Marshes

These typically contain areas of open water that is static or flowing and may be fresh, brackish or saline (Boon and Pringle, 2009). Active decomposition means there is no excessive accumulation of plant material. Marsh wetlands form an ecological continuum with shallow lakes (Sigee, 2004), having standing water present for the entire annual cycle (permanent wetlands) or just part of the annual cycle (seasonal wetlands). Because the water column of

wetlands is normally only 1–2 m in depth, it is not stratified and the photic zone (light penetration) extends to the sediments – promoting growth of benthic and other attached algae.

Wetlands are typically dominated by free floating and rooted macrophytes, which are the major source of carbon fixation. Although growth of algae may be limited due to light interception by macrophyte leaves, leaf and stem surfaces frequently provide a substratum for epiphytic algae – and extensive growths of periphyton may occur. Wetlands tend to be very fragile environments, liable to disturbance by flooding, desiccation (human drainage), eutrophication (agriculture and waste disposal) and increased salinity (coastal wetlands). Algal bioindicators of water quality are particularly important in relation to eutrophication and changes in salinity – such as those occurring in Florida (USA) coastal wetlands.

Case study 3.1   Salinity changes in Florida Wetlands

In recent decades, wetlands in Florida (USA) have been under particular threat due to extensive drainage, with many of the interior marshlands lost to agricultural and urban development. This has resulted in a shrinkage of wetland areas to the coastline periphery. In addition to their reduced area, coastal marshes in SE Florida have also suffered a rapid rise in saltwater encroachment due partly to freshwater drainage, but also to rising sea levels resulting from global warming.

Studies by Gaiser et al. (2005) have been carried out on an area of remnant coastal wetland to quantify algal communities in three major wetland ecosystems – open freshwater marshland, forested freshwater marshland and mangrove saltwater swamps. The study looked particularly at periphyton (present as an epiphyte and on soil sediments) and the use of diatom bioindicator species to monitor changes in salinity within the wetland system.

Effects of salinity   The major microbial community throughout the wetland area occurred as a cohesive periphyton mat, composed of filaments of blue-green algae containing coccoid blue-greens and diatoms. Periphyton biomass, determined as ash-free dry weight, was particularly high (317 g m^{-2}) in open freshwater marshes, falling to values of 5–20 g m^{-2} in mangrove saltwater swamps. Salinity had an over-riding effect on algal community composition throughout the wetlands. The filamentous blue-greens Scytonema and Schizothrix were most abundant in freshwater, while Lyngbya and Microcoleus dominated saline areas. The most diverse algal component within the periphyton mats were the diatoms, with individual species typically confined to either freshwater or saline regions. Dominant species within the separate ecosystems are listed in Table 3.8, with freshwater diatoms predictably having lower salinity optima (2.06–4.20 ppt) compared to saltwater species (11.79–18.38). Salinity tolerance range is also important, and it is interesting to note that dominant diatoms in freshwater swamps had higher salinity optima and tolerance ranges compared to other freshwater diatoms – suggesting that the ability to tolerate limited saltwater contamination may be important. The converse is true for the saltwater swamps, where dominant species were not those with the highest salinity optima.

Table 3.8   Freshwater and Saltwater Wetland Diatoms (Florida, USA): Salinity Optima and Tolerance.

Species	Salinity Optimum[a]	Salinity Tolerance[a]
Diatoms with lowest salinity optima		
Achnanthidium minutissimum	1.80	0.20
Nitzschia nana	1.83	0.72
Navicula subrostella	1.84	0.21
A. Open freshwater marshland		
Nitzschia palea var. *debilis*	2.06	1.24
Encyonema evergladianum	2.25	1.51
Brachysira neoexilis	2.69	1.77
B. Forested freshwater marshland		
Nitzschia semirobusta	2.19	0.86
Fragilaria synegrotesca	3.41	2.68
Mastogloia smithii	4.20	2.75
C. Mangrove Saltwater swamp		
Amphora sp.	11.79	1.57
Achnanthes nitidiformis	17.28	0.51
Tryblionella granulata	18.38	0.92
Diatoms with highest salinity optima		
Caloneis sp.	20.52	1.06
Tryblionella debilis	20.86	1.33
Mastogloia elegans	20.99	1.03

Source: Gaiser *et al.*, 2005. Reproduced with permission of Taylor and Francis.
    Diatoms are listed in relation to
    • Dominant species in three major wetland ecosystems (A, B, C)
    • Three species with lowest salinity optima (Ecosystem A), shaded area – top of table
    • Three species with highest salinity optima (Ecosystem C), shaded area – bottom of table
    [a]Salinity values given as ppt. Optimum and tolerance levels for individual species are derived from environmental analyses (see text).

Predictive model   The salinity optima and tolerance data shown in Table 3.8 were derived from environmental samples. Analysis of diatom species composition and salinity (along with other water parameters) was carried out over a wide range of sites. Salinity optima for individual species were determined from those sites where the species had greatest abundance, and salinity tolerance was recorded as the range of salinities over which the species occurred.

The species-related data were incorporated into a computer model that could predict ambient salinity (in an unknown environment) from diatom community analysis. Environmental salinity at a particular site was determined as the mean of the salinity optima of all species present, weighted for their abundances. Predicted values for salinity based on such diatom calibration models are highly accurate (error <10%). Measurement of salinity using a portable meter has a high degree of accuracy, but values often show considerable variation along a time sequence of readings. One advantage of the use of bioindicator species in this context is that such variability is reduced because the effects of salinity on diatom communities are integrated over a period of time.

3.3.2 Peatlands

Peatlands differ from marshes in having an imbalance between decomposition and accumulation of organic matter, resulting in the long-term deposition of peat. They are of two main types:

Ombrotrophic bogs Fed by precipitation (rain, snow, mist). Typically highly acidic (pH 3.5–4.6), poor in minerals (conductivity <80 iS cm^{-1}) with low Ca concentrations (<2 mg l^{-1}). Dominated by *Sphagnum* moss.

Minerotrophic fens Fed by groundwater. Typically moderately acidic (pH 4.6–7.5), mineral-rich (conductivity >80 iS cm^{-1}) with high Ca concentrations (2–50 mg l^{-1}). Dominated by moss and sedges.

In both cases, the major micro-algae present are desmids (e.g. *Closterium, Cosmarium, Staurastrum*) and diatoms (*Eunotia, Gomphonema, Pinnularia*). Studies by Neustupa *et al.* (2013) on European peatlands have shown that algal species composition and richness are primarily controlled by pH levels. The mean cell size of desmids relates to pH and mineral concentration, with low biovolumes in acidic ombrotrophic bogs, large biovolumes in minerotrophic fens. Cell sizes of desmids can thus be used to monitor environmental processes such as the transition from minerotrophy to ombrotrophy, and acidification. In contrast to desmids, mean cell sizes of diatoms did not relate to the minerotrophic–ombrotrophic gradient.

3.4 Rivers

Until recently, monitoring of river water quality in many countries (Kwandrans *et al.*, 1998) was based mainly on *Escherichia coli* titre (sewage contamination) and chlorophyll-*a* concentration (trophic status). Chlorophyll-*a* classification of the French National Basin Network (RNB), for example, distinguished five water quality levels (Prygiel and Coste, 1996) – normal (chl-*a* concentration ≤ 10 μg ml^{-1}), moderate pollution (10 to ≤60), distinct pollution (60 to ≤120), severe pollution (120 to ≤300) and catastrophic pollution (>300).

The use of microalgae as bioindicators was pioneered by Patrick (Patrick *et al.*, 1954) and has concentrated particularly on benthic organisms, since the rapid transit of phytoplankton with water flow means that these algae have little time to adapt to environmental changes at any point in the river system. In contrast, benthic algae (periphyton and biofilms) are permanently located at particular sites, integrating physical and chemical characteristics over time, and are ideal for monitoring environmental quality. Examples of benthic algae present on rocks and stones within a fast-flowing stream are shown in Fig. 2.23. The use of the periphyton community for biomonitoring normally involves either the entire community or specific taxonomic groups – particularly diatoms and N$_2$-fixing (heterocystous) blue-green algae.

3.4.1 The periphyton community

Analysis of the entire periphyton community clearly gives a broader taxonomic assessment of the benthic algal population compared to diatoms only, but the predominance of filamentous algae makes quantitative analysis difficult. Various authors have used periphyton analysis to characterise water quality, including a study of fluvial lakes by Cattaneo *et al.* (1995; Section 3.2.1). This showed that epiphytic communities could be monitored both in terms of size structure and taxonomic composition, leading to statistical resolution of physical (substrate) and water quality (mercury toxicity) parameters.

3.4.2 River diatoms

Contemporary biomonitoring of river water quality has tended to concentrate on just one periphyton constituent – the diatoms. These have various advantages as bioindicators – including predictable tolerances to environmental variables, widespread occurrence within lotic systems, ease of counting and a species diversity that permits a detailed evaluation of environmental parameters. The major drawbacks to diatoms in this respect are the requirement for complex specimen preparation and the need for expert identification.

Attached diatoms can be found on a variety of substrates including sand, gravel, stones, rock, wood and aquatic macrophytes (Table 2.6). The composition of the communities that develop is in response to water flow, natural water chemistry, eutrophication, toxic pollution and grazing.

Various authors have proposed precise protocols for the collection, specimen preparation and numerical analysis of benthic diatoms to ensure uniformity of water quality assessment (see below). More general aspects of periphyton sampling are discussed in Section 2.10.

Sample collection

The sampling procedure proposed by Round (1993) involves collection of diatom samples from a reach of a river where there is a continuous flow of water over stones. About five small stones (up to 10 cm in diameter) are taken from the river bed, avoiding those covered with green algal or moss growths. The diatom flora can be removed from the stones either in the field or back in the laboratory. As an alternative to natural communities, artificial substrates can be used to collect diatoms at selected sample sites. These overcome the heterogeneity of natural substrata and consequently standardise comparisons between collection sites, but presuppose that the full spectrum of algal species will grow on artificial media. Dela-Cruz et al. (2006) used this approach to sample diatoms in south-eastern Australian rivers, suspending glass slides in a sampling frame 0.5 m below the water surface (see Fig. 2.27). Slides were exposed over a 4-week period to allow adequate recruitment and colonisation of periphytic diatoms before identifying and enumerating the assemblages.

Specimen preparation

The diatoms are then cleaned by acid digestion, and an aliquot of cleaned sample is then mounted on a microscope slide in a suitable high refractive index mounting medium such as Naphrax©. Canada balsam should not be used as it does not have a high enough refractive index to allow resolution of the markings on the diatom frustule.

Cleaning diatoms and mounting them in Naphrax© means that identification and counts are made from permanent prepared slides, rather than from volume or sedimentation chambers (Section 2.5.3). It is normal to use a ×10 eyepiece and a ×100 oil immersion objective lens on the microscope for this purpose. Before counting, it is always desirable to scan the slide at low magnification to determine which are the dominant species.

Numerical analysis

Diatoms on the slide are identified to either genus or species level and a total of 100–500+ counted, depending on the requirements of the analysis being used. Once the counting has been completed and the results recorded in a standardised format, the data may be processed.

In recent times, there has been developed a dual approach to analysis of periphytic diatoms in relation to water quality:

- evaluation of the entire diatom community (Section 3.4.3), often involving multivariate analysis

- determination of numerical indices based on key bioindicator species (Section 3.4.4)

Individual studies have either used these approaches in combination (e.g. Dela-Cruz et al., 2006) or separately.

3.4.3 Evaluation of the diatom community

The term 'diatom community' refers to all the diatom species present within an environmental sample. Species counts can either be expressed directly (number of organisms per unit area of substratum) or as a proportion of the total count. Evaluation of the diatom community in relation to water quality may either involve analysis based on main species or a more complex statistical approach using multivariate techniques.

Main species

Various authors have considered different levels of water quality in terms of distinctive diatom assemblages. Round (1993) proposed the following classification, based on results from a range of British rivers. In this assessment, five major zones (categories) of increasing pollution (inorganic and organic soluble nutrients) were described.

Zone 1: Clean water in the uppermost reaches of a river (low pH) Here the dominant species were the small *Eunotia exigua* and *Achnanthes microcephala,* both of which attached strongly to stone surfaces.

Zone 2: Richer in nutrients and a little higher pH (around 5.6–7.1) This zone was dominated by *Hannaea arcus, Fragilaria capucina* var. *lanceolata* and *A. minutissima. T. flocculosa* and *Peronia fibula* were also common in some instances.

Zone 3: Nutrient rich with a higher pH (6.5–7.3) Dominant diatoms included *A. minutissima* with *Cymbella minuta* in the middle reaches and *Cocconeis placetula, Reimeria sinuate* and *Amphora pediculus* in the lower reaches.

Zone 4: Eutrophic but flora restricted through other inputs Fewer sites in this category were found and more work was considered necessary to precisely typify its flora. However, the major diatom associated with the decline in water quality was *Gomphonema parvulum* together with the absence of species in the *Amphora, Cocconeis, Reimeria* group.

Zone 5: Severely polluted sites, where the diatom flora is very restricted As with zone 4 not many sites in this category have been investigated and more work is needed. The flora was frequently dominated by small species of *Navicula* (e.g. *N. atomus* and *N. pelliculosa*). Detailed identification of these small species can be difficult, especially if only

a light microscope is available. Round (1993) concluded that identification to species level for these was unnecessary and merely to record their presence in large numbers is enough. If the pollution is not extreme there may be associated species present such as *Go. parvulum, Amphora veneta* and *Navicula accomoda.* Other pollution-tolerant species include *Navicula goeppertiana* and *Gomphonema augur.*

Round's diatom assessment for British Rivers is compared with those of other analysts in Table 3.9 for categories of moderate and severe pollution. Although there is substantial similarity in terms of species composition, differences do occur – to some extent reflecting differences in river sizes and geographical location. These differences highlight the problems involved in attempting to establish standard species listings for water quality evaluation.

Multivariate analysis

Multivariate analysis can be used to compare diatom assemblages and to make an objective assessment of species groupings in relation to environmental parameters. Soininen *et al.* (2004) analysed benthic diatom communities in approximately 200 Finnish stream sites to determine the major environmental correlates in boreal running waters. Multivariate statistical analysis was used to define

- diatom assemblage types

- key indicator species that differentiated between the various assemblage groups

- relative contribution of local environmental factors and broader geographical parameters in determining diatom community structure.

The results showed that the ~200 sites sampled could be resolved into major diatom groupings which corresponded to distinctive ecoregions (Table 3.10). These authors concluded that although analysis of diatom communities provides a useful indication of water quality, hydrology and regional factors must

Table 3.9   Comparison of the Diatom Floras Monitored by Lange-Bertalot, Watanabe and Round for Severely Polluted and Moderately Polluted River Sites.

Lange-Bertalot	Watanabe	Round
Severe nutrient pollution: most tolerant diatom species		
Group 1	Saprophilic[a]	Zone 5
Amphora veneta	*Navicula goeppertiana*	*Amphora veneta*
Gomphonema parvulum	Achnanthes minutissima	Gomphonema augur
Navicula accomoda	var *saprophila*	*Navicula accomoda*
Navicula atomus	*Navicula minima*	Small *Navicula*
Navicula goeppertiana	Navicula mutica	Small *Nitzschia*
Navicula saprophila	*Navicula seminula*	
Navicula seminulum	*Nitzschia palea*	
Nitzschia communis		
Nitzschia palea		
Synedra ulna		
Navicula minima		
Navicula frugalis		
Navicula permitis		
Navicula twymaniana		
Navicla umbonata		
Navicula thermalis		
Moderate nutrient pollution: Less-tolerant diatom species		
Group IIa	Eurysaprobic[a]	Zone 4
Achnanthes lanceolata	*Achnanthes lanceolata*	Amphora pediculus
Cymbella ventricosa	Cocconeis placentula	Reimeria sinuata
Diatoma elongatum	Gomphonema parvulum	= Cymbella sinuata
Fragilaria vaucheriae	Navicula atomus	Gomphonema sp.
Fragilaria parvulum	Nitzschia frustulum	
Navicula avenacea		
Navicula gregaria		
Navicula halophila		
Nitzschia amphibian		
Surirella ovalis		
Synedra pulchella		

Source: Adapted from Round, 1993.

[a]Saprophilic (preference for high soluble organic nutrient levels), Eurysaprobic (tolerant of a range of soluble organic nutrient levels). Species present in two or more of the lists area shown in bold.

also be taken into account. The impact of local hydrology was shown by the presence of planktonic diatoms in benthic samples where rivers are connecting lakes (Group D) or have numerous ponds and lakes within the watercourses (Group J).

3.4.4   Human impacts and diatom indices

Diatom indices have been widely used in assessing water quality and in monitoring human impacts on freshwater systems. The latter can be considered

Table 3.10 Benthic Diatom Community Analysis of Finnish Rivers Showing Some of the Major Community Groups.

Group	Ecoregion	Stream water quality	Characteristic taxa
A	Eastern Finland	Polyhumic, acid	*Eunotia rhomboidea* *Eunotia exigua*
B	Eastern Finland, woodland	Oligotrophic, neutral	*Fragilaria construens,* *Gomphonema exilissimum*
C	South Finland, small forest streams	Slightly acid, low conductivity, humic	*Gomphonema gracile* *Achnanthes linearis*
D	South Finland Connect lakes		*Aulacoseira italica[a]* *Rhizosolenia longiseta[a]*
E	Mid-boreal	Mesotrophic, neutral, humic	*Achnanthes bioretii* *Aulacoseira subarctica*
F	North boreal	Oligotrophic, clear water, neutral	*Caloneis tenuis* *Gomphonema clavatum*
H	Arctic-alpine		*Achnanthes kryophila* *Eunotia arcus*
J	South Finland, many small lakes and ponds	Polluted: Eutrophic, high organic content	*Aulacoseira ambigua[a]* *Cyclotella meneghiniana[a]*
K	South Finland	Polluted: Eutrophic – treated sewage, diffuse agricultural loading	Motile biraphid species – e.g. *Surirella brebissonii* *Nitzschia pusilla*

Source: Soininen *et al.*, 2004. Reproduced with permission from John Wiley and Sons Ltd.
[a]Planktonic diatoms. Groups A–K selected from 13 ecoregions.

either in terms of change from an original state or in relation to specific human effects.

Change from a 'natural' community

The use of benthic diatom populations to assess anthropogenic impacts on water quality implies comparison of current conditions to a natural original community, with deviation from this due to human activities such as eutrophication, toxic pollution and changes in hydrology. This concept is implicit in some major research programmes, such as the European Council Water Framework Directive (WFD: European Union, 2000) where the baseline is referred to as 'reference conditions'. In the case of lakes (Section 3.2.2), reference conditions can be determined by on-site extrapolation into the past using sediment analysis. Diatom sediment analysis is not generally appropriate for rivers, however, since there is little sedimentation of phytoplankton, the sediment is liable to disturbance by water flow and the oxygenated conditions minimise preservation of biological material.

An alternative strategy, in the quest for a baseline state, is to locate an equivalent ecological site that has a natural original community – unaffected by human activity. This is not straightforward, however, since local variations in environment and species within a particular ecoregion make it difficult to define what is actually meant by a natural original community. This is evident in the studies of Eloranta and Soininen (2002) on Finnish rivers, for example, where diatom species composition of undisturbed benthic communities varied with river substrate, turbidity and local hydrology (Table 3.11).

In an objective assessment of human impact on natural communities, Tison *et al.* (2005) analysed 836 diatom samples from sites throughout the French hydrosystem using an unsupervised neural network, the self-organising map. In total, 11 different communities were identified, 5 corresponding to

Table 3.11　Diatom Adaptations to Local Environmental Conditions in Finnish Rivers.

Environment	Typical diatoms
Substrate	
Hard substrates	Attached epilithic and epiphytic taxa
Soft substrates	*Pinnularia*, *Navicula* and *Nitzschia*
Turbidity	
Clear waters	*Achnanthes*
Clay-turbid waters	*Surirella ovalis*, *Melosira varians* and *Navicula* spp.
Hydrology	
Lake and pond inflows	*Aulacoseira* spp., *Cyclotella* spp. and *Diatoma tenuis*

Source: Eloranta & Soininen, 2002. Reproduced with permission from John Wiley and Sons Ltd.

natural (undisturbed) conditions and relating to 5 different ecoregions (i.e. river types with similar altitude range and geological characteristics). The six other communities were typical of rivers that had been disturbed by human activity. Although natural variability occurred within individual ecoregions (similar to that noted in Finnish rivers, see above), this was greatly exceeded by the effects of pollution at the different sampling sites. The study aimed to identify diatom indicators for different types of anthropogenic disturbance by comparing benthic diatom communities in natural and disturbed sites.

Use of bioindicators

An alternative approach to analysing human impact in terms of general disturbance is to use diatom indices with indicator species that directly reflect particular environmental changes resulting from human activities. Individual species can be numerically weighted to reflect their degree of environmental specificity. Bioindicators may relate to single parameters such as total-P (Trophic diatom index – TDI – e.g. Hofman, 1996) or reflect more general aspects of water quality, combining organic loading and inorganic nutrient concentration (IPS: index of pollution sensitivity, GDI: generic diatom index). Diatom indicator lists for other variables such as salinity, trophy, nitrogen metabolism types, pH and oxygen requirements have also been published (van Dam *et al.*, 1994).

3.4.5　Calculation of diatom indices

As with lake phytoplankton analyses (Section 3.2.3), the benefit of river diatom indices is that they reduce complex multi-species communities to single numerical values. These provide a very simple quantitative evaluation that can be appreciated and used by biologists and non-biologists alike.

In some cases, river diatom indices have been determined that take into account all recorded species. Stenger-Kovács *et al.* (2014), for example, analysed the biodiversity of benthic diatoms in Hungarian rivers in terms of Shannon index and species evenness. Clear correlations occurred between species number, biodiversity and stream category (based on water chemistry and hydromorphology).

In other cases, analysis has been based on particular bioindicator algae, using either single or multiple taxon assessments.

Single taxon assessment

Indices based on a single species represent a very simple environmental approach.

Kothe's index　This is referred to as the 'species deficiency index' (*F*) and relates the number of cells (N_1) of a particular key species at site 1 to the number (N_x) at site *x*, where:

$$F = \frac{N_1 - N_x}{N_1} \times 100 \qquad (3.1)$$

Table 3.12   Diatom Index and River Water Quality.[a]

Diatom Index (I_d)	River Water Quality
>4.5	No pollution, the best biological quality.
4.0–4.5	Very slight pollution, near normal communities.
3.0–4.0	Toxic pollution at moderate level or nutrient enrichment (eutrophication). Community changes apparent and sensitive species disappearing.
2.0–3.0	Heavy pollution. Only pollution-resistant species abundant or dominant. Sensitive species severely reduced.
1.0–2.0	Severe pollution. Only a few tolerant species survive. Very reduced diversity.

Source: Descy, 1979.
[a]Water quality considered in relation to toxic pollution (e.g. heavy metal content) or nutrient enrichment (eutrophication).

This very simple index is assessing environmental impact on a single species. It assumes that site 1 is the typical one for that river and that changes in the numbers of the selected species will occur at other sites (e.g. a decline with downstream pollution).

Multiple taxon indices

Most diatom indices are based on multiple taxa (genera or species). They are determined either in terms of the presence/absence of key indicator species (e.g. Palmer's index) or are based on the weighted average equation of Zelinka and Marvan (1961):

$$\text{Index} = \frac{\sum_{j=1}^{n} a_j s_j v_j}{\sum_{j=1}^{n} a_j v_j} \qquad (3.2)$$

where a_j is the relative abundance (proportion) of species j in the sample, s_j the pollution sensitivity of species j, v_j the indicator value of species j and n the number of species counted in the sample.

Different indices adapt this equation in different ways, and the performance of a particular index depends on which taxa are used, the number of taxa and the values given for the constants s and v for each taxon. The values of individual indices based on this equation vary from 1 to a maximum value equal to the highest value of s. Commonly used indices include the following.

Palmer's index (Palmer, 1969)   This, applied to river diatoms, is derived from an extensive literature survey and is based on the occurrence of 20 common diatom species, listing them according to their tolerance to organic pollution. The sequence, from least polluted to most polluted waters, includes

Fragilaria capucina → *Achnanthes minutissima* → *Cocconeis placentula* → *Diatoma vulgare* → *Surirella ovata* → *Gomphonema parvulum* → *Synedra ulna* → *Nitzschia palea.*

As with the more general version of Palmer's index (Section 3.2.3, Table 3.5), diatom species are rated on a scale 1–5 (intolerant to tolerant) and the index is calculated by summing up the scores of all relevant taxa present within the sample. Values >20 indicate high levels of organic pollution.

Descy's index (Descy, 1979)   A frequently used index. Values of s_j range from 1 to 5 and v_j from 1 to 3 (Equation 3.2) giving index values from 1 to 5. These can then be related to water quality (Table 3.12; Descy, 1979).

Example. The example of Descy index (I_d) calculation shown in Table 3.13 is taken from the River Semois in Belgium (Round, 1993). The part of the river investigated was heavily polluted with domestic sewage and contained the key diatom species *N. accomoda, Nitzschia palea, Achnanthes lanceolata,*

Table 3.13   Calculation of the Diatom Index (I_d): River Semois.

Diatom	a_j	v_j	$a_j \times v_j$	s_j	$a_j \times v_j \times s_j$	I_d
Achnanthes lanceolata	0.9%	1	0.9	3	2.7	
Gomphonema parvulum	0.5%	1	0.5	2	1.0	
Navicula accomoda	78.7%	3	235.1	1	235.1	
Nitzschia palea	19.7%	2	39.4	1	39.4	
Others	0.2%					
Total			275.9		278.2	278.2/275.9 = **1.0**

Source: Round, 1993.
See text for symbols and I_d calculation (Equation 3.2).

and *Gomphonema parvalum*. The I_d value obtained (1.0) signals heavy pollution, in agreement with the observed sewage contamination.

Some workers have reported that this index tends to give high values and thus underestimates heavy pollution (Leclercq and Manquet, 1987).

CEE index (Descy and Coste, 1991)   This index of general pollution is based on a total of 223 diatom taxa. The index ranges from 0 to 10 (polluted to non-polluted water).

GDI index (generic diatom index) (Coste and Ayphassorho, 1991)   An index of organic and inorganic nutrient pollution, based on 44 diatom genera. Values range from 1 to 5 (polluted to non-polluted water).

IDAP index (Prygiel *et al.*, 1996)   Index based on a combination of 45 genera and 91 diatom species. Values range from 1 to 5 (polluted to non-polluted water).

IPS index (specific pollution index) (Coste, in CEMAGREF, 1982)   This index for pollution sensitivity is based on organic load and nutrient concentrations. Values range from 1 to 5 (polluted to non-polluted water).

TDI (trophic diatom index) (Kelly and Whitton, 1995)   In its original formulation, this index was based on a suite of 86 diatom taxa selected for their indicator value (tolerance to inorganic nutrients)

and ease of identification. The index was determined as the weighted mean sensitivity (WMS: Equation 3.2), with pollution sensitivity values (s_j) from 1 to 5, and indicator values (v_j) from 1 to 3. The value of TDI ranged from 1 (very low nutrient concentrations) to 5 (very high nutrients).

This index has now been modified (Kelly, 2002) to increase the range from 0 to 100, where

$$TDI = (WMS \times 25) - 25 \qquad (3.3)$$

ISL index (index of saprobic load) (Sladecek, 1986)   This index of soluble organic pollution (saprobity) is based on 323 diatom taxa, each with a designated saprobic value. Values range from 4.5 to 0 (polluted to non-polluted water).

3.4.6   Practical applications of diatom indices

The diversity of available and currently used diatom indices presents a complex and confusing picture as to which ones should be used to ensure comparability of studies and consistency of approach. A further potential source of confusion is that some indices (e.g. TDI, see above) have been subsequently modified to give a wider range of values in relation to water quality. The range of available diatom indices raises a number of key practical issues, including

- which index? Ease of use

- comparability between diatom indices

- comparability between diatoms and other bioindicator organisms

- differences between indices in response to changes in water quality

- adoption of a standard approach to use of diatoms as bioindicator organisms

- quality assurance: reliability of analyst assessment

Which index? Ease of use

One practical problem with the use of diatoms as bioindicators is the large number of taxa encountered in environmental samples. This can be overcome in two main ways:

- Use indices based on genera (e.g. GDI) rather than species (e.g. SPI, TDI). Various researchers (e.g. Case study 3.2) have found no significant differences between the two.

- Restrict identification to the most abundant species. Round (1993), for example, used about 20 key species for monitoring rivers in the United Kingdom.

Comparability between indices

Recent projects using diatoms as bioindicators have tended to use groups of indices, to eliminate any irregular conclusions that might have arisen from use of a single index. Different indices have been compared for assessment of water quality within different river systems, where there is variation in trophic status (inorganic nutrients – particularly phosphates and nitrates) and a range of other factors such as salinity, organic pollution (related to biological oxygen demand – BOD) and industrial contamination (metal pollution and acidity).

A summary of studies is shown in Table 3.14, with some detailed examples below. In general, different diatom indices give broadly comparable results. Various investigations (e.g. 3, 4 in Case study 3.2) do indicate, however, that the IPS index is particularly useful for monitoring general changes in water quality. This index best reflects the combined effects of eutrophication, organic pollution and elevated salt concentrations, since it usually integrates all diatom species recorded within the samples.

Case study 3.2   Field studies on different diatom indices (see Table 3.14)

1. *Greek rivers: variation in trophic status, organic and inorganic chemical pollutants.* Iliopoulou-Georgudaki *et al.* (2003) used IPS, Descy (1979) and CEE indices as they were considered more representative of environmental conditions (Table 3.14). The indices gave exactly comparable results.

2. *Selected rivers in England and Scotland.* Ranging from nutrient-poor upland streams to lowland rivers subject to varied eutrophication and contamination with organic pollutants, pumped mine water, heavy metals and agricultural run-off.
    Kelly *et al.* (1995a) assessed water quality using four indices based on diatom genera (GDI) and species (SPI, TDI-P and TDI-NP). The high correlation between indices across the different river sites suggested that any could be used individually for routine monitoring and that diatom recognition to genus rather than species level was adequate.

3. *Metal pollution in lowland river.* De Jonge *et al.* (2008) assessed diatom populations in relation to metal (ZN) contamination and related physicochemical variables. The IPS index best reflected changes in water quality – (pH, conductivity, oxygen concentration, inorganic nutrients) and was the only diatom index that indicated a significant difference between control and contaminated sites.

4. *Rivers of Southern Poland: variation in trophic status and organic pollution.* Kwandrans *et al.* (1998) used the suite of eight diatom indices contained in the 'OMNIDIA' database software to evaluate water quality in this river system. Except for Sladek's index, indices typically showed significant correlations with each other and

also with parameters of water quality – organic load (BOD), oxygen concentration, conductivity, measured ion concentrations and trophic level (NH_4-N, PO_4-P). Two particular indices – IPS and GDI – gave the best environmental resolution in terms of correlation with water quality variables (see Table 3.6) and showing clear differences between the separate river groups.

Although the diatom indicator system emerged as a useful tool to evaluate general water quality, some indices were not able to differentiate between adverse effects of eutrophication (inorganic nutrients) and organic material pollution. Abundance of key indicator species, however, such as *A. minutissima* (highly sensitive towards organic pollution) and *Amphora pediculus* (eutrophic species, sensitive to organic pollution) can be useful in evaluating the type of pollution involved.

Comparison to other bioindicator organisms

Although routine monitoring of all but the deepest rivers has been based, for many years, on macroinvertebrates (e.g. De Pauw and Hawkes, 1993), a range of other bioindicator organisms is available. These include diatoms, other algae, fishes, aquatic and riparian macrophytes. Various authors have made simultaneous comparisons of some or all of these to assess the relative usefulness of particular groups of organisms – with particular emphasis on diatoms and macroinvertebrates. In the examples below (continued from the previous section), diatom communities showed comparable changes to macroinvertebrates in relation to water quality, but (with the exception of the IPS index) were generally less sensitive.

Case study 3.3 Comparisons of diatom indices with other bioindices

1. *Use of different bioindicators for assessing water quality in Greek Rivers.* Iliopoulou-Georgudaki et al. (2003). This study was carried out on river sites with water quality ranging from very poor to very good and used bioindices based on diatoms (see previously) plus four other groups of organisms. Sampling of fish and macrophyte vegetation gave only limited environmental assessment, with macrophytes in particular being incapable of a graded response to varying degrees and kinds of stress. Macroinvertebrates and diatoms recorded the full range of pollution conditions, but indices based on macroinvertebrates were considered to be the most sensitive – responding more readily than diatoms to transient changes to environmental state.

2. *Selected rivers in England and Scotland.* Kelly et al. (1995a) found that their diatom indices were significantly correlated with two benthic macroinvertebrate indices – BMWP (Biological Monitoring Working Party scores) and ASPT (Average Score Per Taxon), both used routinely for water quality monitoring in the United Kingdom.

3. *Metal pollution in a lowland river.* Comparison of macroinvertebrates and diatoms as bioindicators by De Jonge et al. (2008) showed that diatom communities most closely reflected changes in metal concentration, with distinct taxa being associated with low (e.g. *T. flocculosa*), moderate (*Ni. palea*) and high (*Eolimna minima*) zinc levels. In contrast to diatoms, macroinvertebrate communities most closely followed physical–chemical variables and the effects of metal pollution. The combined use of both groups for biomonitoring was considered to be more appropriate than separate use of either.

Table 3.14   Comparability of Diatom Indices in the Assessment of River Water Quality.

Sites	Descy (1979)	CEE	GDI	IDAP	IPS	L&M	SHE	TDI	SLAD	Comments	References
Rivers Alfeios and Pineios in Greece	+	+			+					Indices gave exactly similar quality assessments (bad to very good)	Iliopoulou-Georgudaki et al. (2003)
Rivers in Southern Poland		+	+	+	+	+	+	+	+	Most diatom indices (except SLAD) correlated with each other and water quality	Kwandrans et al. (1998)
English and Scottish rivers			+		+			+		Good correlation between indices	Kelly et al. (1995)
Finnish rivers			+		+			+		Good correlation between indices: rivers ranging from poor to high quality	Eloranta and Soininen (2002)
Metal-contaminated Belgian rivers		+	+	+	+	+	+	+	+	Good correlation with Zn concentrations, but only IPS gave good indication of general water quality	De Jonge et al. (2008)

Shaded area: OMNIDIA database software (Lecointe et al., 1993) containing a range of diatom indices.
CEE (Descy and Coste, 1991), GDI (Coste and Ayphassorho, 1991), IDAP (Prygiel and Coste, 1996), IPS (Coste, in CEMAGREF, 1982), L & M (Leclercq and Maquet, 1987), SHE (Schiefele and Schreiner, 1991), TDI (Kelly and Whitton, 1995), SLAD (Sladacek, 1986).

Response to changes in water quality

One important aspect of algal bioindicators is that they are able to detect a rapid change in water quality. Because of their shorter generation time, diatom communities are potentially able to respond more rapidly than other bioindicator groups (e.g. macroinvertebrates, fish), which integrate water quality over longer time frames.

The time sequence of diatom change can be investigated by transferring diatom biofilms from polluted to non-polluted waters, and record

- how different diatom indices compare over fixed time periods,

- the time needed for different indices to indicate a significant change in water quality.

Biofilm studies by Rimet *et al.* (2005) showed that some indices (e.g. CEE, TDI – high sensitivity) responded more rapidly than others (GDI, ILM, SLA – intermediate sensitivity) to environmental change. All indices showed significant change (integration interval) within 40–60 days of biofilm transfer.

Standardisation of approach

The above studies indicate that benthic diatoms provide the basis for a standard approach to river monitoring, able to be used as an alternative (or addition) to macroinvertebrate sampling. Individual commonly used diatom indices appear to correlate significantly with macroinvertebrate data, and also with each other (see above). Studies by Kelly *et al.* (1995) have also shown that benthic diatom indices do not change significantly either with season or with major flow events (both of which can influence invertebrate populations) – suggesting that diatom indices are robust and that consistent results can be obtained throughout the year.

Standard procedures are important in ensuring comparability of results. These include a unified approach to sampling procedures (Kelly *et al.*, 1998), access to common databases, use of comparable diatom indices and quality assurance.

Access to common databases   Access to web-based data sites, such as the European Diatom Database Initiative (EDDI) and the US geological Survey National water-Quality Assessment program (NAWQA), promotes a standard approach to identification, data acquisition and manipulation. Although the EDDI site has particular application for lake palaeolimnology (Section 3.2.2), other databases specifically on benthic diatoms (Gosselain *et al.*, 2005) may be more applicable to river studies.

Use of comparable diatom indices   The cosmopolitan distribution of diatoms, coupled with the close correlation in results obtained from different indices, suggests that a single unified index should be universally applicable within a particular climate (e.g. temperate) zone.

The use of a single unified index has not been adopted, however, for two main reasons:

- Ecologists tend to prefer the security of using multiple indices, with diatom counts being fed into a database for multiple determinations. The general availability of the 'OMNIDIA' database software (Lecointe *et al.*, 1993), incorporating indices summarised in Table 3.14, has been particularly useful in this respect.

- Evidence suggests that the use of a particular index, coupled with key diatom indicator taxa, will be most usefully applied to the geographic area where it was developed. This is due both to floristic differences between geographic regions and also to environmental differences that modify species responses to water-quality characteristics (Potapova and Charles, 2007).

The indices contained in the 'OMNIDIA' software are based on European diatom communities and clearly apply most directly to European rivers. Application of this software in other parts of the world requires the incorporation of local (endemic) species, as emphasised by Taylor *et al.* (2007) in their studies on South African rivers. It is also important to check that cosmopolitan species have similar ecological preferences in different parts of the world. Dela-Cruz *et al.* (2006), for example, assessed the suitability of 'northern hemisphere' ecological tolerance/preference data for periphytic diatoms in south-eastern Australian rivers before using particular bioindicator indices.

Quality assurance Although techniques for standardised sampling of diatoms (Section 2.10.1), and different types of indices (Section 3.4.5) are well defined, the final environmental assessment ultimately depends on the analyst's ability to accurately record species composition from the environmental samples. Quality assessment of analyst performance is important, since regulatory agencies must be confident that data produced by their staff are relevant and accurate. In Europe, for example, the requirement of water companies to install nutrient removal facilities in certain large sewage treatment works (Urban Wastewater Treatment Directive – European Community, 1991) is determined by assessment of water quality and is highly expensive. The use of benthic diatoms in the implementation of this directive (Kelly, 2002) must be reliable.

In the case of chemical analyses of water quality (including soluble nitrates and phosphates), the situation is relatively simple, since chemical parameters are relatively few, with comparisons that are univariate and amenable to conventional parametric statistics. In contrast to this, biological monitoring is more complex – based on field samples containing many species, all of which may contribute to the assessment. These species counts may be assessed as presence/absence or in terms of relative abundance.

- *Presence/absence*. Invertebrate analyses in the United Kingdom are carried out on this basis, with water quality being determined in relation to the presence/absence of key benthic invertebrate data (Mason, 1996). Reliability of assessment by individual analysts can be measured as a 'quality audit', where the number of taxa 'missed' by an (inexperienced) analyst can be compared to the sample assessment by an experienced auditor.

- *Relative abundance*. A more complicated situation is presented by diatom-based monitoring, where the relative abundance of taxa, rather than their presence or absence, is being recorded. In this situation, comparison of analyses by different analysts can be carried out on the basis of 'similarity measures'. The higher the value of the 'measure' the greater the similarity, rising to a point at which the two sets of data can be considered derived from the same population.

Kelly (2001) has used the 'Bray-Curtis similarity measure' to assess analyst performance, with levels of >60% indicating good agreement between primary analyst and auditor. Evaluation of about 60 comparisons showed that reliability of assessment varied with species diversity, and that samples with large numbers of species had lower levels of similarity compared to those with low numbers. The use of such an audit measure, providing an objective approach to analyst performance, has clear application within regulatory organisations such as Water Authorities.

3.4.7 Nitrogen–fixing blue–green algae

The abundance of benthic nitrogen-fixing blue-green algae can be used for rapid biomonitoring of nitrate levels, as demonstrated by Stancheva *et al.* (2013) in a study on southern Californian streams. This showed that the relative abundance of N_2-fixing heterocystous blue-green algae (*Nostoc* and *Calothrix*) and diatoms (*Epithemia* and *Rhopalodia*) containing blue-green algal symbionts decreased with increasing ambient inorganic N concentrations at the low end of the N gradient. Response thresholds for these N_2-fixers related both to N concentrations (0.075 mg l^{-1} NO_3–N, 0.04 mg l^{-1} NH_4–N) and the N:P ratio (15:1 by weight). The NO_3–N threshold was independently validated by observing nitrogenase gene expression using real-time reverse transcriptase PCR.

3.5 Estuaries

Estuaries are aquatic zones that interface between freshwater rivers and saline seas. As such, they tend to be dominated by saltwater conditions, but also have major freshwater inputs. Algae have been widely used as bioindicators of environmental change (Bortone, 2005) in these highly complex aquatic systems.

3.5.1 Ecosystem complexity

Estuaries are highly complex ecosystems in relation to habitats, hydrology, effects of weather, constituent organisms and human activity.

Habitats

Estuaries can be divided into two main regions – a central river drainage channel (or channels) and a surrounding expanse of mudflats (Fig. 3.8). Major populations of phytoplankton are present within the drainage channel, and an extensive biofilm of euglenoids, diatoms and filamentous blue-greens can occur across the surface of the mudflats (Underwood

Figure 3.8 View across the Mersey Estuary (UK) at low tide, showing the extensive mudflats with main river channel in the distance. Freshwater drainage into the mudflats (foreground), with saltwater flooding at high tides, leads to complex localised variations in salinity and nutrient concentrations. Some epipelic diatoms (present in surface biofilms) have wide tolerances to variations in water quality while others have clear environmental preferences (see text).

and Kromkamp, 1999). In addition, minor drainage channels also discharge freshwater into the mudflats/main channel (Fig. 3.8) and freshwater/saline wetlands are also frequently associated with the main estuarine system.

In many estuaries, water from the catchment area also has a major influence on the local marine environment, flowing as a freshwater plume into the surrounding ocean.

Hydrology

Patterns of water circulation are complex. Water movements within the main channel depend on input from the sea (tidal currents), river (freshwater discharge), surface drainage and ground sources (freshwater discharge). Mixing of freshwater and saltwater within the main channel results in an intermediate zone of brackish water – the 'salt wedge', where dense sea water underlies an upper freshwater layer. The salt wedge is a permanent feature that moves up and down the main channel with advancing and receding tide. Periodic movement water out of the estuary into the surrounding ocean leads to a plume of fresh or brackish water in the bay area.

Weather

Estuaries are markedly affected by local weather events such as droughts, high rainfall (flooding) and winds. During the autumn of 1999, for example, hurricanes inundated North Carolina (USA) with as much as 1 m of rainfall, causing a major flood in the watershed of Pimlico Sound (Paerl *et al.*, 2005). Sediment and nutrient-laden water displaced more than 80% of the Sound's volume, depressed salinity by >70% and accounted for half the annual load of nitrogen in this N-sensitive region.

Estuarine organisms

Many algae living in or around estuaries are typically adapted to either freshwater or saline conditions and have restricted habitats (e.g. Section 3.3 – diatoms in coastal wetlands). Other estuarine organisms have become tolerant of a range of conditions. Diurnal and seasonal changes in tidal flow mean that planktonic organisms in the main water channel are subjected to wide periodic fluctuations in salinity, and biota occurring on the surface of mudflats are exposed to extremes of salinity change, desiccation and irradiation. The typically high nutrient level of river inflow also signals a further difference between freshwater and saline organisms in terms of productivity, with a marked increase in C-fixation from freshwater algae when saline water is displaced (Section 3.5.2).

Algal diversity: the mudflat biofilm Organism diversity within the complex estuarine environment is illustrated by the area of mudflats, where localised drainage and inflow (Fig. 3.8) lead to numerous small-scale salinity and nutrient gradients (Underwood *et al.*, 1998). Epipelic diatoms are the main group of algae inhabiting these intertidal sediments, forming surface biofilms that have high levels of productivity. Algae associated with estuarine sediments are also important in substrate stabilisation (Section 2.7.2) and are in dynamic interchange with planktonic populations (Section 1.1.4).

Although many epipelic diatoms are cosmopolitan in their distribution, indicating broad tolerance of environmental variations, others are more specific. Analysis of diatom distribution within local salinity gradients led Underwood *et al.* (1998) to classify some diatoms as oligohaline (0.5–4% of full salinity: *Navicula gregaria*), mesohaline (5–18%: *Navicula phyllepta*) and polyhaline (18–30%: *Pleurosigma angulatum*, *Plagiotropis vitrea*). Variations along a nutrient gradient indicated some species (*Nitzschia sigma*) typical of high nutrient conditions, while others (*N. phyllepta*, *Pl. angulatum*) were more common of low nutrient levels. Similar studies by Agatz *et al.* (1999) on tidal flat epipelic diatoms also demonstrated nutrient-tolerant (*Ni. sigma*) and nutrient-intolerant (*Achnanthes*, *Amphora*) taxa. The above algae serve as bioindicators for micro-environmental (local) conditions of pore water quality.

Human activity

Human activity has a major and varied effect on estuaries and is a dominant source of stress and change. At least half the world's population resides in estuarine watersheds (Paerl *et al.*, 2005), greatly increasing

sediment and nutrient loads to downstream estuarine and coastal systems. This results in deterioration of water quality and an overall decline in the ecological and economic condition of the coastal zone, including loss of fisheries habitat and resources.

3.5.2 Algae as estuarine bioindicators

Algae have been used to monitor changes in water quality of estuaries and coastal systems in relation to two main aspects – changes in salinity and increases in inorganic nutrient levels (eutrophication). The potential role epipelic diatoms as monitors of local pore water salinity/nutrient concentrations has been noted earlier, and diatom bioindicators of coastal wetland salination discussed in Section 3.3.

Detection of large-scale eutrophication is a more pressing and general problem, since it is a feature of almost all estuaries. Nutrient increases are accelerating with rising levels of human population and have important economic impacts in relation to fisheries. The use of algae as bioindicators of general estuarine eutrophication is considered in relation to two main aspects – monitoring algal populations using pigment concentrations and molecular detection of freshwater species in river plumes.

Eutrophication: analysis of pigment concentrations

Analysis of algal populations as bioindicators in estuarine and related coastal systems typically involves monitoring a large area of water surface. The speed with which environmental changes can occur also means that relatively frequent sampling may be necessary. Because of these requirements, many estuarine monitoring programmes have combined rapid sample analysis (using pigments as diagnostic markers) with automated collection (from fixed sites or by remote sensing).

The combined use of HPLC (pigment separation), spectrophotometry (pigment quantitation) and a matrix factorisation programme such as ChemTax® (calculation of biomass proportions) is described in Section 2.3.3 and provides a speedy and accurate assessment of the algal community. Although

Table 3.15   Algal Bioindicators of Estuarine Pollution: Episodic and Seasonal Events in Southern US Estuaries.

Estuarine System	Source of Eutrophication	Algal Analysis: Results[a]
Neuse River estuary, North Carolina	High rainfall – hurricanes in 1999 Pollution source – agricultural urban and industrial activities	River discharge into estuary – elevated biomass (chl-*a*), increased chlorophytes/diatom populations, seasonal changes – reduced dinoflagellate bloom.
Galveston Bay, Texas	High rainfall – tropical storm in 2000	Freshwater flushed into bay, resulting in reduced salinity, increased DIN plus blooms of cryptophytes and dinoflagellates. Adverse effects on oyster beds
St. Johns River Estuarine System, Florida	High summer rainfall. Accelerating point and diffuse nutrient loading from watershed	Seasonal nutrient enrichment from watershed. Leading to blue-green algal blooms, fish kills, submerged-vegetation loss, wildlife mortalities

Source: Paerl *et al.*, 2005. Reproduced with permission from Taylor and Francis.
[a]HPLC diagnostic photopigment determinations coupled to ChemTax® analysis

pigment analysis only resolves phytoplankton communities to the major group (phylum) level, the results obtained give clear information on the algal response to environmental change and enable these organisms to be used as bioindicators.

The use of HPLC/ChemTax analysis is reported by Paerl *et al.* (2005) for various southern US estuarine systems (Table 3.15), including the Neuse River Estuary (North Carolina), Galveston Bay (Texas) and St. Johns River (Florida). In all of these cases, eutrophication occurs due to flushing of high volumes of nutrient-rich freshwater into the lower reaches of the estuary and into the surrounding ocean. The source of nutrients primarily results from agricultural, urban and industrial activities in the watershed, and flushing occurs due to high rainfall – which may be seasonal (summer wet season) or episodic (hurricane and storm effects).

Neuse River Estuary   This aquatic system is subject to alternations of high river discharge (seasonal and hurricane rainfall) with periods of low discharge (no rainfall). Algal bioindicators indicate phases of eutrophication resulting from high freshwater flow in relation to biomass increase (up to 19 µg chl-*a* l^{-1}) and a switch in phytoplankton populations – promoting growth of chlorophytes and diatoms, but reducing levels of blue-green algae and dinoflagellates. This apparent reversal of the normal effects of eutrophication is due to the conditions of high water

flow (reduced residence time), where algae with rapid growth and short generation time (r-selected organisms) are able to dominate more slow growing algae (K-selected organisms).

Galveston Bay   Eutrophication resulting from tropical storms in 2000 led to increased algal growth in the bay, with HPLC analysis indicating growths particularly of cryptophytes and peridinin-containing dinoflagellates. The location of the algal blooms overlapped with commercial oyster reefs, leading to ingestion of phycoerythrin-containing cryptophytes and a commercially disastrous red/pink coloration of the oysters.

St. Johns River system   This is a 300 mile-long estuarine system composed of lakes, tributaries, riversine segments and springsheds. Eutrophication was indicated by extensive blooms of blue-green algae, which dominated the relatively static parts of the system such as freshwater lakes and various oligohaline (low salinity) sites. These blooms have been associated with fish kills, loss of macrophytes and wildlife mortalities.

Eutrophication: molecular detection of indicator algae

Freshwater plume   Outflow of freshwater from rivers creates a fluctuating zone of fresh (brackish) water within the surrounding marine environment. This plume of water extends into the surrounding ocean and is characterised by reduced salinity, high levels of inorganic nutrients, high levels of suspended material and a distinctive phytoplankton population. The freshwater tends to lie on the surface of the sea-water (forming a 'freshwater lens'), so that the location of the freshwater plume can be considered both in terms of horizontal and vertical distribution.

A good example of this is the Amazon River plume, which extends into the Western Tropical North Atlantic (WTNA) ocean (Foster *et al.*, 2007). The Amazon River represents the world's largest fluvial outflow, releasing an average 1.93×10^5 m^3 s^{-1} of water to the WTNA. The influence of the Amazon River plume is geographically far-reaching, with freshwater lenses containing enhanced nutrients and distinct phytoplankton populations reported as far away as >1600 km from the river mouth (Borstad, 1982). The Amazon plume brings silica, phosphorus, nitrogen and Fe into the WTNA, promoting high levels of phytoplankton productivity within the plume area. This productivity tends to be nitrogen limited, promoting the growth of nitrogen-fixing (diazotrophic) blue-green algae.

Molecular detection The detection of diazotrophic blue-greens can be used to map regions of eutrophication resulting from freshwater outflow.

Studies by Foster *et al.* (2007) on the occurrence of nitrogen-fixing algae in the WTNA targeted the vertical and horizontal distribution of seven diazotrophic populations, including *Trichodesmium* (typical of open ocean environments), three symbiotically associated algae and three unicellular free-living organisms (phylotypes). Algal distributions were examined in terms of the presence of the *nif*H gene, which encodes for the iron protein of the nitrogenase enzyme (responsible for nitrogen fixation) and is highly conserved. The presence of species-specific *nif*H gene sequences was determined using DNA quantitative polymerase chain reaction (QPCR) technology.

The studies showed that one particular algal symbiont (*Richelia*), associated with the diatoms *Hemiaulus hauckii* and *Rhizosolenia clevei*, was distributed within the freshwater lens of the Amazon plume – with abundances of *H. hauckii*-*Richelia nif*H genes being particularly prominent. The QPCR study showed the dominance of *H. hauckii*-*Richelia* symbioses in the Amazon plume waters, implying that these associations had an ecological advantage over other diazotrophs and establishing them as indicator organisms of the plume environment. Free-living unicellular blue-green diazotrophs were particularly abundant in the saline waters outside the plume, where inorganic nutrients were at minimal levels of detection.

4

A Key to the More Frequently Occurring Freshwater Algae

4.1 Introduction to the key

This key allows the user to identify the more frequently encountered algae to genus level with some mention of the more common species within that genus and is constructed on the basis of features that can be observed using a reasonable light microscope. It is not a taxonomic key, but one in which algae are separated on the basis of readily observable morphological features such as shape, motility, cell wall structure and colonial form. Only occasional and very simple laboratory preparations might be required as it is generally assumed that observations will be made on fresh specimens. The main exception to this is with some diatoms, where cleaning is required to remove organic cell material and reveal surface markings on their silica walls (see Section 2.5.2).

Before using the key the reader should familiarise themselves with the main characteristics of each of the major groups (phyla) of algae as described in Chapter 1. The characteristics described there will help understand the points of identification in the specimen being observed.

4.1.1 Using the key

Using the key to identify a particular alga involves a number of stages:

1. *Initial observations with the light microscope.* Before using the key the user should carefully observe the specimen noting size, shape, cell features, both external and internal, and anything else of interest. It is important to observe a number of examples of the alga to be identified to get a good overall picture of its characteristics. This is especially true of cleaned diatoms when views of different aspects of a cell are often needed. Note that it is always best to observe fresh live material as well as preserved, and that (in the case of diatoms) cleaned and mounted material may also be needed to see certain features such as markings on the silica wall surface.

2. *Using the key.* As with all binary keys, two sets of alternative features are presented at each stage and the user must consider which best fits the specimen being observed. Each alternative leads on to a further (numbered) binary choice, with the sequence of choices ultimately reaching the name of the genus to which the specimen is provisionally assigned.

3. Confirmation of genus can be made with reference to the fuller description given and the photographs plus line drawings provided. In some cases, details and illustrations of individual species are also given. Algal photographs are invariably of fresh specimens unless stated otherwise. In many

Freshwater Algae: Identification, Enumeration and Use as Bioindicators, Second Edition. Edward G. Bellinger and David C. Sigee.
© 2015 John Wiley & Sons, Ltd. Published 2015 by John Wiley & Sons, Ltd.

cases, the photographs are taken under bright field microscopy, but other techniques such as phase-contrast, differential interference contrast (DIC) microscopy and negative staining are also used to enhance contrast.

By working through the key, many of the important features of the specimen will have been used so the user will already have in their mind the main features of that organism. Until the user is familiar with the key, there may be occasions when it is difficult to decide which set of alternatives is best and this could lead to incorrect identification. It is always best to note progress through the key by writing down the sequence of numbers, so that if a mistake occurs steps can easily be retraced and an alternative route followed.

4.1.2 Morphological groupings

Because this is a morphological rather than a taxonomic key, algae from different phyla may be juxtaposed in the text. The user will find that most specimens with the following features are found in the following key groups:

- Key nos 5 to 23: branched filamentous forms (non-diatom)

- Key nos 27 to 29: diatom filaments, strands or ribbons

- Key nos 30 to 42: unbranched filament but not blue-greens

- Key nos 43 to 55: filamentous blue-greens

- Key nos 57 to 67: colonial blue-greens

- Key nos 72 to 75: motile colonies

- Key nos 77 to 80: colonial diatoms

- Key nos 81 to 105: other colonial forms (mainly greens)

- Key nos 107 to 125: unicellular flagellates

- Key nos 128 to 143: centric diatoms

- Key nos 144 to 203: pennate diatoms

- Key nos 207 to 212: single-celled green algae

- Key nos 214 to 217: some desmids (also 223)

- Key nos 218 to 222: remaining single-celled green algae

4.2   Key to the main genera and species

1 (a) Plants macroscopic often growing to >20 cm in length. Thallus differentiated into nodes bearing whorls of branches and internodes. Erect in habit and growing anchored to the substrate. Chloroplasts discoid and numerous. Chlorophyta, order Charales 2

(b) Plants microscopic or if visible to the naked eye it is normally because they are present as a mass but still requiring microscopic observation to determine the more detailed morphology 3

2 (1) (a) Plants coarse to touch, frequently encrusted with lime (Common name Stonewort). . . . . *Chara*

(b) Plants not coarse to touch, usually deep green in colour, not encrusted with lime. . . . . *Nitella*

Members of the Charales are macroscopic branched filaments with a main axis of large elongate cells (one per internode) and a whorl of smaller cells forming the branches at each node. The large internodal cells may be several centimetres in length. Both *Chara* and *Nitella* are anchored to the bottom of still (sometimes slow flowing) nutrient-poor clear freshwaters, anchored by means of colourless rhizoids. A few species may occur in brackish waters. They are probably the only true members of the rhizobenthos in freshwaters. ***Chara*** particularly favours hard waters with a pH >7.0 but the pH range of the whole group is between 5.0 and 10.0. They are not present in polluted waters especially where sewage is being discharged. Some species of the Charales impart a garlic-like odour to the water. Charophyta. Plate I.

3 (1) (a) Cells grouped together to form a filament, strand or ribbon (see Glossary for definition of these terms). Sometimes filaments can grow in such profusion as to be visible, en masse, to the naked eye or are visible as multiseriate rows encrusting stones **4**

(b) Cells individual or in groups that may be regular or irregular in shape but not forming a filament, strand or ribbon **56**

4 (3) (a) Cell pigments localized in chloroplasts*. Colour when fresh may be grass green, pale green, golden to brown, olive green or (rarely) bluish or reddish **5**

(b) Cell pigments not localized in chloroplasts. Colour when fresh is frequently blue-green but may be olive green or other colour **43**

5 (4) (a) Filaments branched (false or true), sometimes rarely. Filaments may branch only occasionally (in some cases to generate reproductive structures) so it is important to examine a reasonable length to determine whether branching occurs or not. **6**

(b) Filaments or ribbons unbranched **24**

---

*In some algae, particularly small coccoid species, the chloroplasts may appear to fill the whole cell – giving the appearance of unlocalised pigment. In such cases, it is useful to check this apparent non-localisation over a range of cells and also to look at fresh material to see the true colour of the alga. It is also useful to look for the range of features indicated in Table 1.2. The only group of algae not having chloroplasts are the Cyanophyta and they, for example, do not have flagella – so if the specimen is flagellate it cannot be a cyanophyte and hence must have chloroplasts. Cross-referencing to other features in the table can also help arrive at the correct conclusion.

6 (5) (a) Branches of filament rejoin to form a net . . . . ***Hydrodictyon***

Hydrodictyon (commonly known as the 'water net') forms macroscopic nets that are free floating (occasionally attached) and many centimetres in length. Populations may be large enough to partially block small streams, especially in more nutrient-rich waters. Cells are cylindrical, joined at the ends to form a pentagonal mesh net. Individual adult cells can be several millimetres in length and have a single parietal, net-like chloroplast with several pyrenoids. Widespread in ponds and slow moving ditch water. Can produce very large masses in the summer which may be a nuisance blocking the watercourse. Chlorophyta. Fig. 4.1

(b) Branches of filaments do not rejoin to form a net **7**

7 (6) (a) Each cell is enclosed in a flask-shaped lorica which is narrow at one end and with a wide opening at the other. One or two loricas may arise from the mouth of the one beneath, forming a forked or dendroid series. The algal cells within each lorica are bi-flagellate with usually two brownish chloroplasts . . . . ***Dinobryon***

Dinobryon can be solitary but more frequently occurs in colonies. These are branched or dendroid with each new lorica emerging from the open neck of, and attached by means of a stalk to, the old. Each lorica contains one biflagellate cell attached to the base of the lorica by a fine cytoplasmic thread. Flagella are of unequal lengths (heterokont). There are typically two chloroplasts per cell with chrysolaminarin granules at the base. An eyespot is also present. The lorica may be clear or coloured brown, smooth or ornamented and vary in overall shape depending upon species. The lorica may be up to 100+ μm in length.

Dinobryon is widely recognised as a bacteria-consuming mixotroph (a type of nutrition where both autotrophy and heterotrophy can be used).

Plate I   **1**. *Nitella*. **2.** *Chara.*

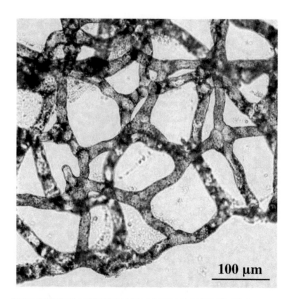

Figure 4.1 *Hydrodictyon*. Detail from edge of tubular net showing 3D intercellular network.

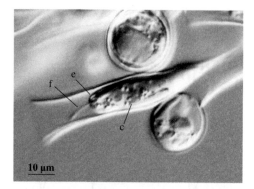

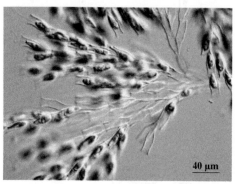

Very widely distributed in the plankton of ponds and lakes from nutrient-enriched to nutrient-poor. It may be present in hard or soft waters. Common in the plankton of cool waters and in mountain regions. In temperate climates, most abundant in summer. Has been associated with odour problems in drinking water (Palmer, 1962). Chrysophyta. Plate II. Fig. 4.2a.

Figure 4.2a *Dinobryon*. Top: Detail of single elongate cell within a flask-shaped lorica, showing apical flagellum (f), red eyespot (e) and live-brown chloroplast (c). Bottom: Low-power view of free-floating dendroid colony. Separate species. Reproduced with permission from R. Matthews.

Epipyxis, a closely-related genus, is a solitary loricate or only forms small colonies. The lorica consists of overlapping scales which are sometimes difficult to see (may need phase-contrast microscopy or staining). Epiphytic on other algae. Plate II. Fig. 4.2b.

(b) Cells not in flask-shaped loricas **8**

8 (7) (a) Filaments without cross walls dividing them into separate cells (siphonaceous or coenocytic), cross walls only appearing when reproductive structures produced. Irregularly branched . . . . **Vaucheria**

Filaments cylindrical, multinucleate – containing numerous disc- to oval-shaped chloroplasts,

Figure 4.2b *Epipyxis*. Colony of cells, with individual sedentary loricae radiating out from a central zone of attachment. Reproduced with permission from R. Matthews.

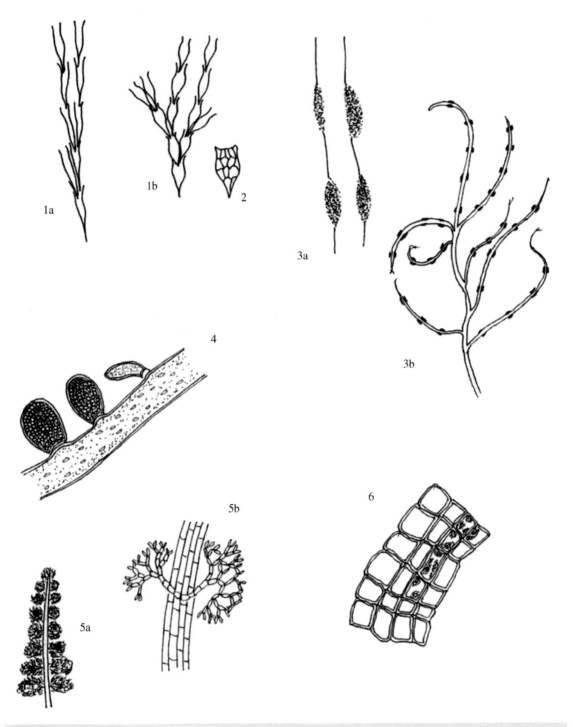

Plate II   **1.** *Dinobryon*: (a) *D. sociale,* (b) *D. divergens*. **2.** *Epipyxis anglica*. **3.** *Lemanea*: (a) growths on higher plant stem, (b) whole alga. **4.** *Vaucheria*. **5.** *Batrachospermum*: (a) whole alga, (b) detail of main stem and branching. **6.** *Hildenbrandia*.

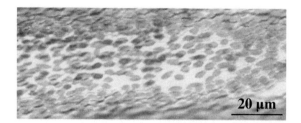

20 μm

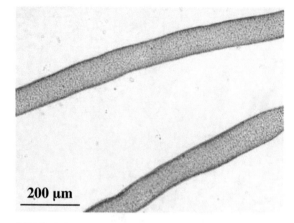

200 μm

Figure 4.3 *Vaucheria*. Top: Detail from surface of filament (DIC image) showing numerous discoid chloroplasts. Bottom: Low power view of filaments showing absence of cell walls.

with or without pyrenoids. The usual storage products are oil and fat, very rarely starch. Branching does occur but is irregular. Filament 20–140 μm wide. The various species are identified mainly by their reproductive structures. Widespread and may form green mats of tangled filaments in shallow freshwaters, on stones and damp soil. May also occur in salt marshes and brackish water muds. Xanthophyta. Plate II. Fig. 4.3.

(b) Filaments with normal cross walls, not siphonaceous **9**

9 (8) (a) Filaments multiseriate, that is, columns of cells in many parallel or radiating rows. Branches may or may not be present **10**

(b) Filaments not multiseriate **13**

(but see also *Coleochaete* which, when older, forms a mass or disc of cells that appear multiseriate).

10 (9) (a) Filaments branched (although the branching can be scarce and difficult to find – hence the need to look along several filaments to confirm the situation) with branches arising in tufts (embedded in mucilage) or as less frequent and more tapered bristles **11**

(b) Filaments encrusting on stones or macrophytes forming single or packed layers of cells, green, red or brown in colour **12**

11 (10) (a) Filaments loosely embedded in copious mucilage. Main filament axis with tufts of branches arranged at regular intervals . . . . *Batrachospermum*

Batrachospermum produces large multibranched filaments in a mucilaginous or gelatinous mass, similar to frog spawn. The overall appearance of the filaments is that they are beaded. There is a main large axis with regular whorls of smaller branches. Chloroplasts lack pyrenoids and are ribbonlike (several per cell), olive brown to reddish or grey-blue in colour. The whorls may be >1 mm in diameter. Widespread and with a characteristic appearance. Found in streams and pools/bogs, often in warmer waters. Rhodophyta. Plate II. Fig. 4.4a.

Audouinella has simple branched filaments with both prostrate and erect branches. Cells cylindrical with parietal chloroplasts. Reproduction by monospores. Common epiphyte. Rhodophyta. Fig. 4.4b.

(b) Plants without regular tufts of branches. Thickened areas along stem. . . . . . *Lemanea*

The thallus of *Lemanea* is cartilaginous and thus has an erect posture being attached at the base. The bristle-like 'stems' can grow

Figure 4.4 (a) *Batrachospermum*. Top: Detail showing a whorl of uniaxial tufts (A) arising from the main axis (B). Bottom: General view of thallus (b) *Audouinella*. Details from an attached colony showing clear red coloration. Top: Cluster of ovoid asexual monospores at the tip of a short lateral branch. Bottom: Simple branched filaments composed of cylindrical cells. Reproduced with permission from R. Matthews.

up to 20 cm in length and show little to frequent branching. When growing in water plants look like stiff tufts of black or brown hair, each hair having irregular thickenings along its length. Chloroplasts are numerous, parietal and disc-shaped. Widespread in moderate- to fast-flowing streams attached to stones. Rhodophyta. Plate II.

12 (10) (a) Cells of filaments stacked in vertical rows to form a pseudoparenchymatous mass. These filaments are found as a thin, flat rose-coloured encrustation on stones. .... *Hildenbrandia*

Thalli of **Hildenbrandia** grow over stones in rivers and streams, frequently in hard waters.

Thalli are crustose, reddish in colour and closely adhere to the substrate. Cells cubical, in vertical rows, each cell up to 16 μm in diameter. Rhodophyta. Plate II.

(b) Filaments attached to a surface, radiating from a central point forming a flattened disc or slightly rounded cushion of cells – some of which have fine hairs or setae which are sheathed at the base .... *Coleochaete*

Coleochaete is uniseriate when younger, subsequently growing as a mass where filaments tend to merge to form a pseudoparenchymatous mass. Branches may be dichotomous or irregular. There may be an erect part of the thallus as well as the flattened

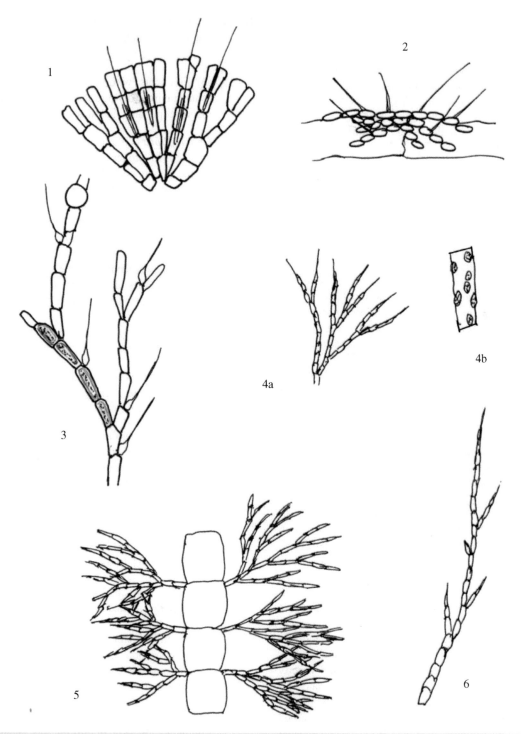

Plate III **1.** *Coleochaete*. **2.** *Aphanochaete*. **3.** *Bulbochaete*. **4.** *Chaetophora*: (a) whole alga, (b) growths on stem of higher plant. **5.** *Draparnaldia*. **6.** *Stigeoclonium*.

disc-like cushion. Cells frequently have long bristles sheathed at their bases. There is a single parietal plate-like chloroplast with one or two pyrenoids. Widespread as an epiphyte on aquatic macrophytes and other surfaces. Chlorophyta. Plate III.

13 (9) (a) Some cells along the filament bear hairs or setae **14**

(b) No hairs or setae present. **18**

14 (13) (a) Hairs or setae having a bulbous base **15**

(b) No bulbous base present with the hairs or setae **16**

15 (14) (a) Filaments growing horizontally or prostrate, epiphytic. Branching irregular or may be absent. Some cells with one to several hairs . . . . ***Aphanochaete***

Aphanochaete is a prostrate creeping epiphyte. Cells longer than wide (sometimes slightly swollen in the centre) often bearing one or more long bristles which are swollen at the base. Chloroplasts are disc-shaped with one or more pyrenoids. Widespread distribution, often reported in nutrient-rich waters on submerged macrophytes and larger filamentous green algae. Chlorophyta. Plate III.

(b) Filaments not prostrate, branched, with cells somewhat wider at the top than the base. Many cells have one or more characteristic colourless chaetae (a terminal hair with a swollen base) arising from the top of the cell. . . . . ***Bulbochaete***

Bulbochaete is a member of the **Oedogoniales.** It has branched filaments and is easily identified by the hairs which have distinctly swollen bases. Chloroplast parietal and net-like with several pyrenoids. Widespread distribution, growing attached to submerged macrophytes or stones. Wide pH and nutrient range. Chlorophyta. Plate III.

16 (14) (a) Filaments enclosed in soft watery mucilage which has no definite shape **17**

(b) Filaments with a firm mucilage surround having a definite shape. Often forming macroscopic attached masses . ***Chaetophora***

Chaetophora filaments are frequently attached to stones or submerged aquatic plants. Filaments can be quite long (10+ cm). Branching more common towards filament apex. End cells of filament/branch with a pointed shape or in the form of a long multicellular hair. The erect system is well developed but the prostrate one is not. Fairly widespread distribution at the edges of shallow flowing or still waters attached to a substrate such as aquatic plants. Chlorophyta. Plate III.

17 (16) (a) Main filament axis composed of a single row of larger cells from which arise tufts of branches composed of much smaller cells, the whole enclosed in soft mucilage . . . . ***Draparnaldia***

Draparnaldia has erect filaments, embedded in soft mucilage, attached to the substratum by means of rhizoids. Main filament cells barrel-shaped (40–100 μm wide) containing a net-like or complete band-shaped parietal chloroplast with several pyrenoids. Branches arise in whorls from main axis. Branch cells smaller with single chloroplast and up to three pyrenoids. Branches have either a blunt terminal cell or form a multicellular hair. Widely distributed, often in phosphorus-poor waters. Chlorophyta. Plate III.

(b) Filament main axis cells not markedly different to branches (except end cells of branches which are thinner and hair-like). Branches not usually born in distinct whorls . . . . ***Stigeoclonium***

Filaments of ***Stigeoclonium*** are usually attached by means of basal cells to a

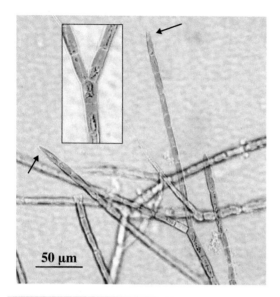

Figure 4.5 *Stigeoclonium*. General view of filaments, with branches tapering to a fine point (arrows). Inset: Mode of branching.

variety of surfaces. Branches can be opposite, alternate but rarely whorled and terminate in a tapering series of cells sometimes forming a terminal multicellular hair. Filaments surrounded by a thin mucilaginous layer. Chloroplasts parietal with one or more pyrenoids. Cells 5–40 μm wide, many times longer than wide. Widespread distribution, usually attached to rocks or stones but may break away and can be found free-floating. Several species, in flowing or still waters. *Stigeoclonium tenue* is common and is often used as an indicator of enriched or organically polluted situations (also reported as tolerant to heavy metal pollution). Chlorophyta Plate III. Fig. 4.5.

18 (13) (a) Branched filaments embedded in mucilage **19**

(b) Branched filaments not embedded in mucilage **21**

19 (18) (a) Plant forms a gelatinous globular cushion. Mucilage may be firm and encrusted with lime in hard water locations **20**

(b) Mucilage covering thin. Usually attached at base and forming trailing branched filaments. Not cushion-like or globular . . . . **Stigeoclonium**

(see also key No. 17)

20 (19) (a) Cushions composed of filaments with branches tapering to a long multicellular hair or having a single rounded cell at apex. Mucilage firm, forming cushions 1–2+ cm wide . . . . **Chaetophora**

(see also key No. 16)

(b) Cushions generally smaller (>5 mm diameter). Filaments with less branches and cells rounded or with swollen ends, often lime encrusted . . . . **Gongrosira**

Gongrosira forms cushion-like colonies that may be lime-encrusted in hard waters. Filaments arise from a basal prostrate group of cells, forming an erect as well as a prostrate system. No hairs present and terminal cells of branches bluntly rounded. Cells cylindrical or swollen, often with thickened walls. Parietal chloroplasts with one to several pyrenoids. Frequent in streams, shallow lake margins or ponds, on stones, often in hard water regions. Cells 4–30 μm wide, 1 to 3 times longer than wide. Chlorophyta. Plate IV.

21 (18) (a) Plants small, erect. Cells 1.5–5 μm diameter, up to ×12 as long as wide. Cell walls thin. Filaments highly branched with the first cross wall of a branch being a small distance from the main axis. Pyrenoids absent . . . . **Microthamnion**

Microthamnion consists of attached delicate branched filaments of the same width throughout. Chloroplasts parietal and pyrenoids absent. Widespread distribution, often in acid organically rich waters and waters with higher iron and/or manganese compounds. Attached to substrate by means

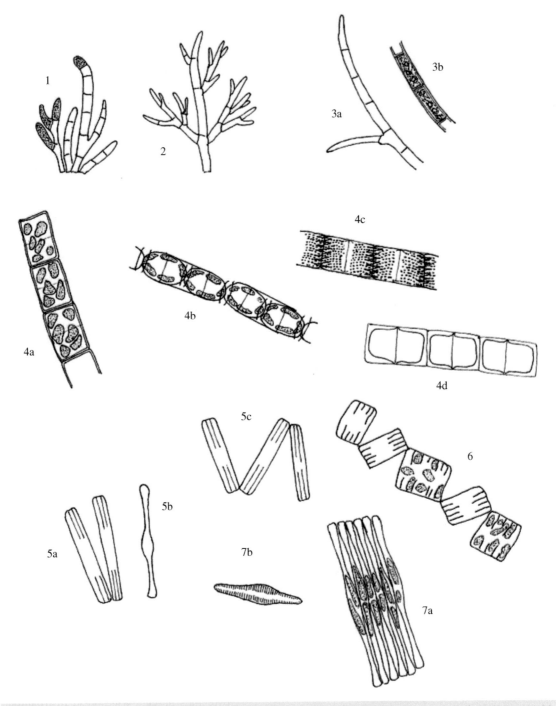

Plate IV **1**. *Gongrosira*. **2**. *Microthamnion*. **3**. *Rhizoclonium*: (a) general view, (b) detail of cell. **4**. *Melosira*: (a) *M. varians*, (b) *M. nummuloides*, (c) *M.* (now *Aulacoseira*) *italica*, (d) *M. dickiei*. **5**. *Tabellaria fenestrata*: (a) girdle view of colony, (b) single cell – valve view, (c) zig-zag colony. **6**. *Tabellaria flocculosa*. **7**. *Fragilaria*: (a) *F. crotonensis*, (b) *F. capucina* – single cell, girdle view.

of a holdfast. Often in association with **Microspora** assemblages. Can also be found on soils. Chlorophyta. Plate IV.

(b) Plants larger. Cells >7 µm in diameter. The first cross walls of each branch occur at the origin of the branch. Pyrenoids present **22**

22 (21) (a) Filaments and branches tapering gradually, over two or three cells, to a fine point . . . . **Stigeoclonium**

(see also key No. 17)

(b) Filaments tapering abruptly, not gradually, with a rounded blunt end cell. **23**

23 (22) (a) Branches, which may be sparse in occurrence, usually short, sometimes consisting of only one cell and almost rhizoidal in appearance . . . . **Rhizoclonium**

Rhizoclonium forms coarse, wiry filaments with short sometimes rhizoidal branches although these are not always present. Cells elongate, with robust walls, 10–40 µm wide, 2–8 times as long as they are wide. Chloroplast net-like with numerous pyrenoids. Common in hard, shallow, waters where it may be found in dense mats. Often found with (and confused with) **Cladophora.** Chlorophyta. Plate IV.

(b) Branches often longer and more robust. Repeated branching may occur, depending upon conditions – but they may be few and far between. . . . . **Cladophora**

Cladophora is typically well branched, but in gently flowing waters branching may be intermittent and difficult to find. In contrast, on a lake shoreline habitat subject to choppy wave action the form may be tufted or bush-like with many branches. **Cladophora** may be free-floating or attached to a substrate by means of a small holdfast. Branches

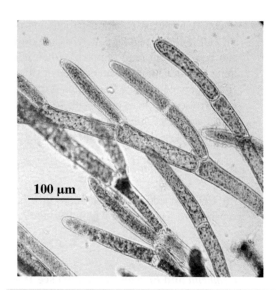

Figure 4.6 *Cladophora.* General view of filaments, showing newly formed branches.

may be alternate or opposite, dichotomous or even trichotomous. Cell walls are robust and the chloroplast net-like (reticulate) and parietal with numerous pyrenoids. Cells 50–150 µm wide, up to ×10 long as wide. Commonly known as 'blanket weed' it can form extensive, coarse-to-touch mats. Frequent in hard or semi-hard waters, and estuarine habitats, especially those enriched with sewage. **Cladophora** can also produce large growths in water treatment filtration systems such as slow sand filters where severe problems of filter clogging may result. Chlorophyta. Figs. 2.28 and 4.6).

(Not to be confused with **Pithophora**, which is common in ponds but whose filaments frequently contain dark akinetes that are swollen and barrel-shaped. Cells 20–100 µm wide, up to 400 µm long. Not illustrated).

24 (5) (a) Cells with a siliceous wall **25**

To determine whether a silica wall is present, it may be necessary to 'clean' the specimen with a suitable oxidising or acidifying agent (see Section 2.5.2). Some indication can be obtained, however, by sharply focussing on the cell wall. If a regular pattern of markings can be made out (e.g. Fig. 1.13), then it is quite probable that the wall is made of silica. If an empty cell can be found then it may be easier to see any markings. With some pennate diatoms, a distinctive gliding movement may be observed in live (fresh) material

(b) Cell wall not made of silica.........**30**

25 (24) (a) Cells embedded in a gelatinous tube but separate from each other within the tube ...

***Frustulia* in part see 203**

***Cymbella* and *Encyonema* in part see 164**

(b) Cells not as above **26**

26 (25) (a) Cells joined together to form a continuous filament not surrounded by extensive mucilage. Filaments and cells circular in cross section **27**

(b) Cells forming a ribbon or chain. Cells elongate and not circular in transverse section **28**

27 (26) The former genus ***Melosira*** consisting of filaments of centric cells has now been subdivided into ***Melosira*** and ***Aulacoseira*** (see (a) and (b) below)

(a) Cell walls with no obvious markings. Cells linked in pairs (may be difficult to see without cleaning) .... ***Melosira***

Cells of ***Melosira*** are rectangular (***M. varians***) or ovoid (***M. nummuloides***) in shape with relatively thin walls or rectangular with quite thick walls (***M. dickiei***). Chloroplasts are small discs or plates and may be golden-brown to dark-brown. ***Melosira varians*** (cells 8–35 µm in diameter, 4–14 µm deep) is common and sometimes abundant in shallow, frequently

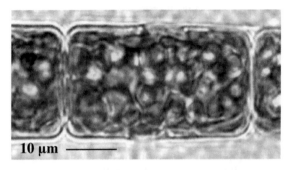

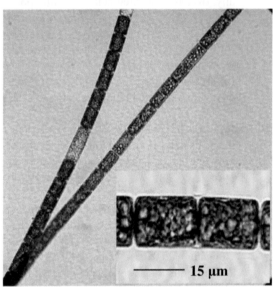

Figure 4.7   *Melosira.* Top: Single cell showing numerous yellow-brown discoid chloroplasts. Bottom: Straight, unbranched filaments typical of this colonial diatom. Inset: Characteristic paired arrangement of cells along filament.

smaller, eutrophic waters. ***Melosira dickiei*** (10–20 µm in diameter, 7–10 µm deep) is found on damp rocks and river banks and ***M. nummuloides*** (9–42 µm in diameter, 10–14 µm deep) is found in brackish or marine environments, often attached to the substratum. ***Melosira varians*** can grow in water treatment filters open to sunlight and cause clogging problems. Bacillariophyta. Plate IV. Fig. 4.7.

(b) Cell walls with granulated markings and often with spines at end (often best seen at end of a filament). Forming a continuous filament and not in pairs.........***Aulacoseira***

In ***Aulacoseira***, the cells are rectangular in shape and are linked together to form long filaments. The surface of the silica cell wall has characteristic markings of rows of dots (punctae) and the ends of the cells bear a ring of spines, one or two of which may be quite long that link the cells together. A sulcus can usually be observed between the mantle and girdle band. Chloroplasts disc-shaped and usually golden-brown. Several species are found in freshwaters. Some are more common in eutrophic lakes, for example, ***A. ambigua*** (cells 4–17 × 5–13 μm in size), ***A. granulata*** (cells 4–30 × 5–24 μm in size), ***A. granulata var. angustissima*** has narrow, long cells with an obvious long spine protruding from the end cell and ***A. italica*** (cells 6–23 × 8–20 μm in size) and others in less enriched waters, for example, ***A. islandica*** (cells 3–28 × 4–21 μm in size) and ***A. subarctica*** (cells 3–15 × 2.5–18 μm in size). ***Aulacoseira granulata*** can cause filter clogging problems in water works (see also ***Melosira***). Bacillariophyta. Figs. 1.3, 4.8 and 4.9.

28 (26) (a) Cells with internal septa or costae (see below) **29**

Costae are silica thickenings on the inside of the valve, orientated at right angles to the apical (long) axis. Septa are silica plates that run across the inside of the cell from the girdle bands partially or wholly dividing the cell into chambers. The septa may protrude for different lengths into the cell.

(b) Cells without internal septa . . . . ***Fragilaria***

In ***Fragilaria***, the pennate or elongate frustules are joined by the central part of their

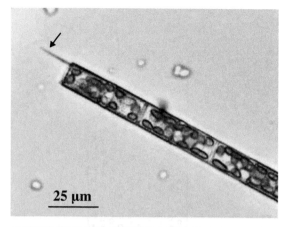

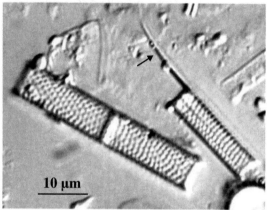

Figure 4.8 *Aulacoseira.* Top: Planktonic filament. Cells with numerous disk-shaped chloroplasts. Bottom: Fragmented filament. Acid digest, lake sediment sample. A terminal spine (arrow) is visible in both cases. Images reproduced with permission from G. Ji (top) and M. Capstick (bottom).

valve faces to form a ribbon-like chain. Valves fusiform in shape, sometimes slightly swollen at the centre (gibbous) or with swollen (capitate) ends or narrow and rectangular in girdle view. Usually seen in girdle view in non-acid-cleaned material. Valve surface with fine striae. Two plate-shaped chloroplasts present. ***Fragilaria crotonensis*** can be found in meso- to eutrophic waters, especially where phosphorus levels are high, sometimes forming blooms. Size,

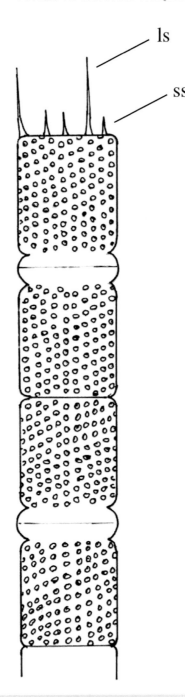

10 μm

Figure 4.9   *Aulacoseira granulata*. Cells have terminal long (ls) and short (ss) spines.

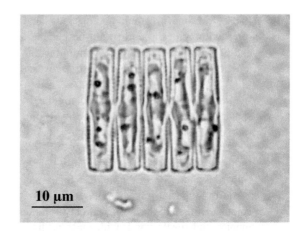

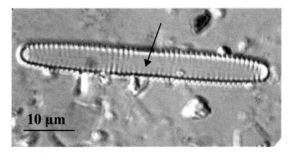

Figure 4.10   *Fragilaria*. Top: Ribbon arrangement of diatom colony. Fixed sample. Bottom: Single cell showing fine transverse striae, with central clear area (arrow). Acid digest, lake sediment sample. Reproduced with permission from M. Capstick.

40–170 μm long, 2–4 μm wide. ***Fragilaria capucina*** is also widespread occurring in lakes and rivers, 10–100 μm long, 3.5–4.5 μm wide. Can impart unwanted odour to drinking waters (Palmer, 1962). Large numbers can cause filter-clogging problems in water treatment works. Bacillariophyta. Plate IV. Fig. 4.10.

29 (28) (a) Cells rectangular or tabular in normal view, sometimes united to form zig-zag colonies. When seen individually and not as part of a chain the valves often show a strongly swollen middle region. Cells without costae but have internal septa which can be clearly seen.................***Tabellaria***

Tabellaria cells are rectangular in girdle view, forming colonies that may be stellate or zig-zag. The stellate colonies can be confused with *Asterionella* but can be told apart by (1) internal septa in *Tabellaria* and (2) more squared-off ends to the cells in *Tabellaria* when compared with the more rounded ends in *Asterionella*. *Tabellaria fenestrata* (33–116 μm long, 4–10 μm wide) and *T. flocculosa* (6–130 μm long, <5 μm wide) are the most frequent species.

Common in the plankton of less nutrient-rich lakes although they can be seen in mildly eutrophic conditions. Different species have been reported to prefer either mildly acid or slightly alkaline conditions (Wehr and Sheath, 2003). Large numbers can clog water treatment filters. Bacillariophyta. Plate IV. Figs. 2.28 and 4.11.

(b) Cells rectangular, may unite to form zig-zag colonies. Cells rod-shaped when seen individually with at the most slightly swollen centres. Transverse costae present within the cells. . . . . *Diatoma*

Diatoma forms zig-zag or ribbon-like colonies or chains. Internal costae are present and these may be seen as ridges or points at the cell margin. Chloroplasts discoid to plate-like, >10 per cell. Cells may be capitate. Several species occur – the more common of which are *D. vulgaris* with cells (8–75 μm long and 7–18 μm wide) having bluntly rounded ends, common in the plankton of moderately eutrophic lakes and rivers and *D. tenuis* (22–120 μm long, >10 m wide) with capitate apices wider than the rest of the cell. Planktonic and epiphytic. Large numbers of planktonic species can cause filter-clogging problems in water treatment works. Bacillariophyta. Plate V. Fig. 4.12.

30 (24) (a) Chloroplasts form a distinct spiral band within the cell . . . . *Spirogyra*

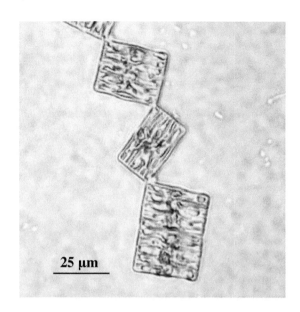

25 μm

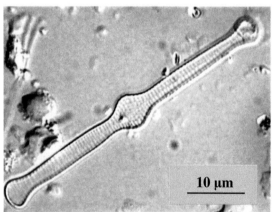

10 μm

Figure 4.11 *Tabellaria.* Top: Zig-zag colony from river biofilm. Fixed preparation. Bottom: Different species – valve view of single cell. Reproduced with permission from M. Capstick. Acid digest, lake sediments.

Spirogyra has cylindrical cells that are joined end to end to form an unbranched filament. Cell walls are firm, with a thin film of mucilage on the outside – often giving them a slimy feel. Chloroplasts (up to 15 per cell) have a helical shape and possess numerous pyrenoids. The various forms of spiral chloroplast in different species are shown in Plate V. The nucleus, often visible in live

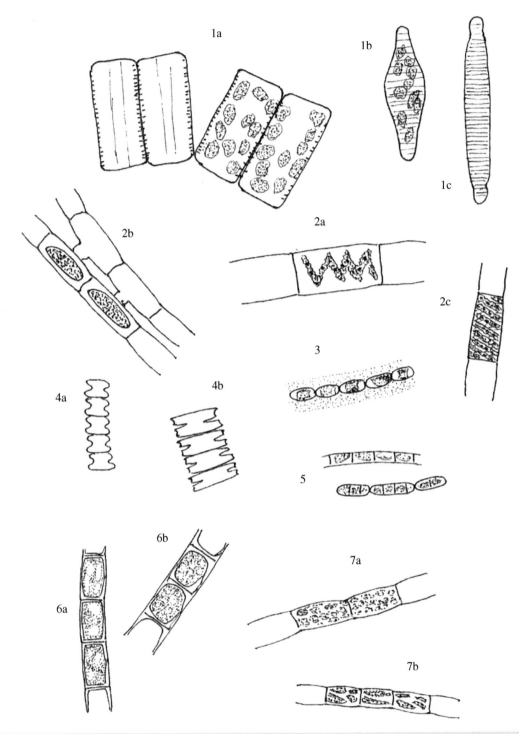

Plate V   **1.** *Diatoma*: (a) *D. vulgaris* girdle view, (b) *D. vulgaris* valve view, (c) *D. anceps* valve view. **2.** *Spirogyra*: (a) *S. protecta*, (b) *S. circumlineata*, (with zygospore) (c) *S. wrightiana*. **3.** *Geminella*. **4.** *Spondylosium*: (a) *S. planum*, (b) *S. rectangulare*. **5.** *Stichococcus*. **6.** *Microspora: (a) M. crassior, (b) M. floccosa*. **7.** *Tribonema*: (a) *T. viride*, (b) *T. minus*.

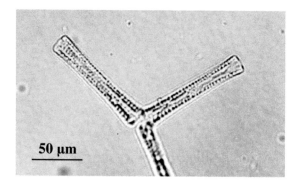

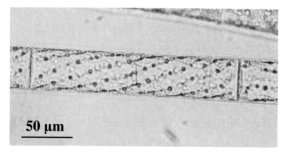

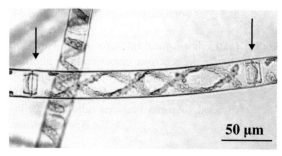

Figure 4.12 *Diatoma*. Top: Star-shaped pattern of cells at edge of planktonic colony. Bottom: Single cell, showing transverse costae. Reproduced with permission from M. Capstick. Acid digest, lake sediment sample.

Figure 4.13 *Spirogyra*. Top: Species with multiple (about 6) spiral chloroplasts and simple cross walls. The chloroplasts have numerous pyrenoids (small dots). Bottom: Species with two chloroplasts and complex (replicate) cross walls (arrows).

material, is in the centre of the cell. Cells are 10–160 μm in diameter, up to 590 μm long. Filaments fragment easily at the cross walls, each fragment growing into a new filament. Sexual reproduction in *Spirogyra* involves conjugation between cells of different filaments and results in the production of a resistant zygote. Common and widely distributed in shallow ponds and ditches, mostly stagnant waters which are neutral to slightly acid although *S. crassa* is confined to hard waters. *Spirogyra* can form dense green masses and can grow in large enough amounts in shallow open water treatment filters to cause blockage. Nearly 400 species have been listed worldwide including commoner species such as *S. condensata*, *S. longata*, *S. neglecta* and *S. tenuissima*. Chlorophyta. Plate V. Fig. 4.13.

(b) Chloroplasts not in the form of a spiral band **31**

31 (30) (a) Filaments are unbranched with the cells embedded in a prominent wide mucilaginous hyaline cylindrical envelope or sheath. Cells are short cylinders with widely rounded ends . . . . ***Geminella***

Cells are 5–25 μm wide, up to twice as long as they are wide, and are surrounded by a thick mucilage sheath. Cells often occur in pairs along the filament. The chloroplast is a parietal plate, roughly saddle-shaped and located near the central region of the cell. A pyrenoid is usually present. Reproduction is by fragmentation of the filament. Occurs in shallow, often slightly acidic, waters where desmids are also present. It can also occur in damp terrestrial habitats. Chlorophyta. Plate V.

(b) Cells not embedded in a prominent mucilaginous envelope................**32**

32 (31) (a) Filament outline with constrictions giving it a toothed appearance............**33**

(b) Filament outline without constrictions and does not have toothed appearance...**34**

33 (32) (a) Cells elliptical in shape in one view but show a deep narrow constriction in alternative view – giving the filament a toothed appearance, with the teeth being smoothly rounded. No gap between adjoining cells. ....*Spondylosium*

Spondylosium is a filamentous desmid. Each cell resembles an angular or rounded (depending upon the species of *Spondylosum*) *Cosmarium* cell with a deep constriction – giving the characteristic toothed appearance to the filament as a whole. Cell walls fairly smooth. There is one chloroplast with pyrenoid in each semicell. Frequent around lake margins in slightly acidic and oligotrophic upland areas although *S. planum* can occur in richer waters. Chlorophyta. Plate V.

(b) Cells angular, rarely elliptical, in one view with a small lens-shaped gap between the cells usually visible. Cells only have a small constriction to the toothed margin, not so pronounced, and the 'teeth' are more angular....................*Desmidium*

Desmidium is a filamentous desmid. The cells can form long twisted filaments enclosed in a gelatinous envelope. When viewed in transverse section, the filaments may be triangular. There is one chloroplast with pyrenoid in each semicell. Cells 26–50 μm wide. Sometimes the space between the cells can be difficult to see. Common in

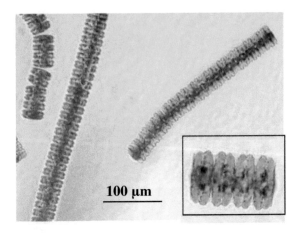

100 μm

Figure 4.14 *Desmidium.* Elongate filaments of this colonial desmid. Inset: Details of four cells, each composed of two semicells.

the vegetation and along the margins of oligotrophic lakes. Chlorophyta. Fig. 4.14.

34 (32) (a) Filaments very short, often only 2 or 3 cells long (more like a short chain than a filament). Cells cylindrical with free ends rounded...................*Stichococcus*

Stichococcus can occur singly or form very reduced filaments, just short chains. Cells 1–16 μm wide and 4–32 μm long. There is a single-lobed chloroplast but no pyrenoid. On damp earth, tree trunks etc. as well as in marine and freshwaters. Chlorophyta. Plate V.

(b) Filaments many cells in length, not typically short.........................**35**

35 (34) (a) Large alga with tubular thallus many cells wide and long, macroscopic in size ....*Enteromorpha*

Enteromorpha has a tube-like thallus that may or may not branch and can grow many centimetres in length. The tube itself consists of a layer one cell thick. The tubes are

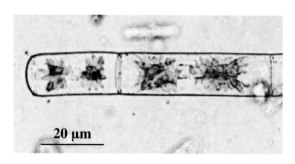

Figure 4.15 *Zygnema*. End of filament showing cells with two typical stellate chloroplasts.

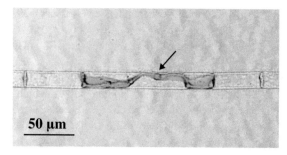

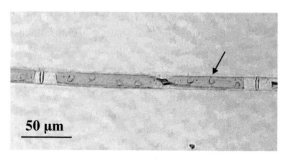

Figure 4.16 *Mougeotia*. Detail of separate cells showing plate-like chloroplast in edge (top) and face (bottom) view. Small refractive pyrenoids can be seen in the chloroplast (arrows).

attached to a substrate by means of rhizoidal branches and an attachment disc. ***Enteromorpha*** has a worldwide distribution, mainly in marine and brackish water habitats – but can stray into freshwater sites. The two most common species are **E. intestinalis** and **E. compressa.** Chlorophyta. (Not illustrated.)

(b) Alga not a large tubular thallus, microscopic **36**

36 (35) (a) Two star-shaped chloroplasts per cell . . . . ***Zygnema***

Cells of **Zygnema** are cylindrical with the two star-shaped chloroplasts separated by a clear area. Each chloroplast has a pyrenoid. The filaments usually have a soft mucilage sheath, are unbranched and not very long. They may be attached to a substrate by means of rhizoids. Cells 16–50 μm in diameter and 2 to 3 times long as wide. As members of the Chlorophyta they store starch. Common in shallow, acidic to alkaline, waters where it can form luxuriant growths – often occurring in a mixed population with other filamentous chlorophytes. Chlorophyta. Fig. 4.15.

(b) Chloroplast either single or more than two per cell **37**

37 (36) (a) One chloroplast per cell in the form of a flat plate arranged along the long axis of the cell. When viewed from one direction the chloroplast fills most of the cell but when viewed from the other it is a thin line down the middle . . . . ***Mougeotia***

The chloroplast of **Mougeotia** is suspended on cytoplasmic strands and can move within the cell depending upon the light. At different times it may be seen face-on, edge-on or twisted. The cells form long unbranched free-floating filaments. Cells 3.5–35 μm in diameter, 5–12 times as long as wide. There are several pyrenoids per cell and when the chloroplast is viewed edge-on the pyrenoids give it a lumpy appearance. Storage product starch. Common in many habitats including lakes, ponds, ditches and streams especially in upland areas. Chlorophyta. Fig. 4.16.

(b) Chloroplast not as above **38**

38 (37) (a) Chloroplast reticulate (see Glossary) **39**

(b) Chloroplast not reticulate **41**

39 (38) (a) Cell walls thick, typically lamellate, made up of two overlapping halves which can break into H-pieces. End cells usually have H-shaped terminations. Pyrenoids absent . . . . ***Microspora***

Microspora forms unbranched filaments, sometimes with a holdfast substratum attachment. Chloroplast is reticulate, not very well defined, and pyrenoids are absent. Storage product starch. Cells 5–30 μm in diameter and one to three times as long as wide. Common but not usually abundant in small freshwater bodies. Some species occur in more acid waters, for example, bogs or even acid-mine drainage. More frequent at cooler times of the year. The morphological characteristics of this genus can vary considerably with environmental conditions, making it difficult sometimes to separate different species. Chlorophyta. Plate V.

(b) Cell wall not thick and lamellate and not composed of H-pieces. Pyrenoids present **40**

40 (39) (a) Cells cylindrical or slightly swollen, with robust walls. Cells 20–80 μm wide, 5–15 times as long as wide . . . . ***Cladophora***

(see also key No. 23)

(b) Cells cylindrical, sometimes slightly swollen at one end. Cell wall firm but not very robust. Filaments unbranched. Some cells along the filament have ring-like transverse lines at the swollen end (cap cells). Cells 10–40 μm wide, 2–5 times as long as wide . . . . ***Oedogonium***

Oedogonium cells form long, unbranched, filaments. The chloroplast is parietal and net-like. Cells are rectangular and longer than

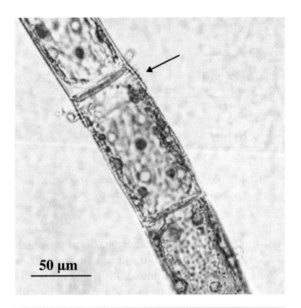

Figure 4.17   *Oedogonium.* The central cell (cap cell) has residual cell wall material, seen as transverse lines (arrow).

wide. The chloroplast is parietal and net-like with several pyrenoids. Oogamous sexual reproduction occurs with large female oogonia and a dwarf male growing close by, producing a thick-walled zygote. Over 200 species have been described and further identification is impossible without the reproductive structures being present. Abundant in still or gently moving waters where they may form dense free-floating, attached or tangled mats. Chlorophyta. Fig. 4.17.

41 (38) (a) Chloroplast a ring or is plate-shaped – extending 1/3 to most way round the cell (at right angles to the long axis). One or more pyrenoids. Starch test positive **42**

(b) Cells cylindrical or slightly barrel-shaped, 2–6 times as long as wide. Chloroplasts one to many, curved discs or plate-shaped with no pyrenoids. Cell walls may fragment into H-pieces. Starch test negative . . . . ***Tribonema***

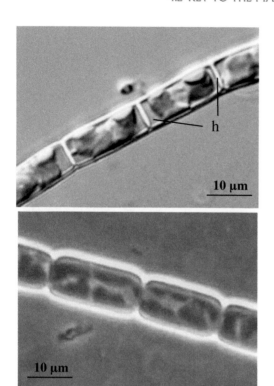

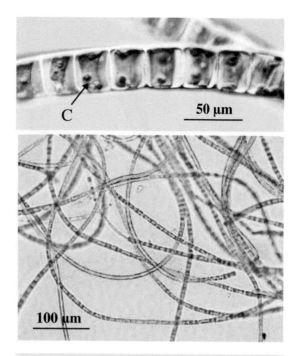

Figure 4.18 *Tribonema*. Top: Detail from filament, showing curved discoid chloroplasts and 'H-piece' end walls (h). Reproduced with permission from R. Matthews. Bottom: Phase–contrast image showing narrowing at cross walls, giving cells a 'barrel shape.' Separate species.

Figure 4.19 *Ulothrix*. Top: Detail from filament showing chloroplast (C) with distinct pyrenoids, wrapping around the inside of the cell. Reproduced with permission from R. Matthews. Bottom: General view of floating mass of filaments, separate sample.

Tribonema forms unbranched filaments. The cell walls are composed of H-pieces (compare with *Microspora*, which it can be distinguished from as it does not store starch). The presence of these H-pieces is best seen at the broken end of a filament, as they are not easily visible within the intact filament. Common in the phytoplankton of lakes and reservoirs especially those rich in organic and humic materials. Can occur at pH levels of up to 9.0 Xanthophyta. Plate V. Fig. 4.18.

42 (41) (a) Chloroplast saddle-shaped extending more than half way around the circumference of the cell . . . . *Ulothrix*

Ulothrix species are composed of unbranched filaments. Cells are cylindrical either longer or shorter than broad depending upon the species, which also vary in cell wall thickness. There is a single annular or saddle-shaped chloroplast (sometimes lobed) with one or more pyrenoids. Commonly forms bright green floating masses in shallow waters particularly at cooler times of the year. It can also be attached to submerged stones or wood or even on damp soil. Common species include *U. zonata* (cells 11–37 μm wide, usually shorter than broad, with fairly thick walls) and *U. aequalis* (cells 13–15 μm wide, 18–30 μm long). Can cause filter blocking in water treatment works. Chlorophyta. Plate VI. Fig. 4.19.

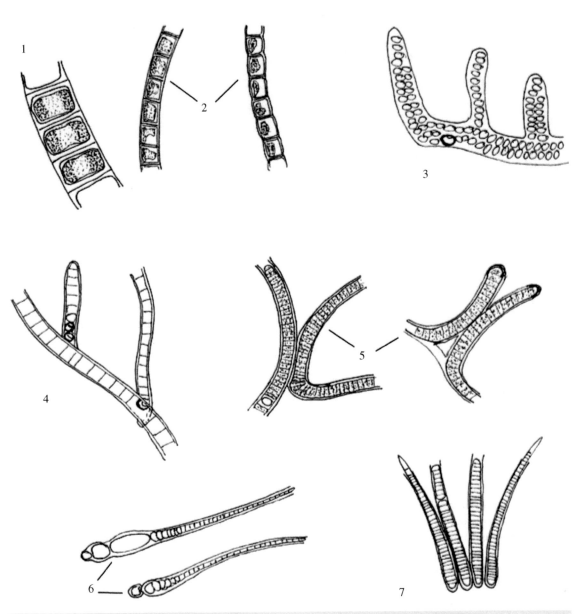

Plate VI   **1.** *Ulothrix.* **2.** *Klebsormidium.* **3.** *Stigonema.* **4.** *Tolypothrix.* **5.** *Scytonema.* **6.** *Calothrix.* **7.** *Homoeothrix.*

(b) Chloroplast extending less than half way round the cell . . . . ***Klebsormidium***

Filaments unbranched. Cells cylindrical (may be slightly barrel-shaped) cells 5–15 μm wide, 1–3 times long as wide. H-shaped pieces sometimes present at cross walls. The chloroplast consists of a parietal band or plate extending part way round the cell. A pyrenoid is usually present. Common on damp soil and in temperate streams, can occur on damp soil or wet rocks. Has been found in acid, metal-rich waters. The filaments fragment easily. For reliable identification, it may be necessary to culture an environmental sample. Chlorophyta. Plate VI.

43 (4) **NB.** In the following section, the terms both *trichome* and *filament* are used when speaking about blue-green algae. The difference between the two terms is defined as follows (John *et al.*, 2002):

A **filament** is a number of cells united in one or more rows to form a chain or thread. In the case of blue-green algae, it refers to species which have a sheath surrounding the cells. A **trichome** is a linear arrangement of cells arranged without a surrounding sheath.

(a) Filaments or trichomes without true branching. False branching may be present **44**

(b) True branching present . . . . . ***Stigonema***

Stigonema grows with branched trichomes up to 50 μm wide, the main axis of which may be multiseriate with the side branches uniseriate. The branched mass is structured enough to be called a colony (John *et al.*, 2002). Young cells at the branch tips can be transformed into hormogonia. Occasional heterocysts may be present. A firm mucilaginous sheath is present which is often stained yellowish-brown in colour. Cells may be quadrate or cylindrical (depending upon

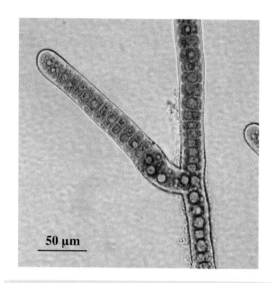

Figure 4.20 *Stigonema.* Blue-green alga, showing uniseriate main axis and side branch (true branching).

species). Grows as a thin mat over stones, rocks, damp soil or damp trees but can break away and become free-floating. Cyanophyta. Plate VI. Fig. 4.20.

44 (43) (a) False branching present **45**

(b) False branching mostly absent (may rarely be present in ***Rivularia*** but this genus is easy to distinguish because of its tapering trichomes) **46**

45 (44) (a) False branches arise singly . . . . ***Tolypothrix***

In ***Tolypothrix*** the trichomes show false branching where the branches are single and are often with a basal heterocyst. A brownish (sometimes colourless) sheath is present. Hormogonia may be produced at the ends. Filaments 6–18 μm in diameter. Usually attached or tangled amongst submerged vegetation but occasionally free-floating, often in more calcareous unpolluted waters. Can occur on damp rocks. Cyanophyta. Plate VI. Fig. 4.21.

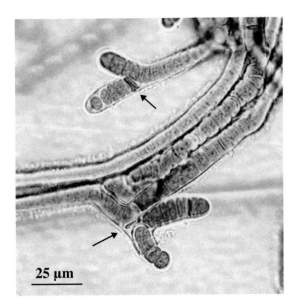

25 μm

Figure 4.21   *Tolypothrix.* Detail from a web of filaments showing the false branching in this alga. Cells at the base of branches (arrows) have a distinctive flattened appearance.

NB. In some genera, false branching is rare but may occur occasionally. This can be true in *Homoeothrix*, *Calothrix*, *Gloeotrichia* and *Rivularia*. In this key, these are separated out on the basis that false branching will not normally be seen.

(b) False   branches   arise   in   pairs
. . . . *Scytonema*

Scytonema has false branches that appear to arise in pairs. This is distinctive for the genus. The branches usually occur between heterocysts. Trichomes sheathed. Cells 5–30 μm wide. Hormogonia may form at the apex of the trichome. Akinetes may be present. Occurs in lakes and on damp rocks and soil, often forming dark tufted mats. Cyanophyta. Plate VI.

46 (44) (a) Heterocysts present (although sometimes infrequent). **47**

(b) Heterocysts absent. **53**

47 (46) (a) Trichomes tapering to a fine point .. . **48**

(b) Trichomes not tapering to a fine point but possibly showing some slight narrowing towards the trichome end **50**

48 (47) (a) Tapered trichomes that are solitary or in small tufts . . . . *Calothrix*

Individual trichomes are tapered, 5–10 μm wide at their base, and have a solitary basal heterocyst. The sheath is firm and often dark straw coloured. False branching may rarely occur. Hormogonia may form towards the end of the filament. Most frequent in hard water streams attached to a substratum, growing as mats or tufts of trichomes over the submerged surface. Cyanophyta. Plate VI.

(b) Tapered trichomes always in a colony, which may be globular or spherical **49**

49 (48) (a) Basal heterocysts usually with an akinete immediately above. Usually planktonic in spherical colonies of radiating trichomes . . . . *Gloeotrichia*

Colonies of *Gloeotrichia* are usually free-floating and spherical. The radiating sheathed trichomes are embedded in soft mucilage. They taper towards the end and have a basal heterocyst. An akinete is frequently present immediately above the heterocyst and is cylindrical in shape. Trichomes 4–10 μm in diameter, colonies up to 5 mm in diameter. Although usually free-floating, the colonies may be attached to a substratum when young before breaking free. Hormogonia may be formed at the trichome ends. Can occur in both brackish and freshwaters. Cyanophyta. Plate VII. Fig. 4.22.

(b) Basal heterocysts present on trichome but no akinetes present. Usually grows as attached globular colonies embedded in firm mucilage . . . . *Rivularia*

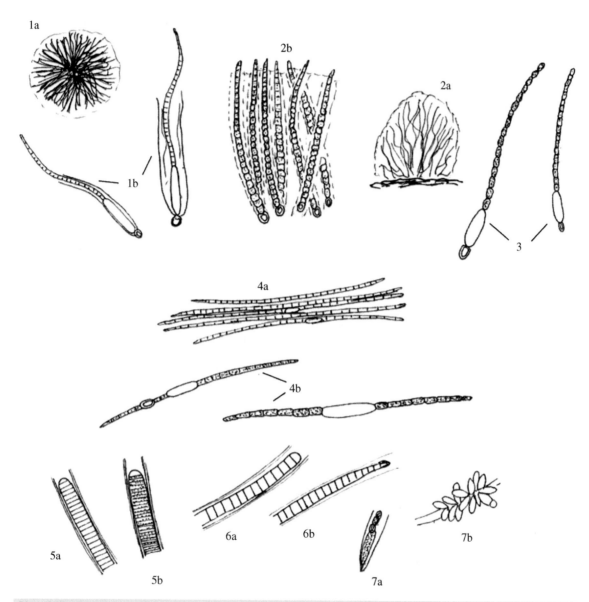

Plate VII **1**. *Gloeotrichia*: (a) colony, (b) filaments. **2**. *Rivularia*: (a) colony, (b) filaments. **3**. *Cylindrospermum*. **4**. *Aphanizomenon*: (a) raft of filaments, (b) single filament. **5**. *Lyngbya*: (a) *L. aestuarii*, (b) *L. major*. **6**. *Phormidium*: (a) *P. lucidum*, (b) *P. autumnale*. **7**. *Chamaesiphon*: (a) single cell with exospores, (b) colony.

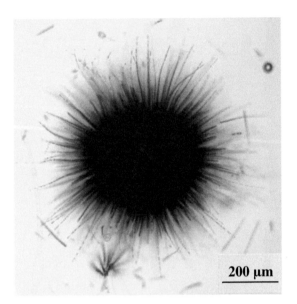

Figure 4.22 *Gloeotrichia.* Single colony showing radiating filaments. Fixed, iodine stained.

Rivularia colonies are sub-spherical to globular and contain numerous filaments that are attached to a substratum. The colony consists of tapering filaments each of which has a basal heterocyst. The tapering trichome often ends in a long hair. False branching, although uncommon, may occur. The trichomes are often radially arranged in firm mucilage. Hormogonia may be produced at the end of the trichome. Most frequently found in hard waters, possibly where there are periods of higher phosphorus concentrations, and colonies may exhibit some calcification. Cyanophyta. Plate VII.

50 (47) (a) Heterocysts terminal, gonidia ellipsoidal to ovate adjoining the heterocyst ....*Cylindrospermum*

Cylindrospermum is characterised by its terminal heterocyst which may be at one or both ends of a trichome. A large akinete may be present, often next to the heterocyst. The trichomes are usually loosely tangled in soft mucilage and grow on damp soil, stones and rocks. Occurs in waters, muds or moist soils that are unpolluted or only slightly eutrophic – often forming dark green patches. Cells 3.5–6 μm in diameter, 4–13 μm long. Cyanophyta. Plate VII.

(b) Heterocysts not terminal but intercalary **51**

51 (50) (a) Each end of the trichome tapers slightly (occasionally slightly curved) and is somewhat elongated. Gonidia, when present, solitary ....*Aphanizomenon*

Trichomes of *Aphanizomenon* may be grouped together like sheaths of wheat to form 'rafts' visible to the naked eye (*A. flos-aquae*) or be solitary (*A. gracile, A. issatschenkoi* and *A. aphanizomenoides*). The trichomes are relatively short, showing a slight taper at each end, with no sheath. Cells are rectangular, 5–6 μm diameter and 8–12 μm long, with slight constrictions at the cross walls. Heterocysts are cylindrical (7 μm diameter, 12–29 μm long) and located in the mid-region of the trichome. Gonidia (about 8 μm diameter, 60–70 μm long) are also cylindrical and in the mid-region but not adjacent to the heterocysts. Common in the plankton especially in nutrient-rich lakes where it can form dense blooms. *Aphanizomenon* is one of the genera of blue-green algae that has been reported to produce toxins in water. Four such toxins have been associated with this genus – saxitoxin, cylindrospermopsin, anatoxin-a and lypopolysaccharides. (Chorus & Bartrum, 1999). Cyanophyta. Plate VII. Fig. 4.23.

(b) End cells of trichome not narrower than the rest so trichome does not taper **52**

52 (51) (a) Trichomes solitary or in a tangled, sometimes coiled, mass ....*Anabaena*

There are many species of *Anabaena*. The trichomes are fairly easy to recognise and

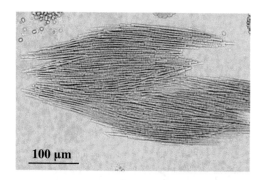

Figure 4.23 *Aphanizomenon*. Top: Typical view of planktonic raft in lake phytoplankton. Bottom: SEM detail of individual filaments.

are of uniform width throughout with a layer of amorphous and frequently inconspicuous mucilage on their surface. The filaments may be straight, curved or coiled depending upon the species. Some species produce gas vacuoles and can form blooms. Cells (3.5–14 μm wide) are rounded or barrel-shaped, giving the filament the appearance of a string of pearls. Heterocysts are spherical and may not be frequent in nitrogen-rich waters. Akinetes (rounded or cylindrical) are produced adjacent to the heterocysts and are frequent in natural populations. Many species can only be identified if the akinetes are present. An extremely common species in the plankton of standing waters where it frequently forms blooms. *Anabaena* can occur in lakes, ponds and ditches. It has been reported (Palmer, 1962) as potentially producing taste and odour in water used for drinking purposes. The occurrence of dense planktonic populations can cause problems of filter clogging when the water passes through a treatment works. *Anabaena* species are also able to produce a range of toxins in water – including microcystins, anatoxin-a, saxitoxins and lipopolysaccharides (Chorus and Bartrum, 1999). Cyanophyta. Figs. 1.3, 1.4, 1.5, 2.7 and 4.24a–4.24c

(b) Trichomes embedded in obvious and extensive firm mucilage............*Nostoc*

Cells of *Nostoc* are similar to *Anabaena* but are embedded in firm, extensive, mucilage which may be leathery in texture and coloured straw or brown. In older colonies, the trichomes tend to be situated towards

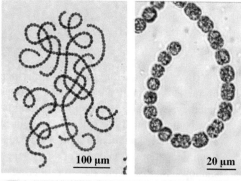

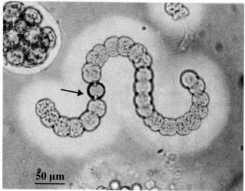

Figure 4.24a *Anabaena*. Top left: Large planktonic colony. Top right: Detail of olive-brown cells in colony. Bottom: Small planktonic colony showing filament embedded in mucilage, and having a single heterocyst (arrow). Indian ink preparation. Comparison of the two colonies shows the colour variation that can occur in this alga.

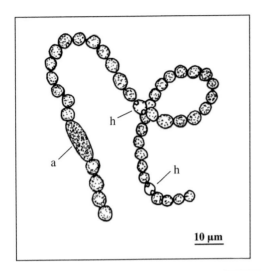

Figure 4.24b *Anabaena circinalis*. Filament with spherical vegetative cells and akinetes (a) not adjacent to heterocysts (h).

the edge. Akinetes may occur in older parts of the colony and are produced between heterocysts (unlike *Anabaena*). Hormogonia may be occasionally produced. The cells are approximately spherical to barrel-shaped 3–6 μm wide. Grows on damp or wet surfaces, shallow waters. Can be free-floating or

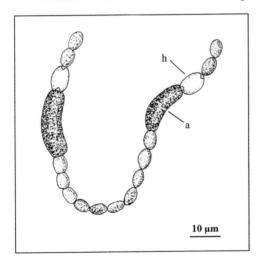

Figure 4.24c *Anabaena flos-aquae*. Single filament with sausage-shaped akinetes (a) adjacent to heterocysts (h).

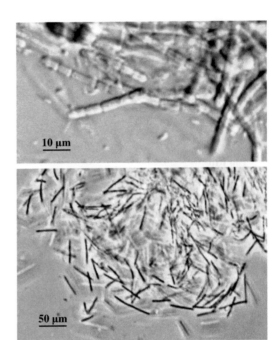

Figure 4.24d *Pseudanabaena*. Top: Details of single filaments, with cylindrical cells deeply constricted at cross walls. Bottom: Edge of colony, showing numerous dispersed short filaments. Reproduced with permission from R. Matthews.

attached. It occurs in rice paddies where it is used to contribute nitrogen to the rice crop. *Nostoc* can produce toxins in freshwaters, for example, microcystins and lipopolysaccharides (Chorus and Bartrum, 1999). Cyanophyta. Fig. 4.25.

The genus *Pseudanabaena* comprises a variety of forms with various common features, including short trichomes capable of motility. Cells are spherical, barrel-shaped or cylindrical with trichomes markedly constricted at the cross-walls. Widespread amongst other algae in ponds and lake margins. Fig. 4.24d.

53 (46) (a) Trichomes form a regularly spirally coiled cylinder in which the individual cells may be difficult to distinguish . . . *Spirulina*

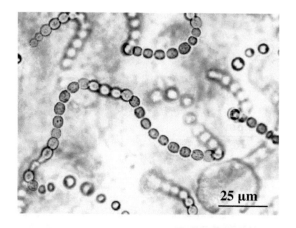

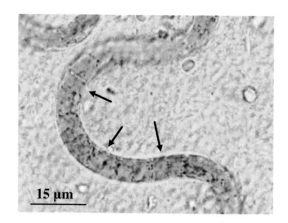

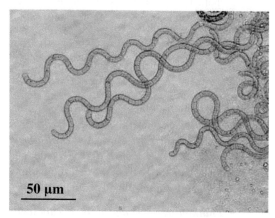

Figure 4.25 *Nostoc.* Top: Detail from colony, with filaments embedded in a mucilaginous matrix. Bottom: Low power view of large globular colony. Indian ink preparation.

Figure 4.26 *Arthrospira.* Top: Detail from spiral filament showing short disk-shaped cells (arrows). Bottom: Low power view of filaments showing typical spiral morphology.

The cylindrical trichomes are twisted into a very regular spiral or helix, with the cell cross walls difficult to see. Colour variable from blue-green to violet. Gas vacuoles may be present. Trichomes (1–8 µm diameter, depending upon species) can be motile if in contact with a surface. ***Spirulina*** (free-floating or attached) is widespread in both freshwaters (including saline lakes, mildly polluted waters) and brackish environments. Cyanophyta.

Arthrospira has cylindrical trichomes (frequently >7 µm diameter) and no heterocysts. Cells are short, with distinct cross walls. Organisms with these features were originally placed within ***Spirulina***, but are now given a distinct genus. Cyanophyta. Fig. 4.26.

(b) Trichomes not forming a regular and definite spiral. The individual cells of the trichome usually more easy to distinguish **54**

54 (53) (a) Trichomes without a sheath ....***Oscillatoria***

Trichomes may be (1) straight or bent, (2) single or in groups, (3) free-floating or attached, (4) short or quite long, (5) capable of a gliding movement or a gentle waving (oscillatory movement), (6) blue-green, olive green, reddish or brownish in colour. Free-floating forms commonly have gas vacuoles. Cells 1–60 μm wide, either longer or usually shorter than broad. The cell cross walls may or may not be narrowed, depending upon species, as may the presence or absence of a calyptra on the end cell of the trichome. The end cell may be rounded or have a characteristic shape. Reproduction occurs by the trichome fragmenting into short hormogonia.

Anagnostidis and Komarek (1985) separate several genera from *Oscillatoria*, including:

- *Tychonema* – with species *T. bornetii* and *T. bourrellyi* (formally *Oscillatoria bornetii* and *O. bourrellyi*) are separated mainly on the basis of the cells containing numerous liquid-filled vacuoles;

- *Planktothrix* – *P. agardhii*, *P. rubescens* and *P. prolifica* (formally *Oscillatoria agardhii*, *O. rubescens* and *O. prolifica*) are all species with narrow cells.

- *Limnothrix* – *L. redekei*, (*Oscillatoria redekei*), which has gas vacuoles localised either side of the cross walls.

Several species are reported to produce taste and odour in water used for drinking purposes (Palmer, 1962). When present in large numbers, it can clog filters used in water treatment processes. Some members of the *Oscillatoria* group are known to produce toxins such as microcystins, anatoxin-a, lipopolysaccharides and aplysiatoxins (Chorus and Bartrum, 1999). Cyanophyta. Plate VIII. Fig. 4.27.

(b) Trichomes surrounded by a sheath .. **55**

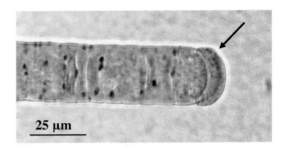

25 μm

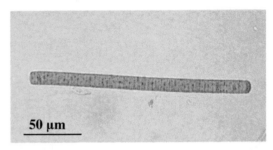

50 μm

Figure 4.27 *Oscillatoria.* Top: Detail showing cap cell (arrow) at the end of filament. Bottom: Single, free floating filament. This species has a distinctive purple coloration.

55 (54) (a) Trichomes single or grouped. Sheath delicate but firm (rarely thick) . . . . *Lyngbya*

Trichomes of *Lyngbya* are a single row of cells enclosed in a fairly firm sheath. Trichomes may be solitary or form a coiled or tangled mass, free-floating or growing on a substrate. The sheaths may become yellowish-brown through staining with chemicals in the water. False branching has been occasionally reported. Cells 1–24 μm diameter. Hormogonia may be formed. Larger species of *Lyngbya* may be confused with *Oscillatoria*, especially if the former has migrated out of its sheath. *Lyngbya* can produce a range of toxins in water such as aplysiatoxin, lyngbyatoxin-a, saxitoxin and lypopolysaccharides (Chorus & Bartrum, 1999). Cyanophyta. Plate VII. Fig. 4.28.

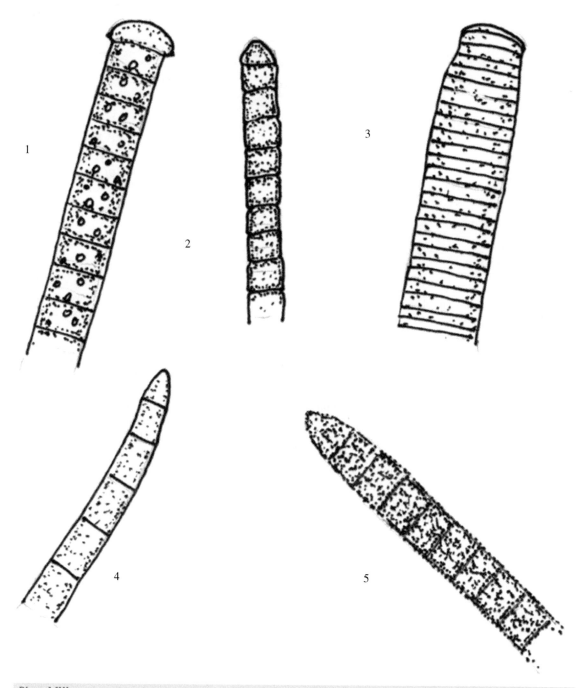

Plate VIII 1. *Oscillatoria rubescens*. 2. *Oscillatoria tenuis*. 3. *Oscillatoria princeps*. 4. *Oscillatoria brevis*. 5. *Oscillatoria agardhii*.

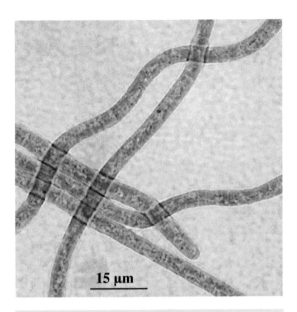

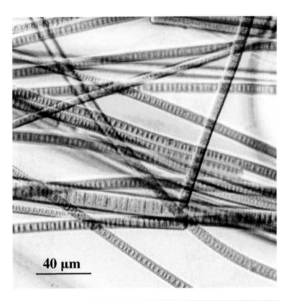

Figure 4.28 *Lyngbya.* General view of tangled filaments. **NB**. Cultured specimen does not show an obvious sheath.

Figure 4.29 *Phormidium.* Detail from web of filaments forming a dense algal mat. Reproduced with permission from R. Matthews.

(b) Filaments interwoven in a sticky gelatinous matrix, sheaths of filaments become indistinct and sticky . . . . *Phormidium*

Phormidium is extremely common on damp soil, stones, rocks etc. or may be free-floating. Trichomes are cylindrical, sometimes tapering slightly towards the ends, with a firm thin sheath. Present within the gelatinous mass, they are capable of movement. Cells, shorter than wide (0.6–12 μm wide, 1.0–10 μm long), may be constricted at the cross-walls. The end cell of the filament is often characteristic of the species. Because of the difficulty in seeing the sheath when preserved it can often be confused with *Oscillatoria.* Cyanophyta. Plate VII. Figs. 2.23, 2.29, and 4.29.

56 (3) (a) Cell pigments not localized in chloroplasts **57**

(b) Cell pigments localized in chloroplasts **68**

NB. some genera, for example, ***Botryococcus***, have colonies that are densely stained brown making it difficult to see cell contents, especially chloroplasts. This can be made easier by gently squashing the colony under a coverslip to spread the cells out thus making it easier to see the contents of individual cells.

57 (56) (a) Epiphytic, unicellular or at the most a few-celled colony in which exospores are produced at the apical end . . . .*Chamaesiphon*

Chamaesiphon is epiphytic on other aquatic plants sometimes forming dense aggregates or colonies, also epilithic. Frequent in flowing streams. Cells normally sausage or club-shaped, sometimes curved, 10–50 μm long and 2–7 μm wide. Cells usually surrounded by a sheath which may be brownish in colour. Exospores are produced at the apex of the mother cell and are spherical 2–9 μm in diameter, 5–70 μm long, and are produced in single or multiple rows. Cyanophyta. Plate VII.

(b) Cells as isolated individuals or form colonies which are not epiphytic **58**

58 (57) (a) Cells arranged in a rectilinear series, often in groups of four, forming a plate one cell thick, often of many cells . . . . ***Merismopedia***

Merismopedia cells are spherical to oval forming a colony of a single layer, shaped as a plate or rectangle within thin structureless mucilage. Cells are arranged in rows, sometimes in groups of four. The cells, often a pale blue in colour, are 0.5–5 μm in diameter and 1–16 μm long. Free-floating or resting on bottom sediments. Also occurs in saline environments. Cyanophyta. Plate IX.

(b) Cells not as above but individual or in spherical, ovate or irregular colonies but 3-dimensional and not a flat plate **59**

59 (58) (a) Cells in distinct colonies which form hollow spheres with cells arranged regularly around the periphery **60**

(b) Cells individual or in colonies that are not hollow at most clathrate (see Glossary) with cells distributed throughout **63**

60 (59) (a) Cells pear-shaped to sub-spherical, sometimes with a mucilaginous sheath. Cells arranged at the ends of branching mucilaginous strands separate from one another and radiating from the colony centre . . . . ***Gomphosphaeria***

Gomphosphaeria is widely-dispersed in lakes, ponds and ditches where it may become the dominant planktonic alga. Cells 1.5–12 μm wide, 2–16 μm long. Cells have distinct gelatinised envelopes and are often arranged in clusters within the colony. Reported as producing taste and odour in waters used for drinking (Palmer, 1962). Cyanophyta. Plate IX. Figs. 1.3 and 4.30.

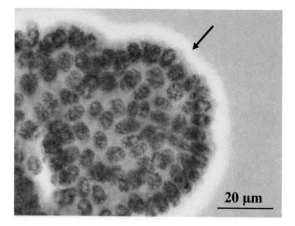

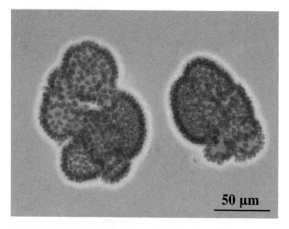

Figure 4.30 *Gomphosphaeria.* Top: Detail from edge of colony showing pear-shaped, granular cells in side view (arrow). Bottom: Typical lobed colonies, which have been slightly compressed to accentuate surface mucilage and to see peripheral cells. Indian ink preparation

(b) Cells spherical, not pear-shaped, in globular to spherical colonies.. . . . . **61**

61 (60) (a) Cells not on the end of mucilaginous stalks within the colony . . ***Coelosphaerium***

Cells of ***Coelosphaerium*** are spherical to sub-spherical or even oval. The cells are not attached to stalks within the colony, are closely-packed, and form a more or

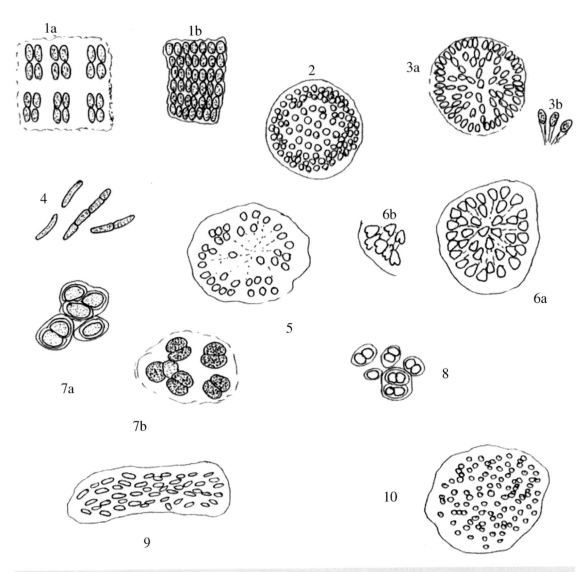

Plate IX   **1.** *Merismopedia*: (a) *M. punctata*, (b) *M. elegans*. **2.** *Coelosphaerium*. **3.** *Woronichinia*: (a) colony, (b) cells with radiating stalks attached. **4.** *Synechococcus*. **5.** *Snowella*. **6.** *Gomphosphaeria*: (a) colony, (b) individual cells. **7.** *Chroococcus*: (a) *C. turgidus*, (b) *C. limneticus*. **8.** *Gloeocapsa*. **9.** *Aphanothece*. **10.** *Aphanocapsa*.

less single layer towards the outside of the colony. Colonies are blue-green in colour and are common in the plankton sometimes reaching bloom proportions. Cells 2–5 μm wide, 2–7 μm long. Gas vacuoles may be present. This genus can potentially produce taste and odours in drinking waters. Colonies 20–180 μm in diameter. Cyanophyta. Plate IX.

(b) Cells with stalks (sometimes not very distinct) radiating from the centre of the colony **62**

62 (61) (a) Cells spherical to slightly elongated. Arranged at the ends of thin branched stalks which are quite easily seen . . . . ***Snowella***

Cells forming a free-floating globular colony embedded in mucilage. The cells are located towards the outside of the colony but do not form a single distinct layer and are located at the ends of distinct branched stalks. Cells 1.5–3.0 μm in diameter, 2–4 μm long. Gas vacuoles uncommon. Colonies up to 80 μm in size. Widespread in the plankton of eutrophic standing waters. Cyanophyta. Plate IX.

(b) Cells spherical to obovoid, stalks indistinct but quite thick near colony centre. Stalks branched, with cell at the end of the sometimes numerous branches . . . . ***Woronichinia***

Woronichinia cells are arranged in the outer region of the colony which may contain several hundred cells. Cells (3.5–5.0 μm diameter, 5–7 μm long) are embedded in fine mucilage. Colonies are free-floating and up to 180 μm in diameter. Common in the plankton of standing waters. Can be difficult to distinguish from Coelosphaerium. Cyanophyta. Plate IX.

63 (59) (a) Cells solitary or forming small colonies of 2–4–8 (rarely 16–32) cells, sometimes but not always embedded in mucilage **64**

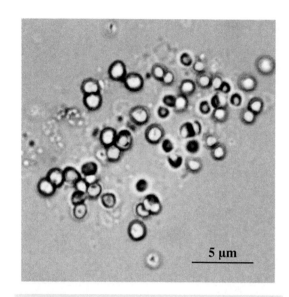

Figure 4.31 *Synechocystis*. Group of loosely associated cells from plankton sample. Reproduced with permission from G. Ji.

(b) Mucilaginous colonies composed of many cells, often in the hundreds **66**

64 (63) (a) Cells spherical or slightly oval **65**

(b) Cells elongate, that is, longer than wide, sometimes cylindrical . . . . . ***Synechococcus***

Synechococcus cells are ovoid to cylindrical, 1–12 μm wide, 2–10 times as long as wide. Cells either solitary or in twos or rarely fours, sometimes in short chains. Mucilage envelope either thin or absent. Frequent in freshwater plankton and on damp surfaces. Also in brackish waters. Cyanophyta. Plate IX. Fig. 4.31 shows ***Synechocystis***, a closely related genus.

65 (64) (a) Cells with a distinct envelope of mucilage which is lamellate and usually thick . . . . ***Gloeocapsa***

Gloeocapsa cells are spherical (1–17 μm diameter) and are surrounded by a sheath

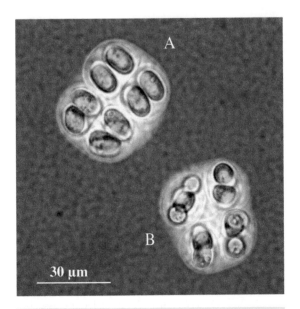

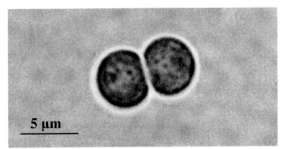

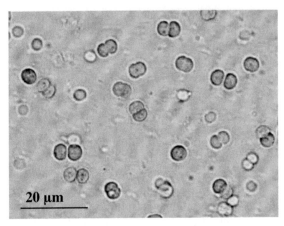

Figure 4.32 *Gloeocapsa*. Two colonies with thick mucilaginous sheaths. The plates of cells can be seen in both face (A) and side (B) view. Indian ink preparation.

Figure 4.33 *Chroococcus*. Top: High contrast image of pair of cells, with granular contents and central pale nucleoid. Bottom: Actively growing culture with cells at various stages of division and separation.

which is usually laminate, can be up to 10 μm thick and can vary in colour. Colonies may grow large enough to see with the naked eye. Abundant especially on wet rocks and tree trunks although sometimes epiphytic. Can produce taste and odour in drinking waters (Palmer, 1962). Cyanophyta. Plate IX. Fig. 4.32.

(b) Cells nearly spherical. After division daughter cells occur in groups of 2–4–8–16 in a gelatinous sheath which is often homogenous with the surrounding mucilage but may be lamellate in some species .... *Chroococcus*

Chroococcus usually forms small groups of cells which can either be free-floating or attached. Cells (2–58 μm diameter) have distinct sheaths which may be reformed after each cell division resulting in a multilayered sheath. Planktonic species tend to have gas vacuoles and cell sheaths that are less

distinct since surface layers are often confluent with the surrounding mucilage. Easily confused with *Gloeocapsa*. Cyanophyta. Plate IX. Fig. 4.33.

66 (63) (a) Large mucilaginous colonies. Cells approximately spherical to globular in shape **67**

(b) Large mucilaginous colonies. Cells cylindrical to elongate in shape .... *Aphanothece*

Aphanothece cells are spherical to ovate or cylindrical and loosely embedded in copious mucilage. Free-floating or sedentary colonies which may be tens of millimetres in size. Cells (1–4 μm wide, 2–8 μm long) have a typical blue-green colour and the smaller species are easily confused with bacteria. Common in the plankton or along the margins of lakes and ponds where it can occur on wet rocks. Cyanophyta. Plate IX. Fig. 1.3.

67 (66) (a) Cells of colony densely crowded within the mucilage . . . . *Microcystis*

Cells of *Microcystis* are spherical to subspherical (very slightly elongate) and are usually gas vacolate. They form large globular to irregular mucilaginous colonies (often containing hundreds of cells) a millimetre or more in size that are planktonic and are often responsible for nuisance water blooms. Colonies may be globular, more elongate or irregular and with holes (clathrate) and the mucilage is distinct and fairly firm. During periods of non-active growth, the colonies may rest on the bottom and in that state may loose their gas vacuoles. Colonies increase in number by fragmentation. Cells 2.5–6 µm diameter. During the actively growing season, it is planktonic and may form blooms in eutrophic waters. Frequently reported as producing taste and odours in drinking water (Palmer, 1962), large planktonic populations can also clog filtration systems in water treatment works. *Microcystis* is well known for producing toxins such as microcystins and lipopolysaccharides in water (Chorus and Bartrum, 1999). Cyanophyta. Figs. 1.3, 2.18, 2.21 and 4.34.

(b) Cells more spaced out within the mucilage so that the colonies appear less dense . . . . *Aphanocapsa*

Aphanocapsa forms globular amorphous gelatinous colonies that are usually freefloating but can be terrestrial. Cells are much more spaced out than in a *Microcystis* colony and also sometimes occur as pairs within the colony as a result of cell division. The mucilage at the outer margin of the colony is not as distinct as in *Microcystis.* Cells

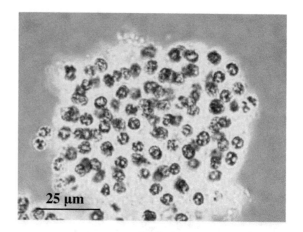

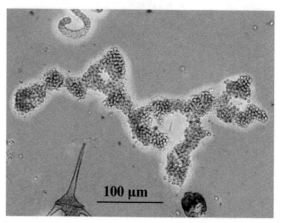

25 µm

100 µm

Figure 4.34 *Microcystis*. Indian ink preparations. Top: Detail from edge of colony showing granular algal cells within copious mucilage Bottom: Typical appearance of entire colony, loosely extending in all directions. The colour of this alga varies from olive brown (top) to fresh blue-green (bottom).

approximately spherical 1–5.5 µm in diameter depending upon species. Common in the plankton. Cyanophyta. Plate IX. Fig. 4.35.

68 (56) (a) Cells arranged in colonies of definite shape **69**

(b) Cells either individual, in pairs or in aggregations with no definite shape **106**

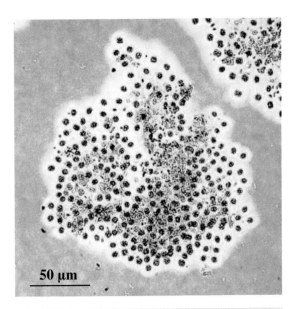

Figure 4.35 *Aphanocapsa.* Low power view of colony. Olive-brown cells of this alga are widely dispersed within the copious mucilage. A range of other algae and bacteria are also present. Indian ink preparation.

69 (68) (a) Cells of the colony with flagella, colony motile **70**

NB. It is important to look at fresh specimens to determine the presence or absence of flagella as they may be difficult to see in preserved samples.

(b) Cells of colony without flagella, colony non-motile **76**

70 (69) (a) Each cell is enclosed in a flask-shaped lorica, narrow and pointed at one end, wide at the other . . . . *Dinobryon*

(see also key No. 7)

(b) Cells not located in a lorica **71**

71 (70) (a) Adjacent cells touching in densely packed colonies. **72**

(b) Cells spaced apart within a colourless mucilage matrix. **73**

72 (71) (a) Chloroplast green and cup-shaped, storage product starch. Cells, embedded in mucilage that obviously extends beyond the cells at the colony edge, have two equal-length flagella . . . . *Pandorina*

Colonies of *Pandorina* are spherical to oval in shape with 8–16 (32) densely packed cells. Individual cells (8–20 μm long) have flattened sides where they touch their neighbour and also slightly flattened apices. On the outside of the cells is a wide band of mucilage through which the flagella protrude. The colony centre is not hollow. Individual cells may divide to form daughter colonies which are later released. Colonies swim with a tumbling motion through the water. Common in the plankton. Reported to impart a fishy odour to drinking water (Palmer, 1962). Chlorophyta. Plate X. Figs. 1.3 and 4.36.

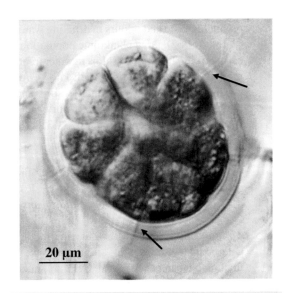

Figure 4.36 *Pandorina.* Colony showing densely packed cells with surface mucilage through which flagella protrude (arrows). Reproduced with permission from R. Matthews.

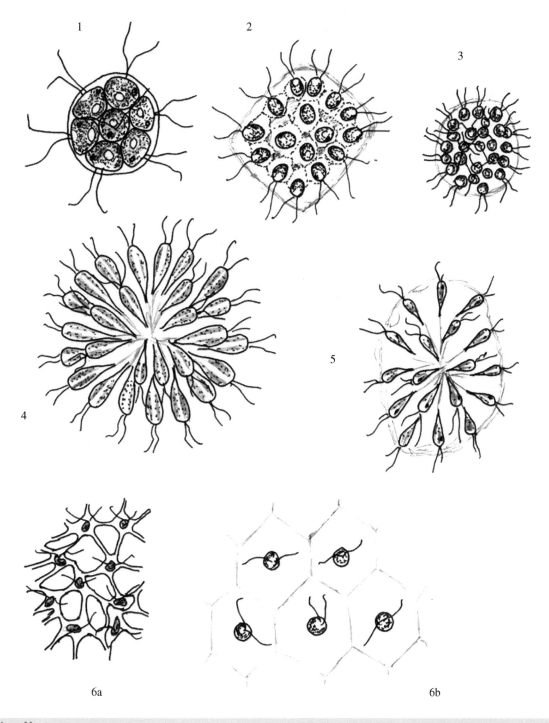

Plate X **1.** *Pandorina.* **2.** *Gonium.* **3.** *Eudorina.* **4.** *Synura.* **5.** *Uroglena.* **6.** *Volvox,* surface detail of colony: (a) *V. globator,* (b) *V. tertius.*

(b) Chloroplasts brown to golden-brown. Storage product leucosin. No obvious wide mucilaginous envelope. Each cell has two flagella of unequal length . . . .***Synura***

Cells of **Synura** have two golden-brown chloroplasts and are more pear-shaped, 7–17 μm wide. Up to 40 closely packed cells per colony. Cells are covered with fine silica scales which are not always obvious. Common in the plankton. Can impart both taste and odour to drinking waters (Palmer, 1962). Chrysophyta. Plate X. Figs. 1.3 and 4.37.

73 (71) (a) Colonies of 64 cells or less **74**

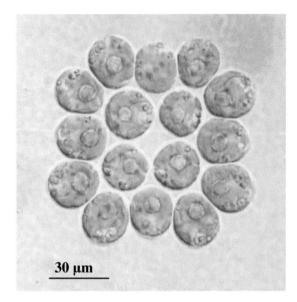

Figure 4.38 *Gonium.* Single colony (one cell deep) showing regular arrangement of cells. Internal detail: prominent nucleus, cup-shaped chloroplast, orange eye-spot, but no flagella are visible in this DIC image.

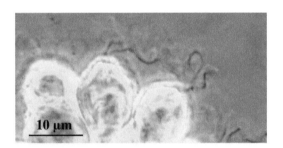

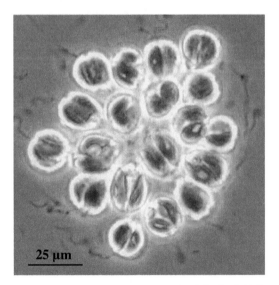

Figure 4.37 *Synura.* Top: Edge of colony (lightly fixed preparation) showing biflagellate golden cells. Bottom: Live colony, phase contrast image.

(b) Colonies with many more than 64 cells, often over 100 . . . . **75**

74 (73) (a) Colony a flat plate of 4–16 (sometimes 32) ovoid to spherical cells. Cell flagellae are all directed within the same plane (i.e. point directly outwards from the side of cells rather than upwards or downwards) . . . . . .***Gonium***

Cells (7–20 μm wide) within a **Gonium** colony are arranged in a flat gelatinous envelope. Each cell has two equal-length flagella. The chloroplast is green and cup-shaped with one or more pyrenoids and an eye-spot. Common in relatively still waters. Can impart both taste and odour to drinking waters (Palmer, 1962). Chlorophyta. Plate X. Fig. 4.38.

(b) Colony globular to elliptical composed of 16–32 (sometimes 64) cells. Cells spherical in shape and arranged near to the surface of the mucilaginous matrix. Each cell has two equal-length flagella which can point in different directions from the colony edge .... ***Eudorina***

Unlike the cells of a ***Pandorina*** colony, those of ***Eudorina*** are spherical and spaced out within the mucilaginous matrix near to the edge, leaving a clear zone at the centre. In immature colonies, they may be more closely packed. The chloroplasts are cup-shaped and green with one or more pyrenoids and an eye-spot. The storage product is starch. Widely distributed in the plankton and can form dense growths. Can impart both taste and odours to drinking waters (Palmer, 1962). Chlorophyta. Plate X. Fig. 4.39.

75 (73) (a) Chloroplasts golden-brown. Cells pear-shaped with thread-like mucilage strands connecting them to the colony centre. Storage product leucosin .... ***Uroglena***

Colonies of ***Uroglena*** can be quite large, up to 1 mm diameter, and are composed of many hundreds of cells. Individual cells (10–22 μm in length) possess one or two chloroplasts and have two, unequal-length, flagella. Cysts are produced which have smooth walls and may not have a collar. Although similar, ***Uroglena*** colonies can be distinguished from ***Volvox*** by their golden-brown chloroplasts, pear-shaped cells and the unequal flagella. Common in the plankton. Reported to be able to impart unwanted odours to drinking waters (Palmer, 1962). Chrysophyta. Plate X.

(b) Chloroplasts green, storage product starch. Cells spherical with interconnecting mucilaginous strands and are situated around the periphery of the colony. Mucilaginous strands do not radiate from the centre of the colony. Each cell with two equal-length flagella .... ***Volvox***

Volvox colonies are large (up to 2.5 mm diameter) and spherical, composed of hundreds or thousands of cells. Asexual repro-

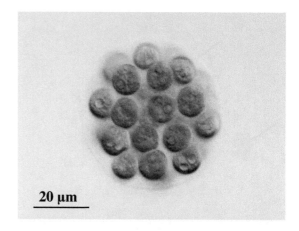

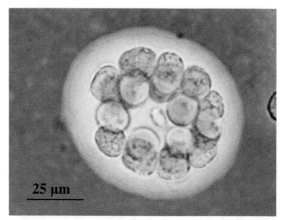

Figure 4.39 *Eudorina*. Top: Cells dispersed (not touching) within the globular colony. Bottom: Colony slightly compressed to show central cavity and mucilaginous matrix. Indian ink preparation. The flagella are not visible in either the DIC (top) or BF (bottom) images.

duction occurs by formation of daughter colonies that develop from special cells within the hollow colony, from which they are released. Sexual reproduction is ooga-mous, with the formation of sperm packets and egg cells from vegetative cells, leading to zygote formation. Common in the plankton. May produce a fishy odour in drinking waters (Palmer, 1962). Chlorophyta. Plate X. Fig. 4.40.

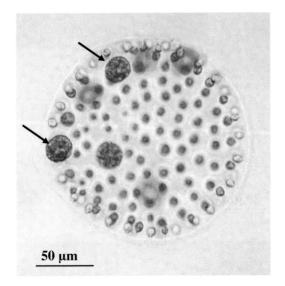

50 μm

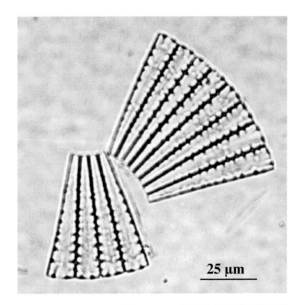

25 μm

Figure 4.41   *Meridion.* Fan-shaped colonies of this colonial diatom. Prepared slide.

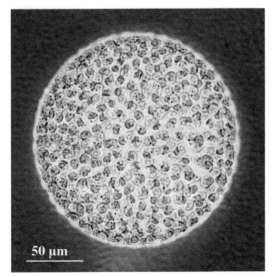

50 μm

Figure 4.40   *Volvox.* Top: Image focussed on the equatorial cells of the spherical colony. Numerous zygotes (arrows) are present in the central cavity. Bottom: Indian ink preparation, showing limits of colony. Flagella not visible in these bright field images.

76 (69) (a) Cells with silica walls which often bear distinct markings. Storage products mainly oils that can often be seen as small globules within the cell, not starch **77**

(b) Cells do not have silica walls. Main storage product starch. **81**

77 (76) (a) Cells cuneate in shape, so when joined together form a fan-shaped colony . . . . *Meridion*

The cells, or frustules, of ***Meridion*** are heteropolar in girdle view and isobilateral in valve view. Within the frustule are thickened transparent costae and on the frustule surface fine parallel striae. A pseudoraphe is present on both valves. Individual cells (24–45 μm long) are usually united to form fan-shaped colonies. If growth of the colony is vigorous then a flat spiral-shaped colony develops. Grows attached to surfaces in shallow waters. Bacillariophyta. Plate XI. Fig. 4.41.

(b) Cells form stellate or zig-zag colonies **78**

78 (77) (a) Frustules swollen at each end to form a knob, the inner one at the hub of the

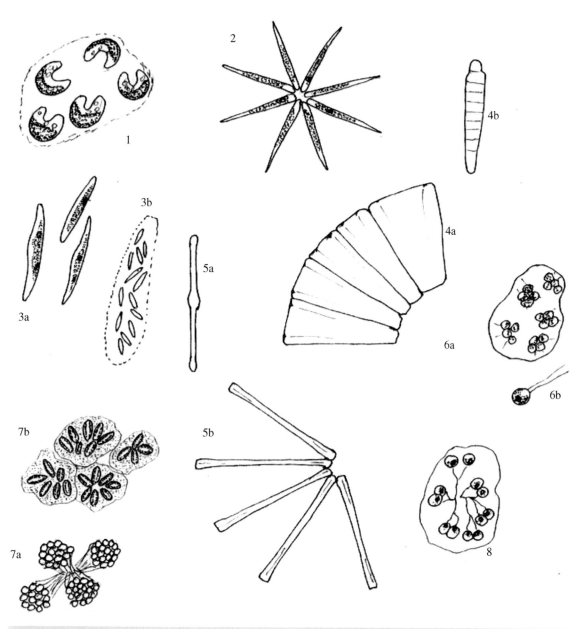

Plate XI   **1.** *Kirchneriella.* **2.** *Actinastrum.* **3.** *Elakatothrix*: (a) individual cells, (b) colony, **4.** *Meridion*: (a) colony, (b) single cell, valve view. **5.** *Asterionella*: (a) single cell, girdle view, (b) colony. **6.** *Tetraspora*: (a) colony, (b) zoospore. **7.** *Botryococcus*: (a) general view of colony, (b) detail of cells. **8.** *Dictyosphaerium.*

star-shaped colony being slightly larger than the outer one . . . . ***Asterionella***

Asterionella frustules are long (40–130 μm), straight and narrow – up to 20 times as long as wide. The narrow central part of the frustule does have fine transverse striae but these are difficult to see. There are usually two or more chloroplasts per cell. ***Asterionella*** can be abundant in the plankton of lakes especially in the spring (spring diatom bloom) and to a lesser extent in the summer and autumn. Can impart a geranium or spicy odour to drinking waters (Palmer, 1962). Large growths of ***Asterionella*** can cause problems of filter clogging in water treatment works supplied by lakes or reservoirs. Bacillariophyta. Plate XI. Figs. 1.16 and 4.42.

(b) Frustules not swollen at each end, colonies either stellate or zig-zag **79**

79 (78) (a) Frustules with neither internal septa nor costae . . . . ***Synedra***

Synedra has elongate, linear, isopolar valves. There is a narrow pseudoraphe and fine transverse striae. Colonies of ***Synedra*** may be stellate or short chains but it can also be present as single cells or as an attached epiphyte. Widely distributed and may be abundant. Can produce both unwanted taste and odour problems in drinking waters and, when in large numbers, clog filters (Palmer, 1962). Bacillariophyta. Plate XXVI. Figs. 2.23 and 4.43. **(see also key No. 191)**

(b) Frustules have either internal septa or costae **80**

80 (79) (a) Frustules with internal longitudinal septa but no costae. There are small polar and larger central swellings when seen in valve view . . . . ***Tabellaria***

(see also key No. 29)

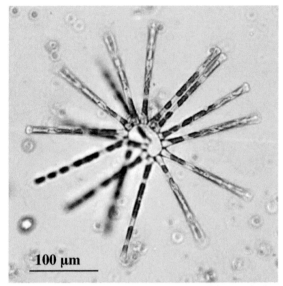

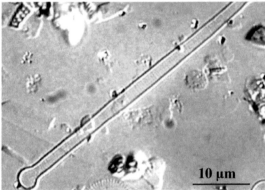

Figure 4.42 *Asterionella*. Top: Stellate planktonic colony, double spiral. Cells with 5–10 brownish chloroplasts. Bottom: Acid digest, lake sediment sample. Single cell showing frustule with rounded ends. Images reproduced with permission from G. Ji (top) and M. Capstick (bottom).

NB. Although the swollen centre can be clearly seen in valve view if you are observing a stellate colony you see the cells in girdle view and the swelling is not visible. ***Tabellaria*** cells, unlike those of ***Asterionella*** with whose colonies they may be confused, do not have swollen and rounded ends to the cells. However, shorter cells and colonies of ***Asterionella*** do appear (sometimes less than a

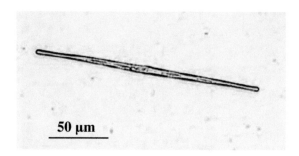

Figure 4.43 *Synedra.* Pennate diatom. Elongate fusiform cell of a common planktonic species.

quarter the length of normal cells) and these do not have such obvious swellings at the poles which can give rise to confusion.

(b) Frustules without internal septa but with thick internal costae. Valve ends may be swollen but no median swelling present. Cells almost elliptical (albeit elongated in some species) in valve view. Bacillariophyta. ....*Diatoma*

(see also key No. 29)

81 (76) (a) Cells elongate, cigar-shaped, without silica walls, radially arranged in a star-shaped colony with the cells attached to each other at one end only ....*Actinastrum*

Cells of **Actinastrum** join together to form 4–8–16 celled star-shaped colonies. Each cell (10–25 μm long, 3–6 μm wide) has a single chloroplast with pyrenoid. Common in the plankton of lakes and may occur in rivers, ponds and boggy waters. Chlorophyta. Plate XI. Fig. 4.44.

(b) Cells not elongate like a cigar, often more spherical, cubical, crescent-shaped or at the most like a short cylinder. Can form plate-like colonies. Stellate colonies not formed **82**

82 (81) (a) Cells of colonies arranged within a definite mucilaginous envelope.............**83**

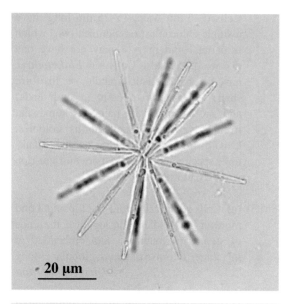

Figure 4.44 *Actinastrum.* Single colony, with cells radiating out from a central point.

(b) Cells not within a definite mucilaginous envelope **91**

83 (82) (a) Cells crescent-shaped or fusiform ... **84**

(b) Cells spherical, ovoid or other shape **85**

84 (83) (a) Cells crescent-shaped (lunate), irregularly arranged in small mucilaginous colonies...................**Kirchneriella**

Kirchneriella cells have a single parietal chloroplast with a pyrenoid. Lunate or crescent-shaped cells are often so curved that their ends almost touch, although some may be sigmoid or slightly spiral. Cells (3–8 μm wide, 10–14 μm long) are often enclosed in mucilage and irregularly arranged in groups of 2, 4, 8 (up to 32). Free-floating and widely distributed in the plankton. Chlorophyta. Plate XI.

(b) Cells fusiform or wedge-shaped ovals ....*Elakatothrix*

Cells (3–6 μm wide, 15–25 μm long) have a single chloroplast (sometimes two) which is parietal and may or may not have one or two pyrenoids. Cells of **Elakatothrix** have a characteristic spindle or fusiform shape (tapering to a point at both ends) and are arranged in pairs or somewhat irregularly within mucilaginous colonies. Usually free-floating, an occasional planktonic species especially in more nutrient-rich waters. Chlorophyta. Plate XI.

85 (83) (a) Cells spherical, ovoid or ellipsoidal and more or less radially arranged at the ends of strands of mucilage, and embedded in mucilage, to form a radiating globular colony . . . . *Dictyosphaerium*

Dictyosphaerium coenobia are composed of 4 to 64 cells which are spherical to subreniform and are connected by threads radiating from the colony centre. The whole is embedded in mucilage. Cells spherical to ovoid, 3–10 μm in diameter, with parietal cup-shaped chloroplasts with pyrenoids. Frequent in the plankton of lakes and ponds and slow moving waters. Can be found amongst submerged macrophytes or submerged objects. May impart an odour to drinking waters (Palmer, 1962). Chlorophyta. Plate XI. Fig. 4.45.

(b) Cells not attached to mucilage strands that are radially arranged **86**

86 (85) (a) Cell of colony very densely packed, often appearing dark brown to black in colour. Individual cells can be difficult to distinguish . . . . **Botryococcus**

Cells of **Botryococcus**, difficult to see because they are so densely packed, may be visualised by gently squashing the colony under the microscope cover slip. The colonies may be pale but are often quite dark and look like a piece of organic debris. Cells, roughly spherical in shape (3–9 μm wide, 6–

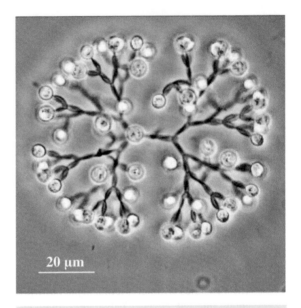

Figure 4.45 *Dictyosphaerium*. Phase contrast image of single colony. Cells are attached to fine mucilaginous threads radiating out from the centre of the colony. Reproduced with permission from R. Matthews.

10 μm long), are embedded in a tough oily mucilage. The chloroplast is parietal with a pyrenoid. Widespread in the plankton but not usually abundant. Chlorophyta. Plate XI.

(b) Cells of colony not so densely packed and colonies not brown in colour **87**

87 (86) (a) Cells with pseudocilia (see Glossary) . . . . *Tetraspora*

Tetraspora forms large irregular gelatinous, often balloon-like, colonies that can be visible to the naked eye. The cells (6–12 μm diameter) are in groups of 2 or 4 within the mucilage. Each cell usually has two fine pseudocilia and they are arranged around the edge of the colony and may extend beyond the mucilage. The chloroplast is cupshaped and a pyrenoid is present. Common in the plankton of shallow waters, ponds and ditches or associated with submerged macrophytes. Chlorophyta. Plate XI.

(b) Cells without pseudocilia.......... **88**

88 (87) (a) Mucilaginous surrounds to the cells show marked stratification or layering ... **89**

(b) Mucilaginous surrounds to cells more or less homogenous, no marked stratification **90**

89 (88) (a) Chloroplast dense and star- shaped .... *Asterococcus*

Cells of ***Asterococcus*** are spherical or globular, up to 40 μm in diameter. A single dense star-shaped chloroplast with arms radiating from the centre is present together with a centrally located pyrenoid. Small groups (1–16) of cells are surrounded by thick stratified or lamellate hyaline mucilage. Free-floating, common in softer shallow waters usually amongst other algae and moss. Chlorophyta. Plate XII.

(b) Chloroplast cup-shaped, not star-shaped, and nearly fills the whole cell . . ***Gloeocystis***

Gloeocystis cells (4–23 μm diameter) are spherical and usually in groups of 4–8–16 surrounded by stratified or lamellate mucilage but may also be solitary. There is a single cup-shaped parietal chloroplast with pyrenoid and often many starch grains. Found in more stagnant ponds and ditches as part of the tychoplankton. Can, if present in large numbers, impart unwanted odour to drinking waters (Palmer, 1962). Chlorophyta. Plate XII.

90 (88) (a) Chloroplast cup-shaped with a single pyrenoid .... ***Palmella***

Palmella cells are spherical or rounded cylinders, 3–15 μm wide, rather like Chlamydomonas in the way they are organised. They are surrounded by an indefinite mucilaginous mass which may be coloured greenish to reddish and are found on damp surfaces. Some

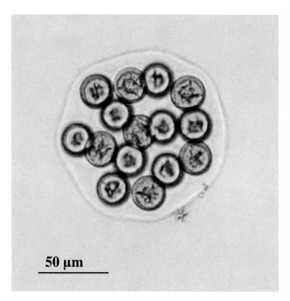

50 μm

Figure 4.46 *Sphaerocystis*. Free-floating colony of cells dispersed throughout a globule of mucilage.

other species of algae have palmelloid stages in their life cycle which gives rise to confusion. The jelly-like masses can cause blockage of filters in water treatment processes. Chlorophyta. Plate XII.

(b) Chloroplast parietal with none to many pyrenoids .... ***Sphaerocystis***

Cells 7–20 μm wide, spherical. Colonies globular and free-floating with cells embedded in stuctureless mucilage. Reproduction occurs by division of the mother, generating 8–16 daughter cells which form characteristic microcolonies within the parent colony. Common and sometimes abundant in the plankton of lakes and also found on damp soils. Chlorophyta. Plate XII. Fig. 4.46.

91 (82) (a) Colonies either spherical, oval or a flat disc **92**

(b) Colonies not as above **101**

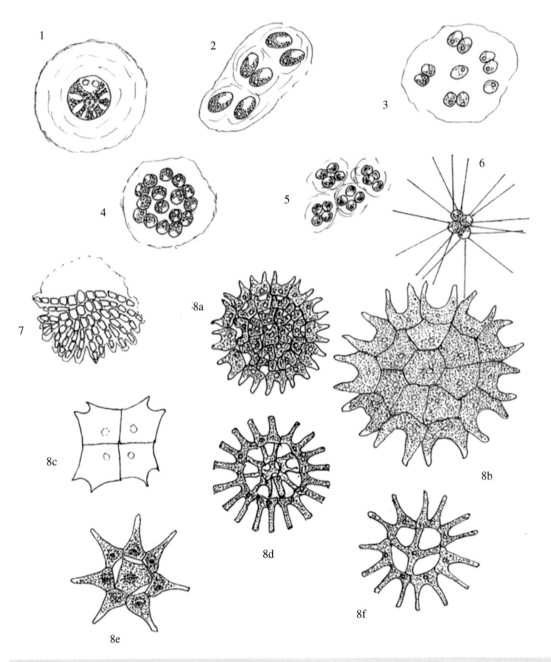

Plate XII   **1.** *Asterococcus.* **2.** *Gloeocystis.* **3.** *Palmella.* **4.** *Sphaerocystis.* **5.** *Westella.* **6.** *Micractinium.* **7.** *Protoderma.*
8. *Pediastrum:* (a) *P. boryanum* v. *cornutum,* (b) *P. boryanum,* (c) *P. tetras,* (d) *P. duplex,* (e) *P. simplex,* (f) *P. simplex* v.
gracillimum.

92 (91) (a) Cells of colony form a flat disc......**93**

(b) Cells forming a spherical or ovoid colony or globular groups of 4–16 cells **96**

93 (92) (a) Colony free-floating.......**Pediastrum**

Pediastrum forms characteristic flat circular, plate-like colonies which are common in nutrient-rich lakes, ponds and slow flowing rivers. The cell walls are often quite tough and persist for some time after the contents have disappeared. Can, if present in large numbers, impart unwanted odour to drinking waters (Palmer, 1962). Chlorophyta. There are many species, differing in overall colony appearance and the shape of the marginal cells – examples shown in Plate XII. Figs. 2.18 and 4.47.

(b) Colony attached to a surface.......**94**

94 (93) (a) At least some of the cells bear setae (see Glossary)............................**95**

(b) Cells without setae ....**Protoderma**

Protoderma forms a flat cushion or disc one cell thick (sometimes becomes a little thicker with age) that is composed of irregularly branched partially radiating filaments. This is more obvious towards the edge of the disc. Cells quadrate to cylindrical up to 15 μm in length. Chloroplasts are parietal with a pyrenoid. Commonly attached to submerged aquatics. Chlorophyta. Plate XII.

95 (94) (a) Setae with sheaths, cells up to 40 μm in length......................**Coleochaete**

(see also key No. 12)

(b) Setae not sheathed, frequent. Cells up to 30 μm long.................**Chaetopeltis**

Colonies composed of radiating rectangular cells forming a flat disc (up to 1

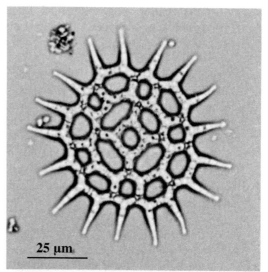

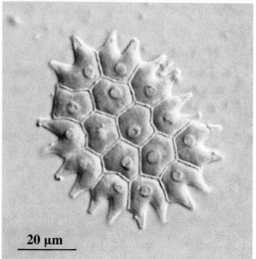

Figure 4.47 *Pediastrum.* Top: Open colony with fretwork of cells. Reproduced with permission from G. Ji. Bottom: Closed plate-like colony, with prominent cell walls and nuclei (DIC image). Separate species. In both cases, the colony is one cell thick – with distinctive peripheral and central cells.

mm in diameter) on submerged aquatic plants. Chloroplast cup-shaped. Some cells with pseudocilia. Chlorophyta. (Not illustrated.)

96 (92) (a) Cells bearing long spines and forming free-floating colonies of 4–16 cells
. . . . ***Micractinium***

Cells (3–7 μm diameter) more or less spherical, forming small clusters or colonies, 4
being the most frequent number but can be
many more. Each cell with 1–5 fine tapering
spines, 10–35 μm long. Chloroplasts parietal
with a single pyrenoid. Occurs frequently in
the plankton of lakes, ponds and larger rivers,
especially if enriched with nutrients. Chlorophyta. Plate XII. Fig. 4.72.

(b) Cells without spines **97**

97 (96) (a) Cells globose to spherical forming a
tightly packed hollow sphere of up to 100
cells **98**

(b) Cells oval or with finger-like processes
extending outwards, more loosely arranged.
Oval cells may have thickened poles or may
be within an encasing envelope **99**

98 (97) (a) Up to 100 cells per colony, loosely held
together by the mother cell walls from the
previous generation . . . . ***Westella***

Westella cells form a free-floating colony
loosely connected by the original mother cell
walls. Cells 3–9 μm in diameter, chloroplast
parietal with or without a pyrenoid. Planktonic in ponds and lakes.

NB. Some authors regard ***Westella*** as syn-
onymous with ***Dictyosphaerium.*** Chlorophyta. Plate XII.

(b) Hollow spherical colonies of up to 64
closely joined cells arranged in a regular
sphere rather like a football . . . . ***Coelastrum***

Cells of ***Coelastrum*** are spherical, 8–
30 μm in diameter with a parietal chloroplast and pyrenoid. Found in the plankton of eutrophic and mesotrophic lakes and

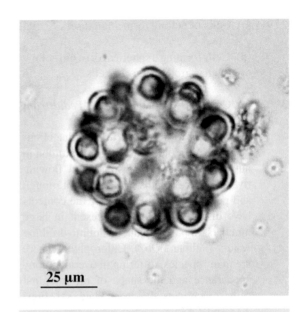

25 μm

Figure 4.48 *Coelastrum.* Hollow colony of closely
joined, regularly arranged cells. Reproduced with permission from G. Ji.

slow flowing rivers. Chlorophyta. Plate XIII.
Fig. 4.48.

99 (97) (a) Cells with finger-like or spike-like processes extending outwards (1–4 per cell).
Cells kidney or pear-shaped and joined to
each other by a mucilaginous protrubance so
as to form a radiating colony . . . . ***Sorastrum***

Sorastrum is related to ***Pediastrum*** but the
cells form a more spherical colony of 8–64
cells. Chloroplast parietal with a pyrenoid.
The outer facing surface of the cell bears
1–4 finger- or spine-like processes 4–15 μm
long. Uncommon, found in similar locations
to ***Pediastrum.*** Chlorophyta. Plate XIII.

(b) Cells ovoid to spherical and without
finger-like processes or spines **100**

100 (99) (a) Cells (4–50 μm long) usually oval and
often having a clearly observable polar
swelling or nodule at each end. Mother cell
wall, when present, more or less entire and

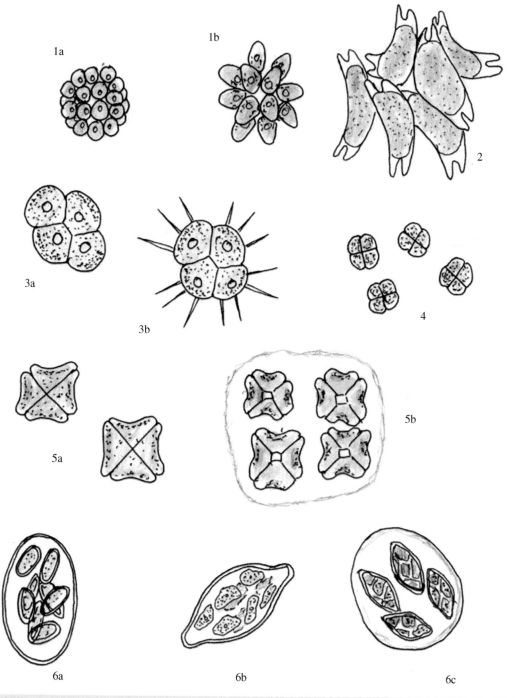

Plate XIII **1.** *Coelastrum*: (a) *C. microporum*, (b) *C. asteroideum*. **2.** *Sorastrum*. **3.** *Tetrastrum*: (a) *T. triangulare*, (b) *T. staurogeniforme*. **4.** *Pleurococcus*. **5.** *Crucigenia*: (a) *C. tetrapedia*, (b) *C. fenestrata*. **6.** *Oocystis*: (a) *O. elliptica*, (b) *O. solitaria*, (c) *O. natans*.

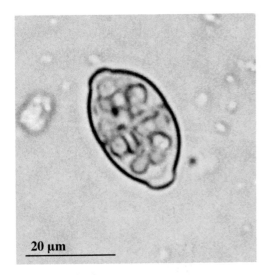

20 µm

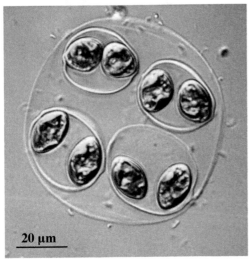

20 µm

Figure 4.49 *Oocystis*. Top: Joined pair of cells, showing disk-shaped chloroplasts and a polar swelling at each end of the colony. Bottom: Pairs of daughter cells, enclosed within two sets of mother cell walls. Reproduced with permission from R. Matthews.

surrounds 2–4–8 daughter cells. Chlorophyta Plate XIII. Fig. 4.49. . . . . . *Oocystis*

(see also key No. 209)

(b) Cells (3–9 µm wide) normally spherical, without polar nodules. Mother cell

walls fragmented. Up to 10+ cells per colony, usually arranged in groups of 4 or 8. . . . . *Westella*

(see also key No. 98)

101 (91) (a) Cells roughly cylindrical (or sausage-shaped, ellipsoidal or fusiform) and usually more than 2 times long as wide. Colonies of usually single, but sometimes double alternating rows of 4–16 (32) cells that are joined along the long axis. . . . . *Scenedesmus*

Scenedesmus is a very common and sometimes abundant genus, especially found in eutrophic and hypertrophic waters. Some species bear spines, others ridges and others no ornamentation. Cell sizes vary greatly from species to species. When abundant may impart unwanted odours to drinking waters (Palmer, 1962). Chlorophyta. Plate XIV – showing some of the various coenobium arrangements and cell ornamentations for different species. Fig. 4.50.

NB. Some authorities re-classify spine-bearing members in the genus *Desmodesmus*.

(b) Cells as long as wide or spherical to very short cylinders joined end to end **102**

102 (101) (a) Cells very short cylinders (2–6 µm wide, 3–20 µm long) joined end to end in groups of 2,3,4 etc. . . . . . *Stichococcus*

(see also key No. 34)

(b) Cells as long as wide, spherical or globose in shape. **103**

103 (102) (a) Cells with spines or projections **104**

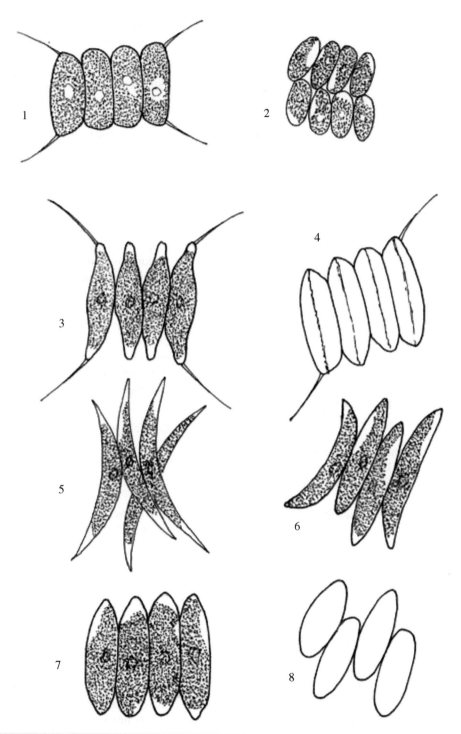

Plate XIV Species of *Scenedesmus*. **1**. *S. communis* (synonym *S. quadricauda*). **2**. *S. arcuatus*. **3**. *S. opoliensis*. **4**. *S. armatus* var. *bicaudatus*. **5**. *S. acuminatus*. **6**. *S. dimorphus*. **7**. *S. obliquus*. **8**. *S. obtusus*.

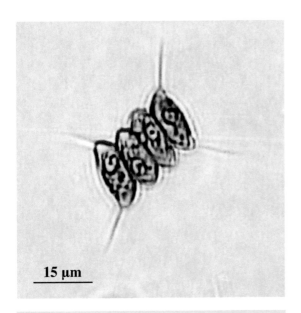

15 μm

Figure 4.50 *Scenedesmus*. Four-cell colony. This species (*S. opoliensis*) has two prominent spines on each outer cell.

(b) Cells without spines or projections
. . . . **105**

104 (103) (a) Cells spherical, in groups of 4–16, each bearing long tapering spines (many times as long as the cell) . . *Micractinium*

(see also key No. 96)

(b) Cells angular and usually in groups of four but may be solitary. Spines usually very fine and short (not more than 2 times cell diameter) . . . . *Tetrastrum*

Cells of **Tetrastrum** are 3–7 μm wide (excluding the spines) with a cup-shaped chloroplast with or without a pyrenoid. Planktonic but not usually abundant. Chlorophyta. Plate XIII.

105 (103) (a) Cells growing on a damp substrate, aerial, globular or angular in shape when

touching and forming definite clusters
. . . . *Pleurococcus*

Pleurococcus (synonyms **Apatococcus** and **Desmococcus**) has cells (5–20 μm wide) which are globular to angular and usually in dense masses, sometimes forming short filaments. The chloroplast is large, parietal and lobed, with or without a pyrenoid. Frequently found on tree bark, fences and other aerial habitats. Chlorophyta. Plate XIII.

(b) Cells aquatic, oval or triangular in shape, usually in groups of four sometimes sticking to each other by means of a thin mucilaginous surround . . . . . . *Crucigenia*

Crucigenia colonies consist of four cells arranged in a cross so that a gap may be present at the colony centre. Cells have a parietal chloroplast with a small pyrenoid, 5–10 μm wide. Frequent in the plankton. Chlorophyta. Plate XIII.

106 (68) (a) Cells with flagella, motile **107**

(b) Cells without flagella, non-motile **126**

NB. Some groups of diatoms are able to move with a gliding action and do not have flagella.

107 (106) (a) Cells covered with delicate scales bearing long bristles or spines which, at a first glance, resemble many flagella but are incapable of movement. One single flagellum occurs at the apical end and is used for locomotion. . . . . *Mallomonas*

Mallomonas is unicellular and free swimming. Its cells are an elongated oval shape, 8–100 μm long and up to 30 μm wide. They are covered with numerous fine

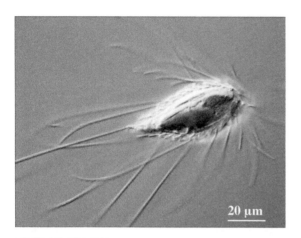

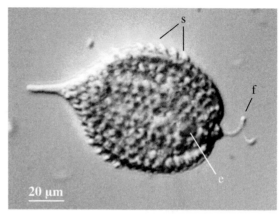

Figure 4.51a *Mallomonas.* Single motile cell (flagellum not visible) showing a surface covering of silica scales, many of which bear elongate trailing bristles. Reproduced with permission from R. Matthews.

Figure 4.51b *Phacus.* Typical flattened pear-shaped cell with a single emergent flagellum (f), a red eyespot (e), a rigid pellicle covered in spines (s) and an elongate posterior tail. Reproduced with permission from R. Matthews.

overlapping plates made of silica. These plates (which may not be obvious in live material) bear spines so that the whole organism is covered with these long spines. The spines may become detached from the plates, especially on preservation. There are two golden-brown parietal chloroplasts. The food reserve is leucosin. Common in the plankton of standing waters. ***Mallomonas*** can impart unwanted odours to drinking water (Palmer, 1962). Chrysophyta. Plate XV, Fig. 4.51a.

(b) Cells without spine-bearing scales
.... **108**

108 (107) (a) Cells with one, two or four flagella attached at one end only **109**

(b) Cells with two flagella attached other than at anterior end **123**

NB. In dinoflagellates two flagella arise at the centre of the cell. One trails backwards

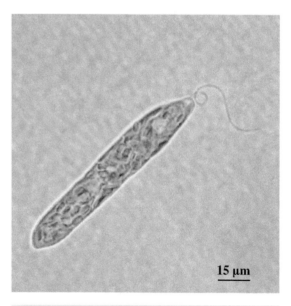

Figure 4.51c *Euglena.* Single cell with discoid chloroplasts and prominent apical flagellum.

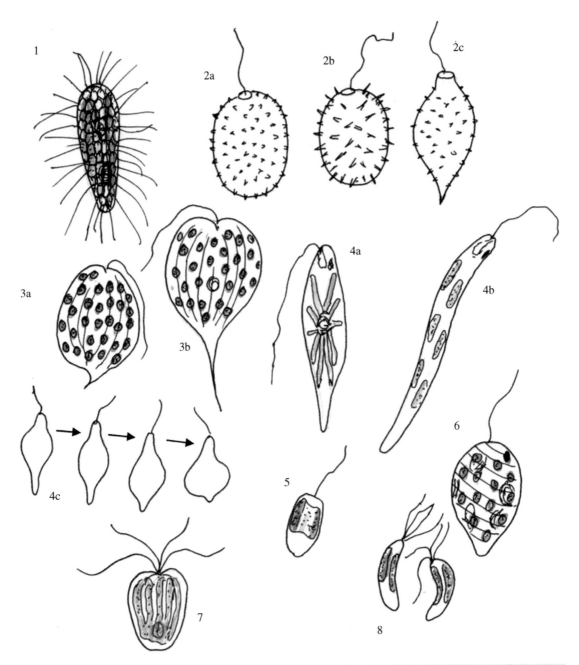

Plate XV   **1**. *Mallomonas*. **2.** *Trachelomonas*: (a) *T. hispida*, (b) *T. superba*, (c) *T. caudata*, **3**. *Phacus*: (a) *P. triqueter,* (b) *P. longicauda*. **4**. *Euglena*: (a) *E. viridis*, (b) *E. mutabilis*, (c). Euglenoid movement. **5**. *Chromulina*. **6**. *Lepocinclis*. **7**. *Pyramimonas*. **8**. *Spermatozopsis*.

and is relatively easy to see and the other lies in a furrow around the centre of the cell and is more difficult to visualise (see Fig. 4.56).

109 (108) (a) Cells with a single emergent flagellum **110**

(b) Cells with two or more emergent flagella **114**

110 (109) (a) Cells enclosed in a brownish test (see Glossary) of various shapes with the flagellum emerging from an anterior aperture .... *Trachelomonas*

Trachelomonas is free swimming and unicellular. The cells, which are *Euglena*-like, are enclosed in a roundish (although there are many variations on this shape) test, lorica or theca which may range from pale to dark brown in colour and is usually opaque. The surface of the test may be smooth, granulate or spiny. The single flagellum emerges from a round aperture at the anterior end. A red eyespot is present. There are two to many disc-shaped chloroplasts. Small paramylum bodies may be present. Found mainly in shallow waters, ditches and ponds, often as a tychoplankter, but may occur in larger water bodies. May be abundant enough to form blooms colouring the water brownish. Many species separated on the basis of the shape and ornamentation of the test. Euglenophyta. Plate XV.

(b) Cells not enclosed in a test **111**

111 (110) (a) Cells with pronounced dorsiventral flattening (leaf-like in shape) often with part of the cell twisted .... *Phacus*

Phacus is solitary and free swimming. Cells are markedly flattened and may be twisted along their length. Wide at

the anterior with single emergent flagellum but pointed at the posterior end, the point being of variable length. There are numerous disc-shaped chloroplasts without pyrenoids, and an eyespot (may not be obvious). Disc-shaped paramylum bodies (one to many) are present. The pellicle is rigid and longitudinal striations and granules may be visible on the surface. Common in ponds and other still waters, tychoplanktonic, especially where enriched with organic matter such as swamps. Numerous species. Euglenophyta. Plate XV. Fig. 4.51b.

(b) Cells not flattened or leaf-like ... **112**

112 (111) (a) Cells roughly cylindrical or fusiform, often showing metaboly (see Glossary) .... *Euglena*

Euglena cells are solitary and free swimming. They are usually elongated but may be spindle-shaped or twisted. The periplast may be flexible or firm and when firm may, in some species, show spiral markings or rows of granules. Chloroplasts one to many and variously shaped, disc-shaped or stellate to band-like, depending upon the species. Sometimes pyrenoids are present. The emergent flagellum arises from an apical reservoir. A prominent eyespot is present. Storage product is paramylum and numerous granules of variable shape may be present. Numerous species, but identification to species level difficult. Common to abundant in ponds and other shallow waters. Some species found in acid waters. If present in large numbers, *Euglena* can impart both taste and odour to drinking waters (Palmer, 1962). Euglenophyta. Plate XV. Figs. 1.6 and 4.51c.

(b) Cells ovoid to pyriform and do not show metaboly **113**

113 (112) (a) Cells ovoid, periplast firm with pointed tail . . . . ***Lepocinclis***

Lepocinclis is found in ponds and small water bodies, often alongside other euglenoids. Cells (10–20 μm wide, 25–40 μm long) are circular in section, spherical to spindle-shaped, sometimes with a tapering tail. The pellicle is rigid with spirally arranged striations on the surface. Cells have numerous discoid chloroplasts, an anterior eyespot, two annular-shaped paramylum granules and a single long flagellum (up to twice the length of the cell). Euglenophyta. Plate XV.

(b) Cells oval, without pointed tail, may show slight metaboly . . . . ***Chromulina***

Widespread in cleaner waters often amongst vegetation, often in the neuston or plankton. Cells (6–7 μm wide, 9–14 μm long) are ovoid with one or two curved plate-like chloroplasts, a leucosin body located at the posterior end and a single long flagellum. Chrysophyta. Plate XV.

Fig. 4.52c shows ***Chrysochromulina***. This is a member of the Haptophyta – unlike ***Chromulina*** which is a Chrysophyte. Haptophytes usually have two flagella and a haptonema (see Glossary). ***Chrysochromulina*** cells are spherical with two chloroplasts. Widely distributed.

114 (109) (a) Cells with four flagella **115**

(b) Cells with two flagella **117**

115 (114) (a) Cells with a wide crescent shape . . . . ***Spermatozopsis***

Only one species found, ***Spermatozopsis exsultans***. Cells (7–12 μm long, 3–4 μm wide) are crescent-shaped or curved, sometimes S-shaped. Cells have four flagella, a parietal chloroplast, an eyespot

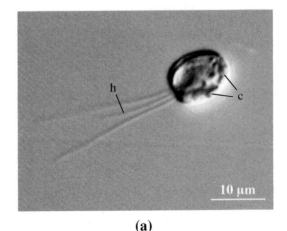

(a)

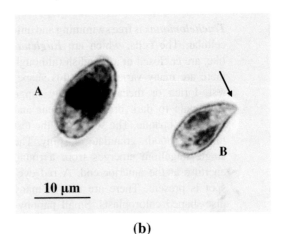

(b)

Figure 4.52 (a) *Chrysochromulina*. Single ovoid cell with two equal flagella, a central long haptonema (h) and parietal olive-green chloroplasts (c). Reproduced with permission from R. Matthews. (b) *Rhodomonas*. Cells appear oval in face view (A) and comma-shaped in side view (B), with a distinct posterior hyaline tail (arrow). Flagella not visible in this iodine-stained preparation.

but no pyrenoids. Quite widely distributed although never abundant – in small ponds, nutrient-rich pools, puddles or slow flowing waters. Chlorophyta. Plate XV.

(b) Cells ovoid, sometimes with a truncated anterior end **116**

116 (115) (a) Cell strawberry-shaped, four lobes forming an indentation at the anterior end from which the flagella arise .... *Pyramimonas*

Cell shape is characteristically strawberry or sub-pyramidal. Cells (12–16 μm wide, 20–30 μm long) are naked (no wall) and have a cup-shaped chloroplast which is often four lobed towards the anterior end. A pyrenoid is present towards the base of the cell. Widespread in ponds and still waters, frequently at colder times of the year although it can occur all the year round. Prasinophyta. Plate XV.

(b) Cells oval or occasionally heart-shaped but not lobed .... *Carteria*

Carteria cells (9–20 μm diameter) are virtually identical to those of *Chlamydomonas* except that they may be pear-shaped or ellipsoidal and have four, not two, flagella. The cup-shaped chloroplast usually lacks pyrenoids but occasionally one may be present. Frequent and occasionally abundant in still waters and habitats similar to *Chlamydomonas*. Less widespread than *Chlamydomonas*. Chlorophyta. Plate XVI.

117 (114) (a) Cells fusiform in shape .... *Chlorogonium*

Cells of *Chlorogonium* are elongate (fusiform or spindle-shaped) tapering at both ends, with two flagella arising from the narrow apical end. Cells (2–15 μm wide, up to 80 μm long) have a rigid cell wall, a chloroplast which fills most of the cell and an anterior eyespot. Pyrenoids are usually present. Can occur in large numbers in small water bodies, especially those rich in humic materials. Chlorophyta. Plate XVI.

(b) Cells not fusiform in shape **118**

118 (117) (a) Cells with anterior end flattened obliquely (cut off at an angle in lateral view, as in Plate XVI). Flagella, of slightly unequal length, arise from the oblique face near to the front end. **119**

(b) Cells with anterior end rounded or flattened transversely. Flagella inserted at the end (apex) of the cell and of equal length **121**

119 (118) (a) Cells obovoid but strongly curved towards the narrowed rear end to form a hyaline tail .... *Rhodomonas*

Rhodomonas cells are small (8–13 μm long, 3–8 μm wide) and fairly delicate so are often overlooked even though the alga is widespread. The single reddish-golden chloroplast (contains phycoerythrin) has a large pyrenoid to one side with a definite starch sheath. A hyaline tail with a basal granule is present in **R. lacustris var. nannoplanctica**, otherwise the end just tapers. Two slightly unequal flagella are present. Widespread in plankton. Some reports suggest that it is able to survive under low-light conditions but this is not a prerequisite in its ecology (Wehr and Sheath, 2003). Cryptophyta. Plate XVI. Fig. 4.52b

(b) Cells not strongly curved. Tail, if present, not hyaline **120**

120 (119) (a) Gullet absent, chloroplast blue to blue-green or reddish. .... *Chroomonas*

Chroomonas and *Cryptomonas* are very similar except that the colour of the chloroplast is markedly different in *Chroomonas*. There are two unequal flagella. One chloroplast (rarely two) which is blue-green in colour as it contains phycocyanin. A single pyrenoid is present. Cells (7–18 μm long, 4–8 μm wide) are also usually somewhat smaller than *Cryptomonas*.

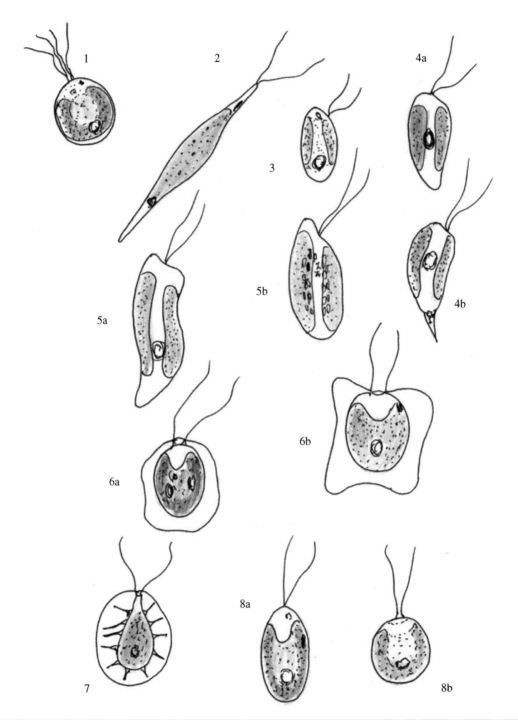

Plate XVI   1. *Carteria*. 2. *Chlorogonium*. 3. *Chroomonas*. 4. *Rhodomonas*: (a) *R. lacustris*, (b) *R. nanoplanktica*. 5. *Cryptomonas*: (a) *C. curvata*, (b) *C. ovata*. 6. *Pteromonas*: (a) *P. aequiciliata*, (b) *P. aculeata*. 7. *Haematococcus*. 8. *Chlamydomonas*: (a) *C. reinhardtii*, (b) *C. globosa*.

Planktonic, in ponds and shallow waters. Cryptophyta. Plate XVI.

(b) Gullet present, chloroplasts brown to olive green . . . . ***Cryptomonas***

Cells of **Cryptomonas** (up to 80 μm in length, 26 μm wide) are slipper- to bean-shaped with some dorsiventral flattening. They have two nearly equal flagella, one or two reddish-brown chloroplasts (with up to four pyrenoids) and a conspicuous gullet extending as a longitudinal furrow from the anterior end. Free swimming in the plankton. Widespread and sometimes in large numbers, when it may impart an odour to drinking water (Palmer, 1962). Cryptophyta. Plate XVI. Fig. 4.53.

121 (118) (a) Chloroplast in centre of cell, suspended there by strands of cytoplasm . . . . ***Haematococcus***

Cells of **Haematococcus** (sometimes called **Sphaerella**) are oval to spherical, 8–30 μm in diameter, and have a thick mucilaginous wall. The green pigments in the chloroplast are often masked with red haematochrome so that the cells, particularly in the encysted state, appear red. The cup-shaped chloroplast has one to four pyrenoids and appears attached to the cell wall by cytoplasmic strands. Occurs in small water bodies and snow as well as bird baths which it may colour red. Chlorophyta. Plate XVI. Fig. 4.54.

(b) Chloroplast not attached to the cell wall by means of cytoplasmic strands **122**

122 (121) (a) Cells with hyaline wing around the cell but most prominent on either side . . . . ***Pteromonas***

Pteromonas cells have two flagella emerging from the anterior end, with wing-

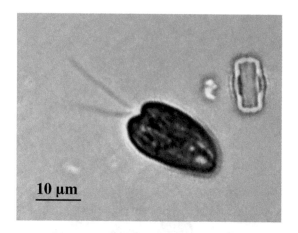

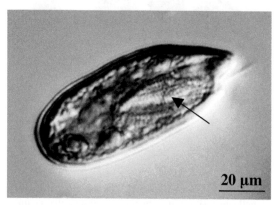

Figure 4.53 *Cryptomonas*. Top: Small reddish brown, slipper-shaped biflagellate cell. Reproduced with permission from G. Ji. Bottom: Large olive-green cell showing conspicuous anterior gullet (arrow). Reproduced with permission from R. Matthews. Separate species.

like extensions of the cell wall giving the cells an angular appearance. The cells are rounded but flattened in side view. A cup-shaped chloroplast is present with one to many pyrenoids. **Pteromonas** is widespread in its occurrence but generally prefers nutrient-rich waters, either ponds or slow rivers. Plate XVI.

(b) Cells without hyaline wing . . . . **Chlamydomonas**

Chlamydomonas is a very large genus with numerous species. Widely

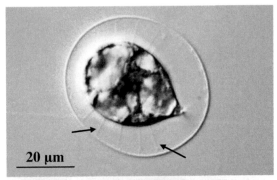

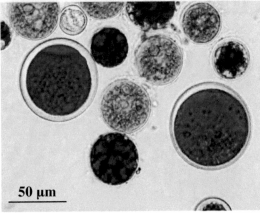

Figure 4.54 *Haematococcus*. Top: Single motile cell (flagella not visible) showing broad hyaline wall containing fine protoplasmic extensions (arrows). The central pear-shaped protoplast has large vacuoles and granules of red pigment (astaxanthin). Reproduced with permission from R. Matthews. Bottom: Non-motile cells, ranging from immature (green) to fully mature cysts filled with astaxanthin.

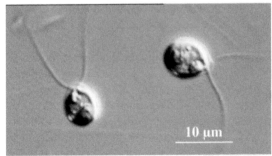

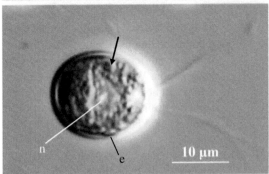

Figure 4.55 *Chlamydomonas*. Top: Two small biflagellate cells. Bottom: Large biflagellate cell showing central nucleus (n) and cup-shaped parietal chloroplast (arrow) with eyespot (e). DIC images from a single lake sample. Reproduced with permission from R. Matthews.

distributed in a range of habitats, especially small pools and ditches, often nutrient-rich. The chloroplast is large and cup-shaped, filling much of the cell, with one or more pyrenoids. Two flagella are present inserted at the anterior end, which also has a prominent eyespot. Non-motile palmelloid stages are known to occur. If present in large numbers it can impart odour to drinking water (Palmer, 1962). Chlorophyta. Plate XVI. Fig. 4.55.

123 (108) (a) Cells with a long anterior horn and two or three posterior horns . . . . *Ceratium*

Ceratium cells are quite large and brownish in colour with two or three posterior 'horns'. The shape, with the horns, is characteristic. There is a narrow but prominent transverse furrow across the middle of the cell around which a flagellum occurs. A second flagellum trails backwards from the mid-region. The cell surface is covered by a number of plates of specific shape and arrangement. The numerous chloroplasts are discoid and golden-brown in colour. Characteristic thick-walled resistant cysts are produced towards the end of the growing season. *Ceratium* commonly imparts a fishy odour and bitter taste to drinking waters (Palmer, 1962). Pyrrophyta. Plate XVII (showing *Ceratium hirundinella* and *C. cornutum*). Figs. 1.3, 1.9, 1.10, 2.7, and 4.56.

(b) Cells otherwise shaped **124**

124 (123) (a) Cell wall thin and difficult to see (except in empty cells) or completely absent **125**

(b) Cell wall thick, cellulose plates covering the cell easily seen. Transverse furrow encircling the central area of the cell . . . . *Peridinium*

Peridinium cells are normally ovoid in outline and free swimming. The cell is covered with angular plates arranged in a specific order with an epitheca and hypotheca of approximately equal size. There are numerous brown chloroplasts. A widespread genus found in a range of freshwater habitats. Reported to impart both taste and odour to drinking waters (Palmer, 1962). Pyrrophyta. Plate XVII. Figs. 1.8 and 4.57.

125 (124) (a) Cells without plates on walls . . . . *Gymnodinium*

Cells (7–80 μm wide, 8–118 μm long) approximately oval but sometimes flattened dorsiventrally. A median transverse groove is present. Epicone (above median groove) and hypocone (below median groove) about equal in size. The cell wall is thin and smooth. Pyrrophyta. Plate XVII.

(b) Cells with thin walls with plates but these are delicate and difficult to see unless cell is empty . . . . *Glenodinium*

Cells (13–48 μm wide, 25–50 μm long) approximately oval in shape, sometimes dorsiventrally flattened with groove around the mid-region bearing one of the

two flagella. Cell wall thin and sometimes lightly ornamented with about 20 plates. Numerous yellow-brown chloroplasts. Widespread in ponds and swamps. Also found in saline situations. Epitheca (above the central groove) and hypotheca (below the central groove) of similar size. Pyrrophyta. Plate XVII.

Woloszynskia is a genus with epicone and hypocone of nearly equal size and with up to 100 delicate thin plates over the cell surface. An eyespot may be present. Plate XVII.

126 (106) (a) Cells isolated or in groups, wall siliceous decorated with grooves or dots (punctae) which form a definite pattern or other definite markings on the surface. Storage products (mainly lipids and droplets) may be visible within the cell. Chloroplasts one to many, yellow-green to golden-brown in colour **127**

(b) Cell wall not made of silica or decorated with patterned dots or bars. No conspicuous lipid droplets present **204**

127 (126) (a) Cells circular in outline in valve view.[1] Decorations usually arranged in radial rows or radial segments, sometimes quite faint; or in the form of large processes (ocelli, see Glossary).[2] Cells often solitary but can occur in loose chains **128**

[1]**NB.** *Campylodiscus* may appear circular in one view but is saddle-shaped in another. This is a distorted form of a pennate diatom, not a centric one, so will key out through section (b).

[2]**NB.** Plate XVIII shows some typical examples of the markings on the valve faces of centric diatoms to indicate the types of patterns seen, not all of which, however, are included – see also Plates XIX and XXI.

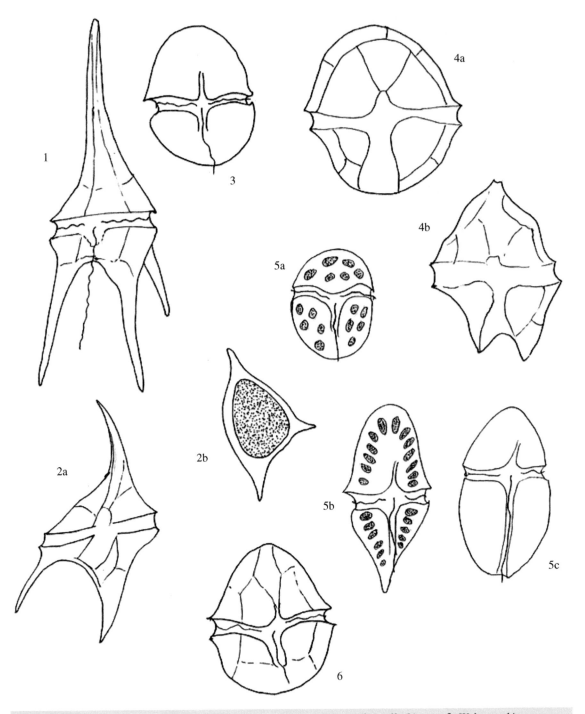

Plate XVII   1. *Ceratium hirundinella*. 2. *Ceratium cornutum*: (a) vegetative cell, (b) cyst. 3. *Woloszynskia coronata*. 4. *Peridinium*: (a) *P. volzii*, (b) *P. limbatum*. 5. *Gymnodinium*: (a) *G. chiastosporum*, (b) *G. fuscum*, (c) *G. aeruginosum*. 6. *Glenodinium cinctum*.

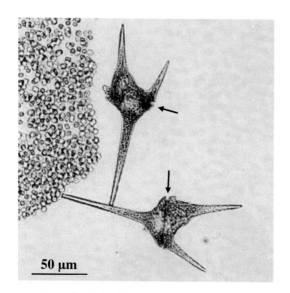

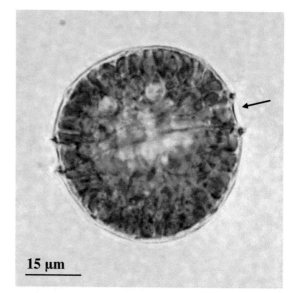

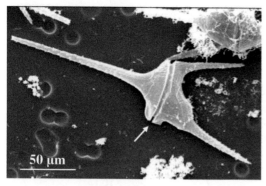

Figure 4.56 *Ceratium.* Top: Two cells in fresh phytoplankton sample. Bottom: Scanning electron microscope image showing typical dinoflagellate plated surface (more clearly visible in Fig. 4.57). Cells have a clear transverse furrow (arrow).

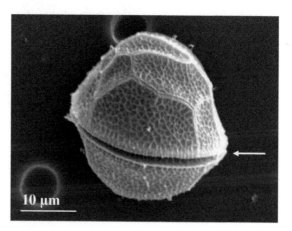

Figure 4.57 *Peridinium.* Top: Living cell containing numerous brownish chloroplasts. Bottom: SEM image showing surface outlines of thecal plates. Both images show a clear transverse furrow (arrow).

(b) Cells elongate, cigar-, boat-, crescent-shaped or a distorted version of any of these. Decorations arranged bilaterally, although this is not always obvious in cells having a crescent or distorted shape..**139**

NB. The structure of the diatom wall is complex (see Chapter 1) and some features may be difficult to see using light microscopy alone. In many cases, scanning electron microscopy may be required for positive identification. Since this facility will not be available to many users of this text, features used in the key will be generally limited to those visible using a good quality light microscope. It is always best to observe both live and cleaned diatom material (see Section

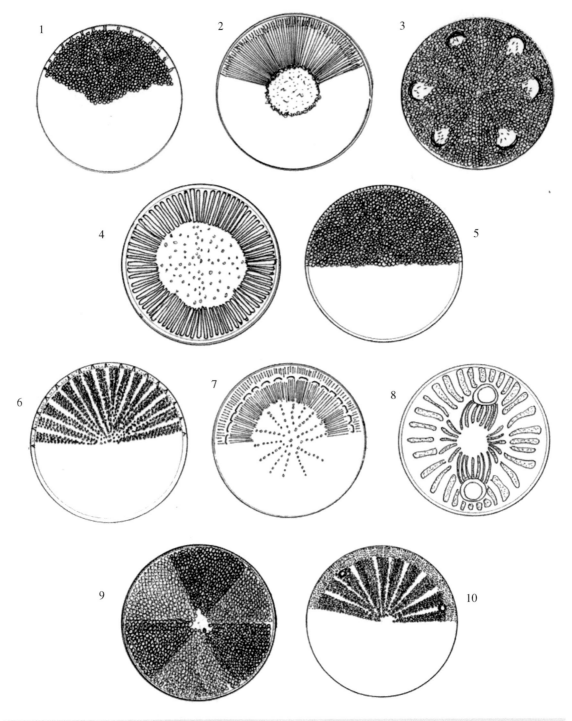

Plate XVIII   **1.** *Thalassiosira.* **2.** *Hyalodiscus.* **3.** *Aulacodiscus.* **4.** *Cyclotella.* **5.** *Coscinodiscus.* **6.** *Stephanodiscus.* **7.** *Cyclostephanus.* **8.** *Auliscus.* **9.** *Actinoptychus.* **10.** *Actinocyclus.*

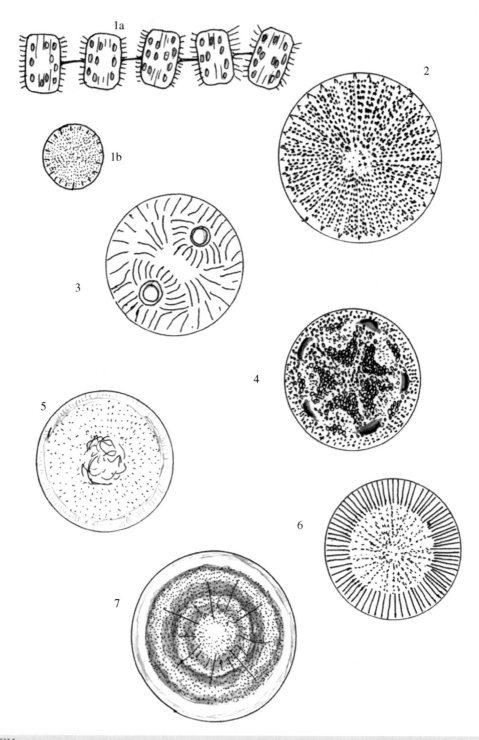

Plate XIX **1.** *Thalassiosira:* (a) Chain of cells, girdle view, (b) single cell, valve view. **2.** *Stephanodiscus.* **3.** *Auliscus.* **4.** *Aulacodiscus.* **5.** *Hyalodiscus.* **6.** *Cyclotella.* **7.** *Actinocyclus.*

2.5.2) so that the cell contents and external mucilage can be seen together with any markings on the silica frustules. Readers are strongly recommended to consult the more detailed literature mentioned in the bibliography.

128 (127) (a) Valve with clear (hyaline) central area devoid of regular markings and 5–7 wide hyaline rays extending to the cell margin . . . . ***Asteromphalus***

Asteromphalus is a widely distributed marine and estuarine species common in cool temperate waters. The valves have a relatively flat surface and between the hyaline rays, there may be fine or coarse areolae (depending upon the species). Cell diameter 25–100 μm. Numerous discoid chloroplasts. Bacillariophyta. Plate XXI.

(b) Valve without wide hyaline rays . **129**

129 (128) (a) Markings on valve surface divided into (usually) six segments that are alternately raised or level giving an alternating light and dark appearance . . . . ***Actinoptychus***

Actinoptychus is a cosmopolitan marine, coastal and estuarine species common at most times of the year. Markings on the valve are characteristic, with segments are covered in coarse areolation. Chloroplasts plate-like or irregular. Cells 20–80 μm in diameter. Bacillariophyta. Plates XVIII and XXI.

(b) Valve markings not as above **130**

130 (129) (a) Margin or rim of valve with spines **131**

NB. ***Coscinodiscus*** (138) may have spines which may be difficult to see using a light microscope.

(b) No spines present at rim or margin of valve **134**

NB. Some centric diatoms secrete chitin fibres or threads that protrude from the valve margin and may superficially look like spines. These may function to bind cells together in loose chains or act as a deterrent to grazing animals.

131 (130) (a) Cells usually in chains joined together by mucilage strands originating from the valve surface centre or embedded in mucilage. Valve face gently undulating, with punctae in a radiating mesh-like pattern . . . . ***Thalassiosira***

Cells of ***Thalassiosira*** can be solitary but are more frequently joined to form loose chains or may even be embedded in mucilage. Valve face circular, fairly flat to gently undulating, and covered with quite coarse punctae which may be radial or arranged in arcs. Small spines may be visible at valve margin. Chloroplasts small discs and numerous. Very common and widely distributed in marine and estuarine habitats. Bacillariophyta. Plates XVIII and XIX.

(b) Cells solitary, or if in loose chains not embedded in mucilage or with mucilaginous strands joining from valve face to valve face **132**

132 (131) (a) Valves with small central area bearing 2–5 pores. Distinct striae radiate out from this central area to the valve margin. Usually linked to form short chains by means of small spines. Valve faces then held close together. . . . . ***Orthoseira***

Cells short cylinders with numerous discoid chloroplasts. Found on damp rocks, in wet moss or on stream banks, usually in upland (sometimes alkaline) areas. Bacillariophyta. Plate XXI.

(b) Cells not as above............**133**

133 (132) (a) Valves large (35–130 μm) and more heavily silicified. Many radially arranged ribs and grooves on valve surface originating from a clear central area.....................*Ellerbeckia*

Ellerbeckia cells are large diameter cylinders sometimes joined in short chains. Numerous small disc-shaped chloroplasts. Clear differentiation between the markings on the valve face and the mantle, which has a distinctive hatched pattern. Previously included in *Melosira*. Found in damp habitats including sandy and other sediments. Only in plankton if sediment stirred up. Bacillariophyta. Plates XX and XXI.

(b) Cells usually solitary but sometimes forming loose chains. Valve surface with radiating rows of punctae (single towards the centre but often double or more towards the edge) separated by clearly defined ribs or clear areas. Central area punctate but rows not as clearly defined. Valve surface concentrically undulate ....*Stephanodiscus*

Cells of *Stephanodiscus* are disc-shaped and slightly barrel-shaped in girdle view. The valve face is undulate with the centre either raised or lower than the margin. The valve margin has a ring of short spines and in some small species long delicate chitin threads that allow cells to link together to form loose chains. Numerous discoid chloroplasts sometimes appearing to lie around the margin of the cell. Common in freshwaters, especially eutrophic ones, planktonic. *Stephanodiscus* has been reported (Palmer, 1962) as producing geranium and fishy odours and tastes in drinking waters. High populations

can cause blockage of filters. Bacillariophyta. Plate XIX. Fig. 4.58.

Cyclostephanus shows features that are intermediate between *Stephanodiscus* and *Cyclotella*.

134 (130) (a) Valve face with two or more circular processes (ocelli) on the surface – clearly visible as obvious large structures compared with normal punctae..........**135**

(b) No obvious large circular structures on the valve surface **136**

135 (134) (a) Cells widely elliptical or subcircular in valve view. Two large ocelli on valve surface which is also decorated with lines around the margin and radiating from the central area ....*Auliscus*

Auliscus cells are frequently attached to sand grains, stones or rocks in inshore and estuarine habitats. Although bottom living they may be found in the plankton. Cells 40–80 μm in diameter. The shape of the cells and the large ocelli are characteristic of the species. Bacillariophyta. Plate XIX. Fig. 4.59.

(b) Cells circular in valve view. More than 2 marginal processes. Valve surface flat at centre but raised as it goes towards the marginal processes – which may have a furrow running back towards the centre. Valve surface areolate .....*Aulacodiscus*

A marine and brackish water species associated with sediments where it can be very common. The processes on the valve surface are variable in number but normally range between 3 and 8. Cell diameter 60–200 μm. Bacillariophyta. Plate XIX. Fig. 4.60.

136 (134) (a) Valve surface apparently almost devoid of markings and with large

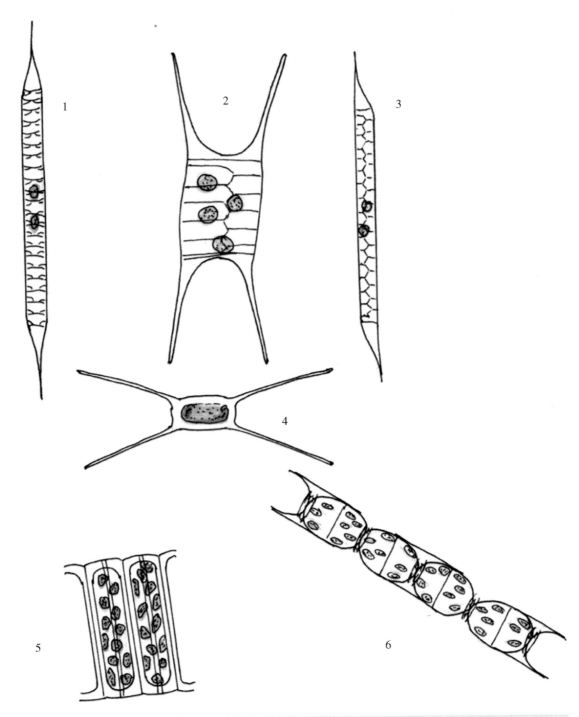

Plate XX   **1**. *Rhizosolenia.* **2**. *Acanthoceras.* **3**. *Urosolenia.* **4**. *Chaetoceros.* **5**. *Ellerbeckia.* **6**. *Melosira nummuloides.*

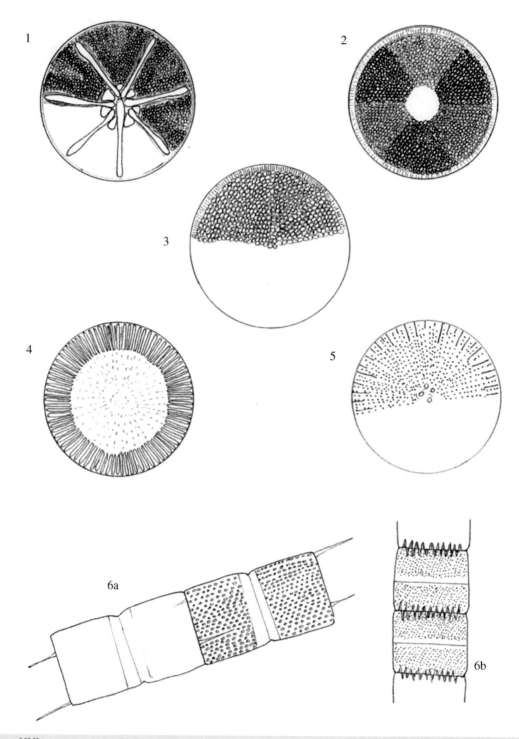

Plate XXI **1**. *Asteromphalus*. **2**. *Actinoptychus*. **3**. *Coscinodiscus*. **4**. *Ellerbeckia*. **5**. *Orthoseira*. **6**. *Aulacoseira*: (a) *A. granulata*, (b) *A. italica*.

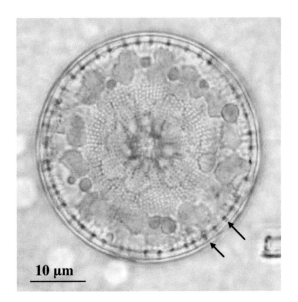

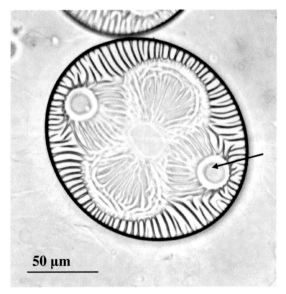

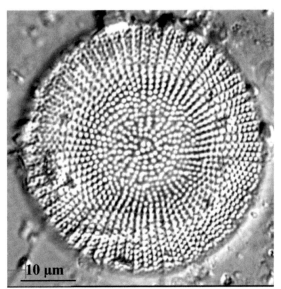

Figure 4.59 *Auliscus*. Surface view of frustule, showing ocelli (arrow). Prepared slide.

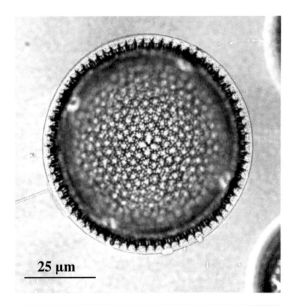

Figure 4.58 *Stephanodiscus*. Centric diatom. Top: Live planktonic cell, with numerous discoid chloroplasts and a frustule with clear marginal spines (arrows). Bottom: Acid digest, lake sediment sample – showing radiating rows of fine punctae.

Figure 4.60 *Aulacodiscus*. Valve view showing central areolate pattern of frustule. Acid digest, lake sediment sample.

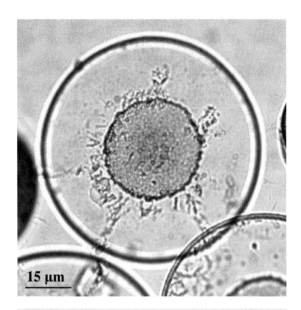

Figure 4.61 *Hyalodiscus.* Diatom with peripheral clear area on valve surface. Prepared slide.

concentric area around the centre
.... ***Hyalodiscus***

Hyalodiscus cells are either sessile, in short chains or as individuals in the marine plankton. Valves circular and strongly convex with the centres slightly flattened. Because of the markedly convex nature of the valve face, only a small area can be in focus under a light microscope at one time, hence the apparent lack of markings. In fact the whole valve surface is covered with areolae arranged in radial rows. Cells (12–115 μm diameter) have numerous rod-shaped chloroplasts. Bacillariophyta. Plate XIX. Fig. 4.61.

(b) Valve surface with obvious markings over whole or a large part of the area. **137**

137 (136) (a) Valve surface with two distinct areas of markings. The middle area is punctate and the outer striate or ribbed
.... ***Cyclotella***

Cells of ***Cyclotella*** are disc-shaped with circular-shaped valves having a slightly undulate surface. Valve margins without spines but in some species small tubules are present that could be mistaken as spines. The central area is irregularly punctate and distinct from the concentric outer area, which is regularly striate. Cells frequently solitary but may be attached in chains by mucilaginous threads. There are several discoid chloroplasts usually arranged around the cell margin. Widespread in lakes, rivers, marine and brackish water habitats. Has been reported to produce geranium and fishy odours and tastes in drinking water (Palmer, 1962). Two of the more common species in cool temperate waters are *C. menegheniana* with cells 10–30 μm in diameter and *C. kuetzingiana* with cells 10–40 μm in diameter. The former has 40–50 rows of radiating striae and the latter up to 90 radiating rows. Can be present as large numbers in reservoirs and lakes and may cause problems of filter blocking in water treatment works. SEM has revealed other features on the valve surface not easily seen using light microscopy, resulting in some workers dividing the genus into specific sub-groups. Bacillariophyta, Plate XIX. Fig. 4.62.

(b) Valve markings not in two distinct zones of punctae and striae **138**

138 (137) (a) Valve surface flat to slightly convex, circular or slightly elliptical in outline. Areolae in straight parallel lines (tangential to the radius) of decreasing length giving the surface a segmented cross-hatched appearance.. .... ***Actinocyclus***

Valves of ***Actinocyclus*** are circular to slightly subcircular. There is a small central hyaline area with irregular punctae. The valve face is marked with areolae in regular parallel rows of decreasing length

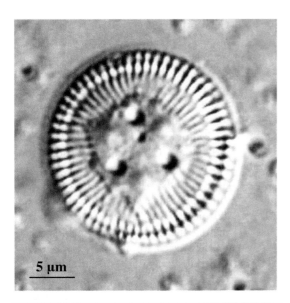

5 μm

Figure 4.62 *Cyclotella.* Valve view of diatom, with clear separation of central and peripheral areas. Acid digest, lake sediment sample. Reproduced with permission from M. Capstick.

lying parallel to the main radial ones. Cells (20–170 μm diameter) contain numerous plate-like chloroplasts. Common in coastal and estuarine plankton and as an epiphyte. Bacillariophyta. Plate XIX.

(b) Valve surface gently undulate. Coarse punctate markings over surface forming radial rows or arcs. Punctae circular or hexagonal . . . . *Coscinodiscus*

The valve markings on *Coscinodiscus* are generally coarse and often hexagonal in shape although not always so. They cover the whole valve face except for a small central hyaline area which may be present together with a rosette of larger punctae (these may be present without the hyaline area). The valve margin may have small spines although these are seldom visible under a light microscope. Cells (up to 300 μm diameter) contain numerous plate-like chloroplasts. Free-floating and abundant in the plankton. Mostly salt water

with only one British freshwater species, *C. lacustris.* Bacillariophyta. Plate XXI.

139 (127) (a) Valves with at least one long spine. The spines are as long as, or are longer than, the cell **140**

(b) Valves without long spines **144**

140 (139) (a) Cells (many times as long as wide) forming a typical filament, with one long spine visible at the end . . . . *Aulacoseira granulata v. angustissima*

(see also key No. 27)

(b) Cells not as above **141**

141 (140) (a) Cells with two long spines per cell, one at each end **142**

(b) Cells with four long spines per cell, two at each end **143**

142 (141) (a) Cells long and thin, up to 200 μm × 10 μm. Spines arise from the centre of the valve end . . . . *Rhizosolenia*

Rhizosolenia cells may be solitary or may form chains. Numerous small plate-like/disc-shaped chloroplasts. Frustule wall only lightly silicified so can be overlooked especially in acid cleaned specimens when the frustule components often disintegrate. Planktonic in lakes and slow flowing rivers. Bacillariophyta. Plate XX.

(b) Cells long, but not as long as above, 150 × 20 μm. Spines arise from opposite corners . . . . *Urosolenia*

Cells are most frequently solitary with many disc-shaped chloroplasts. A planktonic species in usually large and deep lakes which are enriched and of a slightly alkaline nature. Bacillariophyta. Plate XX.

143 (141) (a) Cells not markedly elongate but square to rectangular. Frustules have four spines, one arising from each corner and either projecting in the direction of the long axis of the cell or at a shallow angle. Many girdle bands visible. Four discoid to plate-like chloroplasts per cell. ....*Acanthoceros*

Acanthoceros cells (up to 40 × 80 μm in size) are normally solitary with one to many plate-like/disc-shaped chloroplasts. Found in the plankton of enriched lakes usually in slightly alkaline conditions. Bacillariophyta. Plate XX.

(b) Cells square to shallow rectangular in shape, up to 15 × 30 μm in size, containing 1 or 2 cup-shaped chloroplasts. Few girdle bands visible. Spines project at 45° angle from each corner. .... *Chaetoceros*

Chaetoceros cells are usually joined together into filaments by the interlocking of the spines. Found in waters with a high conductivity, especially in coastal regions, and in brackish waters. Bacillariophyta. Plate XX.

144 (139) (a) Cells cuneate, heteropolar and forming fan-shaped colonies .... *Meridion*

(see also key No. 77)

(b) Cells not as above **145**

145 (144) (a) Cells with either costae or septa present **146**

(b) Cells with neither costae nor septa present **147**

146 (145) (a) Valves swollen in the middle and with slightly swollen poles. Isopolar ....*Tabellaria*

(see also key No. 29)

Tabellaria forms zig-zag, almost linear or sometimes stellate colonies. Cells rectangular to oblong in shape and usually seen in girdle view – making the prominent septa easy to see. In valve view they are elongate, slightly capitate with a slightly inflated mid-region. Chloroplasts elongate and lying between the septa. *Tabellaria fenestrata* (33–116 μm long, 4–10 μm wide) and *T. flocculosa* (6–130 μm long, <5 μm wide) are the most frequent species. Widespread in plankton (oligotrophic/mesotrophic waters) or attached to stones or submerged vegetation. Reported to produce fishy tastes in drinking water if present in large numbers (Palmer, 1962). Bacillariophyta Plate IV. Figs. 2.28 and 4.11.

(b) Valves isopolar with costae but no septa ....*Diatoma*

(see also key No. 29)

Diatoma cells often form ribbon or zig-zag–shaped (occasionally stellate) colonies. In girdle view, the valves are rectangular to oblong with small plate-like or discoid chloroplasts. There are distinct thickened transverse bars (costae) across the valves. Valve ends sometimes capitate, rostrate or bluntly rounded. Cells up to 120 μm long. Epiphytic or planktonic, in littoral regions of lakes or slow flowing rivers. Can impart an odour to drinking water (Palmer, 1962). Bacillariophyta. Plate V. Fig. 4.12.

Williams (1985) separates two subgenera. One with uniseriate rows of striae, prominent internal ribs and scattered spines near polar areas. Chloroplasts plate-like or discoid. A second subgenus, *Odontidium*, with a more diffuse area of markings and lobed chloroplasts is also recognised.

147 (145) (a) Cells triradiate in valve view with a small chloroplast near to the base of each arm . . . . *Centronella*

Centronella has characteristically triradiate- shaped cells with a small chloroplast at the base of each arm. It occasionally occurs in the plankton of meso- and eutrophic lakes although it is quite rare. Bacillariophyta. (Not illustrated.)

(b) Cells not triradiate **148**

148 (147) (a) Cells S-shaped (sigmoid) in outline **149**

(b) Cells not S-shaped in outline. **151**

149 (148) (a) Cells sigmoid in outline, wide in the central region and narrowing towards the apices. Usually quite large (>70 μm long). Two large plate-like chloroplasts, one either side of the longitudinal axis . . . . *Gyrosigma*

Gyrosigma cells are sigmoid in shape (although some species are only slightly so) with rounded poles and S-shaped raphe. The striae on the valve surface are both parallel to the raphe and transverse. They are generally fine but may be coarser in some species. Central areas round or cylindrical. Two plate-like chloroplasts are present lying either side of the girdle. Pyrenoids may be present. *Gyrosigma acuminatum* and *G. attenuatum* are widespread species. Found in flowing or standing waters on sediments and rocks (epipelic) usually in waters with high electrolyte content.

Pleurosigma and *Gyrosigma* are very similar but *Gyrosigma* has markings in lines from the apical and transapical directions and is generally a fresh to brackish water species whereas *Pleurosigma* has its markings diagonally and is predominantly marine. Bacillariophyta. Plate XXII. Fig. 4.63 (*Pleurosigma* illustrated).

(b) Cells narrow and sigmoid, only slightly tapering towards the ends. Longitudinal margins of frustule more or less parallel. One chloroplast or if two one either side of the centre **150**

150 (149) (a) Cells long (up to 200 μm) and narrow (6–9 μm). Single long plate-like chloroplast. Around the cell margin may be seen a line of dots. . . . . *Stenopterobia*

Stenopterobia occurs in oligotrophic, often upland, pools but is rarely abundant. Cells are sigmoid in some species (e.g. *S. sigmatella*), linear in others (e.g. *S. delicatissima*). Cells have one chloroplast which is divided into two parts. As with other members of the Surirellales, the raphe is supported on a raised wing around the margin of the cell. Bacillariophyta. Plate XXII (linear form illustrated).

(b) Cells long but wider (>10 μm wide). Two plate-like chloroplasts, one above and one below the central area. . . . . *Nitzschia sigmoidea*

As with all members of the genus *Nitzschia*, *N. sigmoidea* has its raphe raised on a keel along one margin. The keel supports show up along that margin as a series of dots (carinal dots). *Nitzschia sigmoidea* cells are large, 90- over 200 μm long, 8–14 μm wide. Two plate-like chloroplasts are present, one above and one below the central area. Common and widespread in meso- to eutrophic waters. A benthic species. Bacillariophyta. Plate XXII.

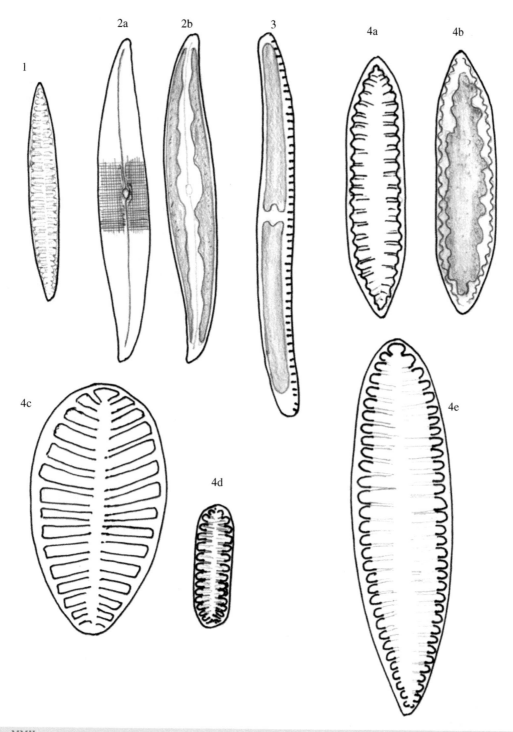

Plate XXII 1. *Stenopterobia*. 2. *Gyrosigma*: (a) frustule markings, (b) chloroplasts. 3. *Nitzschia*. 4. *Surirella*: (a) *S. linearis* – frustule markings, (b) *S. linearis* – chloroplast, (c) *S. robusta*, (d) *S. minuta*, (e) *S. elegans*.

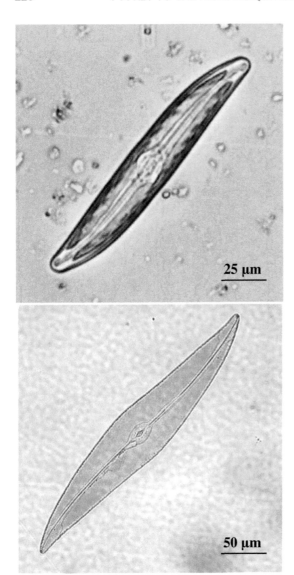

25 μm

50 μm

Figure 4.63 *Gyrosigma.* A solitary diatom with sigmoid valves. Top: Live cell, with two elongate chloroplasts. Reproduced with permission from G. Ji. Bottom: Prepared slide, separate species.

NB. Although *N. sigmoidea* is the most common of the sigmoid *Nitzschia* species others do occur, for example, *N. flexa*, *N. vermicularis*, *N. acula*, *N. filiformis* and *N. clausii*.

151 (148) (a) Cells circular in outline in one view but saddle-shaped in another. Striae radiate on valve surface . . . . . *Campylodiscus*

NB. It is important to look at a number of specimens (or turn a single specimen over) to get the range of views needed to see the saddle-shaped appearance.

Cells of *Campylodiscus* are quite large (60–200 μm diameter) and approximately circular in shape (valve view) or saddle-shaped (girdle view). One large lobed chloroplast is present. Widespread in oligo to eutrophic waters and in brackish and marine habitats. Epipelic. *Campylodiscus hibernicus* and *C. noricus* are the most frequent freshwater species. Bacillariophyta. Plate XXIII.

(b) Cells not saddle-shaped in any view
. . . . **152**

152 (151) (a) Cells widely elliptical/oval, one valve with true raphe, the other with a pseudoraphe. Cells isopolar. One C-shaped chloroplast . . . . *Cocconeis*

Cells of *Cocconeis* are isobilateral and nearly elliptical in shape. Fine or coarse striations may be visible on the valve surface running from the raphe/pseudoraphe to the cell margin. The raphe-bearing valve slightly concave. Epiphytic and epilithic, found on rocks and plants, attached by the concave raphe-bearing valve. Very widespread and common. Two common species, *C. pediculus* (11–45 μm long, 9–30 μm wide) with slightly more rhomboid cells and *C. placentula* (10–70 μm long, 8–35 μm wide). Bacillariophyta. Plate XXIII. Figs. 2.28, and 4.64.

(b) Cells more narrowly elliptical, oval or elongate. Heteropolar or isopolar. One or more chloroplasts, not C-shaped **153**

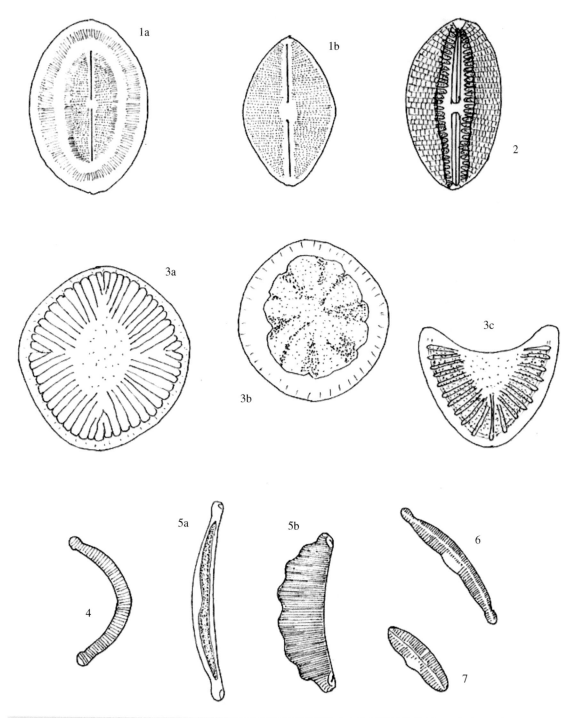

Plate XXIII **1.** *Cocconeis*: (a) *C. placentula* (showing frustule markings/chloroplast), (b) *C. pedicula.* **2.** *Diploneis.* **3.** *Campylodiscus*: (a) valve view with frustule markings, (b) girdle view with chloroplast. (c) side view to show saddle-shape **4.** *Semiorbis.* **5.** *Eunotia*: (a) *E. arcus* (showing chloroplast), (b) *E. serra* (with markings). **6.** *Hannaea.* **7.** *Reimeria.*

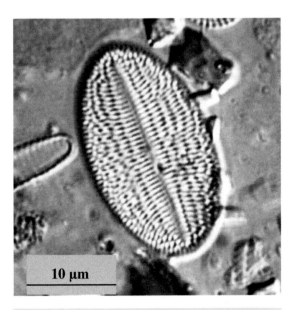

Figure 4.64 *Cocconeis*. Valve view of this sedimentary diatom. Acid digest, lake sediment sample. Reproduced with permission of M. Capstick.

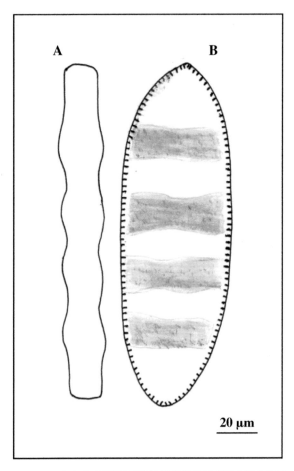

Figure 4.65 *Cymatopleura elliptica*. (A) girdle view; (B) valve view.

153 (152) (a) Cells not as broadly elliptical with bluntly rounded ends, both valves with a true raphe which lies within a long thickened ridge. Two chloroplasts with lobed margins, one either side of the long apical axis . . . . ***Diploneis***

The raphe is situated between prominent silica ridges. Striae on the valve surface run from the raphe to the margin. Chloroplasts may contain a pyrenoid. Cells 20–130 μm long, 10–60 μm wide. Widespread in mainly oligotrophic waters on bottom deposits, some species brackish or marine. Bacillariophyta. Plate XXIII.

(b) Cells without thickened ridges along most of the length of the raphe **154**

154 (153) (a) Cells with a prominent transverse undulate surface in girdle view and banded transverse shaded areas in valve view. Isopolar . . . . ***Cymatopleura***

Cells are elliptical often with a central 'waist'. The valve surface is characteristically undulate, particularly visible in girdle view, giving it a transverse banded appearance. The raphe is on a raised ridge running around the valve margin. A single large lobed chloroplast is present. Found on bottom sediments, commonly epipelic. Widespread and common. Cells 30 to >200 μm in length, 20–60 μm wide. Bacillariophyta. Fig. 4.65.

(b) Cells without undulate surface. Lanceolate to oval in shape, sometimes egg-shaped or narrowing towards the centre (slipper-shaped). Iso- or heteropolar.

.... **155**

155 (154) (a) Cells with permanent rib-like markings around the margin **156**

(b) Cells not as above **157**

156 (155) (a) Cells isopolar ..... *Surirella* **group a**

(b) Cells heteropolar .. *Surirella* **group b**

Surirella cells (20–400 μm long, 10–150 μm wide) have rounded to slightly pointed ends that may be iso- or heteropolar. The raphe runs around the margin of the cell on a wing supported by ribs which give the characteristic markings on the frustule surface. There are two large lobed and plate-like chloroplasts but as each lies beneath the valve face only one is visible in valve view. Widespread on sediments and rocks. Some species found in brackish and marine habitats. *Surirella* **group a** includes *S. angusta* (isopolar), while **group b** includes *S. tenera* and *S. robusta* (heteropolar, with widely rounded apices at one end and more narrow, slightly more pointed apices at the other). Bacillariophyta. Plate XXII.

NB. Species of *Surirella* that are not so widely oval but are more linear to elliptical, may be keyed out elsewhere.

157 (155) (a) Cells crescent-shaped (like the segment of an orange), with one edge strongly convex and the other either concave or nearly straight, isopolar **158**

(b) Cells not crescent-shaped or as above
.... **165**

158 (157) (a) Cells strongly crescent-shaped or distinctly curved in valve view (the more common view) **159**

(b) Cells more mildly crescent-shaped with both margins showing more gentle convex curves **161**

159 (158) (a) Cells strongly lunate (moon-shaped) with parallel sides, raphe difficult to see other than in girdle view and then only towards the end of the valve. Valve apices strongly arcuate. Conspicuous transverse ridges on the valve face extending into short spines at the cell margin. Two plate-like chloroplasts .... *Semiorbis*

Semiorbis cells (20–40 μm long, 3.0–5.5 μm wide) have rounded apices, a reduced raphe (not conspicuous) and can occur singly or in short filament-like groups. Usually found in upland areas, sometimes in coloured waters, and is not particularly common. Bacillariophyta. Plate XXIII.

(b) Cells not lunate. **160**

160 (159) (a) Cells with swollen area on concave or ventral side **163**

(b) Cells not as above **161**

161 (160) (a) Cells gently curved into a crescent or nearly straight and with bluntly rounded apices. Raphe visible towards the poles of the cell .... *Eunotia*

Eunotia cells (17–220 μm long, 1.5–7.0 μm wide) not as strongly curved as *Semiorbis*. Sometimes with a slightly undulate convex margin but often a smooth curve. Apices bluntly rounded or capitate/reflexed. Raphe only visible towards the cell apices where the polar nodules are obvious. Two chloroplasts lying side by side (often only visible as two in valve view). Lives attached to other surfaces, particularly plants, by means of mucilage pads. Can be single or in ribbon-like strands. Found in slow flowing or standing waters often poor in nutrients and acidic. Bacillariophyta. Plate XXIII.

(b) Cells not as above **162**

162 (161) (a) Cells crescent-shaped with a straight ventral surface in valve view, oval in girdle view, with squared-off apices. Cell apices rounded, hyaline area between markings on central dorsal surface in some species . . . .*Amphora*

Amphora cells usually lie so that they are seen in girdle view. The valves are strongly arched so that each raphe, which is curved, appears on the same side. There is an H-shaped chloroplast. Cells (30–90 μm long, 20–40 μm wide) broadly oval in girdle view with rounded apices. *Amphora* is a benthic diatom, primarily marine but also widespread in freshwaters. Perhaps more common in harder richer waters. Often found as an epiphyte and usually with its concave face against the substrate. Bacillariophyta. Plate XXIV.

(b) Cells not shaped as above and without hyaline area between markings on dorsal surface **164**

163 (160) (a) Cells with no true raphe. Up to 150 μm long, 4–8 μm wide . . . .*Hannaea*

Cells curved (banana-shaped) with a concave ventral surface that has a swollen area in its centre. Cells (5–150 μm long, 2–50 μm wide) with slightly capitate or rostrate ends, containing two lobed chloroplasts (only one of which may be visible). Either solitary or in small groups on stones and rocks. Single common species, *H. arcus*. Found in cooler upland streams, sometimes slightly acidic and in larger lake margins. Bacillariophyta. Plate XXIII.

(b) Cells with a true raphe (9–40 μm long, 3–9 μm wide) . . . .*Reimeria*

Cells of *Reimeria* are dorsiventral in shape (see Glossary) with rounded (slightly rostrate) apices, and have an obvious swollen area in the centre of the concave margin. Cells (9–40 μm long, 3–9 μm wide) contain a single-lobed chloroplast. Widespread on damp rocks, in mosses and in rivers. Bacillariophyta. Plate XXIII.

164 (162) (a) Cells with strongly curved dorsal surface and a slightly convex, straight or slightly inflated ventral surface (see Plate XXIV). Central ends of raphe bent upwards (towards dorsal surface) and polar ends bent downwards (towards the ventral surface) . . . . *Encyonema*

Valves strongly arched on the dorsal margin and nearly straight on the ventral margin giving them the appearance of an orange segment (see also *Cymbella*). The raphe is closer to the ventral margin and approximately parallel to it. Frustule surface with very fine striae not easily seen with a light microscope. Cells (20–58 μm long, 3–30 μm wide) contain an H-shaped chloroplast with a pyrenoid and are solitary or enclosed in a mucilage tube. A widespread genus in rivers on rocks and other substrata. In older texts this genus in usually included with *Cymbella*. Bacillariophyta. Plate XXIV. Fig. 4.66.

(b) Cells with gently curved dorsal surface. Central raphe ends bent downwards (towards the ventral surface) and polar ends bent upwards (towards the dorsal surface) . . . . *Cymbella*

Cymbella cells have a convex dorsal margin and a straight, concave or slightly convex ventral margin. Valve surface with striae at right angles to the raphe, which is usually central with the outer ends turned upwards and the central ends turned downwards towards the ventral margin. Cells

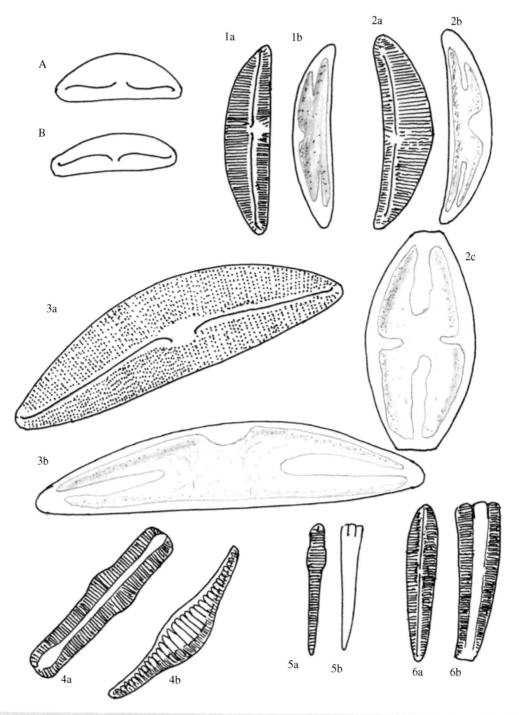

Plate XXIV A/B: general valve view – different orientations of raphe **1.** *Encyonema*: Type A raphe (a) frustule markings, (b) chloroplast. **2.** *Amphora*: (a) valve view – frustule markings, (b) valve view – chloroplast, (c) girdle view – chloroplast. **3.** *Cymbella*: Type B raphe (a) frustule markings, (b) chloroplast. **4.** *Rhopalodia*: (a) valve view, (b) girdle view. **5.** *Peronia*: (a) valve view, (b) girdle view. **6.** *Rhoicosphenia*: (a) valve view, (b) girdle view.

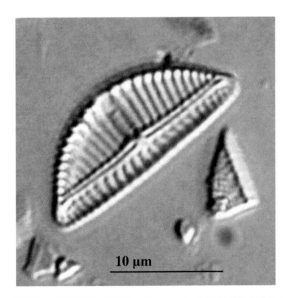

Figure 4.66 *Encyonema*. Valve view of this asymmetric diatom. Acid digest, lake sediment sample. Reproduced with permission from M. Capstick.

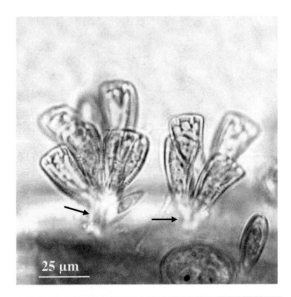

Figure 4.67 *Rhoicosphenia*. Epiphytic clusters of this sedentary diatom with cells attached by stalks (arrows). Fixed preparation.

(10–260 μm long, 4–50 μm wide) have a single H-shaped chloroplast with a single central pyrenoid. Cells can be free-floating or attached by means of a mucilage pad to a solid substratum. Can grow in open water treatment filters and reach large enough numbers to cause filter-blocking problems. Bacillariophyta. Plate XXIV.

165 (157) (a) Cells forming star-shaped colonies
. . . . *Asterionella*

(see also key No. 78)

(b) Cells not forming star-shaped colonies **166**

166 (165) (a) Cells wedge-shaped and curved along their apical axis in girdle view, more club-shaped in valve view. Single H-shaped chloroplast . . . . ***Rhoicosphenia***

Cells (12–75 μm long) often attached by a pad of mucilage at their narrow

end to plants and other substrata. Heterovalvar with the lower valve having a fully developed raphe with central nodules and the upper valve with an extremely reduced raphe in the form of short slits near the poles (only visible using electron microscopy). Abundant and widespread in richer and even brackish waters. Bacillariophyta. Plate XXIV. Figs. 2.29 and 4.67.

(b) Cells not as above **167**

167 (166) (a) Cells heteropolar, wedge-shaped in one view but shaped like an Egyptian mummy in other view (see Plate XXV) sometimes many times as long as wide
. . . . **168**

(b) Cells not as above, isopolar and not shaped like an Egyptian mummy **170**

168 (167) (a) Cells narrow, much longer than wide, tapering slightly to an acute lower pole, bluntly rounded and slightly capitate upper pole . . . . ***Peronia***

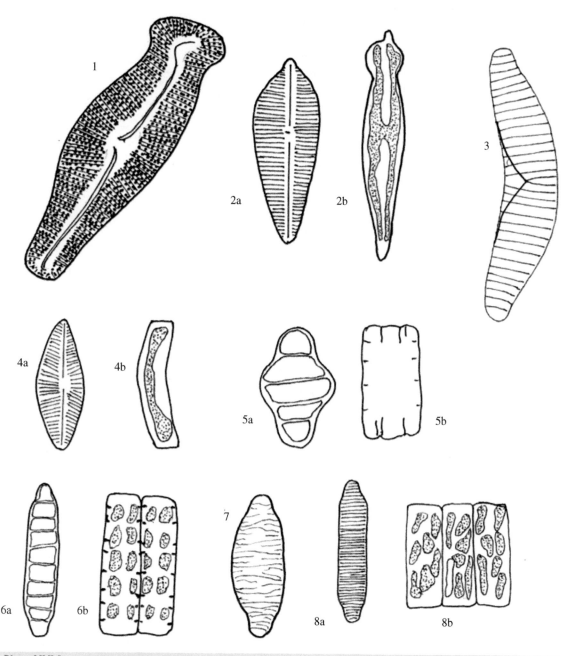

Plate XXV 1. *Didymosphenia*. 2. *Gomphonema*: (a) *G. augur* frustule markings, (b) *G. acuminatum* chloroplast arrangement. 3. *Epithemia*. 4. *Achnanthes*: (a) valve view, (b) girdle view, chloroplast. 5. *Tetracyclus*: (a) valve view, (b) girdle view. 6. *Diatoma hyemalis*: (a) valve view, (b) girdle view. 7. *Diatoma vulgaris*. 8. *Fragilariaforma*: (a) valve view. (b) girdle view.

Cells of **Peronia** (15–70 μm long, 2.5–5.0 μm wide) are relatively straight and heteropolar, with a subcapitate or bluntly rounded apical end and a tapering, narrowly rounded, base end. One valve has an obvious raphe but the other only has a rudimentary one. The valve surface is finely striated, with small spines around the edge. There are two chloroplasts, one either side of the centre. Grows attached to a substratum by means of a mucilage pad. Found in acidic oligotrophic waters. Bacillariophyta. Plate XXIV.

(b) Cells wider with either bluntly rounded, capitate or rostrate upper pole **169**

169 (168) (a) Cells large (>60 μm) with markedly capitate upper and lower poles, rather like an Egyptian mummy in shape ....*Didymosphenia*

Didymosphenia cells (60–140 μm long, 25–43 μm wide) have markedly capitate apices, with the apical end being larger than the base. The central cell area is inflated. The valve face has conspicuous striae with a clear central area which has 2–4 prominent stigmata on one side. A single H-shaped chloroplast is present with a central pyrenoid. Occurs on damp rocks and other substrata, attached by a mucilage stalk. Can occur in large numbers, when the mucilaginous stalks may cause an obstruction to water flow. Bacillariophyta. Plate XXV.

(b) Cells bluntly rounded or rostrate, upper pole at most slightly capitate .... *Gomphonema*

Gomphonema cells are heteropolar in valve view but cuneate in girdle view. Some species are only slightly heteropolar (*G. angustatum*) whilst others are more strongly heteropolar, somewhat like an

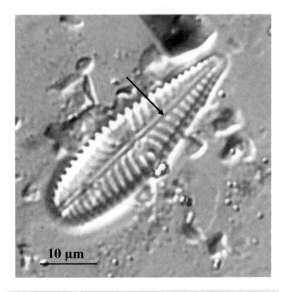

Figure 4.68 *Gomphonema*. Club-shaped diatom with prominent central raphe (arrow). Acid digest, lake sediment sample. Reproduced with permission from M. Capstick.

Egyptian mummy (**G. truncatum** and **G. acuminatum**). The valve surface has fine to obvious striae, parallel in some species or radiate in others. There is a single stigma in the clear central area. A single H-shaped chloroplast is present with a central pyrenoid. Usually attached to a surface by a single or branched mucilaginous stalk. A widespread species in a range of waters. Some species are reported as being sensitive to pollution (**G. subtile**). Bacillariophyta. Plate XXV. Figs. 2.29 and 4.68.

170 (167) (a) Cells isopolar, linear-lanceolate to elongate-elliptical in valve view, bent or curved in girdle view............... **171**

(b) Cells isopolar, lanceolate with rounded or protracted apices but not curved or bent in girdle view........ **176**

171 (170) (a) Cells with distinct costae extending from margin to margin. Single much-lobed chloroplast **172**

(b) Costae absent. Cells either with one C-shaped/plate-like or two elongate/plate-like chloroplasts **173**

172 (171) (a) Cells with arched dorsal surface and concave ventral surface. Raphe characteristically V-shaped at centre of concave surface (see Plate XXV) .... ***Epithemia***

Epithemia is a widely distributed epiphyte found in freshwaters and sometimes in brackish conditions but rarely in more acid waters. Some species found in habitats where phosphorus is more readily available. The cells are solitary with a strongly dorsiventral shape, and show distinct transverse striae. The single chloroplast has distinctly lobed margins. Cells (8–200 μm long, 4–35 μm wide) often contain cyanophycean endosymbionts that play a role in nitrogen fixation (DeYoe et al., 1992). Bacillariophyta. Plate XXV.

(b) Raphe close to dorsal surface but not V-shaped .... ***Rhopalodia***

Rhopalodia cells are solitary, either attached or free-floating, in freshwater or brackish waters. They are dorsiventral in shape often having turned-down apices. Cells (40–150 μm long, 7–12 μm wide) have a single-lobed chloroplast and show coarse transverse striae on the valve surface. The raphe curves along the margin of the cell in a keel but can be difficult to see. Bacillariophyta. Plate XXIV.

173 (171) (a) Cells with a strongly curved appearance but with margins/sides parallel. Bluntly rounded ends, raphe hardly visible. Transverse ridges on the frustule surface ending in short spines .... ***Semiorbis***

(see also key No. 159)

(b) No prominent transverse ridges or short spines. Cells slightly but not strongly curved **174**

174 (173) (a) Cells gently curved with a small swelling at centre of concave surface. Slightly capitate ends. Costae and septa absent .... ***Hannaea***

(see also key No. 163)

(b) Cells gently curved or bent at centre in girdle view. Elliptical to oval or nearly rectangular in valve view **175**

175 (174) (a) Cells gently curved, no small central swelling. Dorsal or convex margin sometimes undulate. Ends of raphe visible towards poles in valve view . ***Eunotia***

(see also key No. 161)

(b) Cells with raphe clearly visible on one valve. Other valve with pseudoraphe. Cells rectangular to elliptical in valve view and bent (genuflexed) in girdle view .... ***Achnanthes***

Cells of ***Achnanthes*** are heterovalvar, one valve bearing a true raphe and the other a pseudoraphe. Cell apices rounded or variable. Cells (5–35 μm long, 3–10 μm wide) contain a single plate-like chloroplast and are either single or form short chains. A widespread and common alga attached to a range of surfaces in flowing waters. Many species in marine or brackish waters. Recent revisions of the old genus ***Achnanthes*** has resulted in a number of new genera being separated out including ***Achnanthidium***, ***Eucocconeis***, ***Planothidium***, ***Kolbesia*** and ***Psammothidium***. ***Achnanthes minutissima***, which is very common, has now been reclassified as ***Achnanthidium minutissimum***. Bacillariophyta. Plate XXV. Fig. 4.69.

176 (170) (a) Cells linked to form colonies or chains **177**

(b) Cells mostly solitary **191**

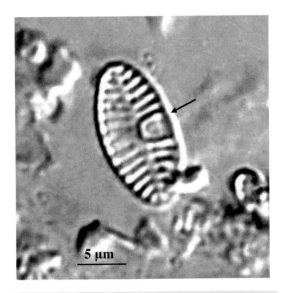

Figure 4.69   *Planothidium.* Acid digest, lake sediment sample. Face view of valve with central pseudoraphe and horseshoe-shaped marking (arrow). Reproduced with permission from M. Capstick.

This section of the key (177–191) contains colonial species which can be found as single cells if the colony breaks up. It is important to look at as many examples of a species in a sample to determine whether it is normally solitary or colonial.

177 (176) (a) Cells joined at corners to form star-shaped or zig-zag colonies **178**

(b) Cells joined along valve face to form ribbons or chains rather than being joined at corners **182**

178 (177) (a) Colonies zig-zag in shape **179**

(b) Colonies star-shaped or like the radiating spokes of a wheel **181**

179 (178) (a) Cells without transverse costae or septa. In valve view, cells have a swollen central area . . . . *Tabellaria*

(see also key No. 29)

(b) Cells usually with strong septa and less clearly defined costae **180**

180 (179) (a) Cells with transverse costae, no septa, elongate and rectangular in girdle view. In valve view with parallel sides and bluntly to slightly curved and rounded ends . . . . *Diatoma*

(see also key No. 29)

(b) Cells with both costae and septa present . . . . *Tetracyclus*

Tetracyclus cells are small and form zig-zag colonies. They are rectangular in girdle view and oval to elliptical in valve view. Short septa are present as are prominent transverse costae. Cells (4–30 μm long, 3–12 μm wide) contain several disc-shaped chloroplasts. Found in both the plankton and sediments of lakes as well as amongst mossses. Bacillariophyta. Plate XXV.

181 (178) (a) Colonies star-shaped, cells many times as long as wide, rectangular in shape (squared off ends in girdle view), septate, valves swollen in centre . . . . *Tabellaria*

(see also key Nos. 29 and 146)

Tabellaria can form zig-zag or star-shaped colonies which are commonly seen in girdle view. Valves 12–140 μm long, 5–16 μm wide. *T. fenestrata* **var. asterionelloides** can be confused with **Asterionella** but the cell ends are not swollen. There are distinct septa within the cell, which has several plate-like chloroplasts. Widespread and common in the plankton of neutral to slightly acidic lakes. Some species can be attached to a substratum. Can produce odour in drinking waters and block filters if present in large numbers. Bacillariophyta.

(b) Colonies star-shaped but valves not as above. Valve apices wider and more rounded, slightly heteropolar with the inner pole a little larger than the outer one. Non-septate and cells hardly or not swollen in the centre. Several plate-like chloroplasts. . . . . *Asterionella*

(see also key No. 78)

Widespread and common in the plankton of meso- and eutrophic lakes. Bacillariophyta.

182 (177) (a) Valves with costae **183**

(b) Valves without costae **186**

183 (182) (a) Valves with costae and septa. Cells square to rectangular in girdle view . . . . *Tetracyclus*

(see also key No. 180)

(b) Cells without septa **184**

184 (183) (a) Cells square to rectangular in girdle view, often linked into loose filaments by means of mucilage pads . . . . *Diatoma mesodon*

Diatoma mesodon cells (10–40 μm long, 6–14 μm wide) are rectangular in girdle view, being wider than the cells are long. In valve view, the cells are slightly elliptical or lanceolate. Costae are present within the valve. The chloroplasts are small and disc-shaped or plate-like. It occurs in upland flowing waters. Bacillariophyta. (Not illustrated.)

(b) Cells not square in shape but more elongate. Costae extend near to apices **185**

185 (184) (a) Cells linear to lanceolate with bluntly rounded apices (30–100 μm long, 7–13 μm wide) . . . . *Diatoma hyemalis*

Cells contain many small and plate-like or discoid chloroplasts, and are usually closely attached to one another to form a chain or ribbon. Found in lakes and ponds, frequently in the littoral zone. Bacillariophyta Plate XXV.

(b) Cells 12–85 μm long, 4–9 μm wide. Cell apices subcapitate . *Meridion anceps*

Meridion anceps cells are more rectangular than those of *M. circulare* (which are cuneate) and are linear to long elliptical in girdle view. They also do not tend to form fan-shaped colonies. Cell apices capitate to subcapitate. Found in more upland waters low in nutrients. Bacillariophyta. (Not illustrated.)

186 (182) (a) Cells elongate, >30 μm long **187**

(b) Cells shorter and wider **189**

187 (186) (a) Cells narrow but slightly swollen in centre. Two chloroplasts per cell **188**

(b) Cells wider, rectangular in girdle view, several disc-shaped chloroplasts . . . . *Fragilariaforma*

Cells of *Fragilariaforma* are closely linked along their long axes to form chains, ribbons or sometimes zig-zag colonies. Cells more linear in valve view with subcapitate or rostrate apices, and contain many small discoid chloroplasts. No transverse costae present. Found in oligotrophic waters, especially streams, lake margins and bogs. Bacillariophyta. Plate XXV.

188 (187) (a) Cells with slightly swollen central region where adjacent cells touch . . . . *Fragilaria crotonensis*

(see also key No. 28)

(b) Cells rectangular, touching along entire length .... *Fragilaria capucina*

(see also key No. 28)

189 (186) (a) Cells rectangular in shape with bluntly rounded corners and with a single chloroplast. Valve wall thickened towards the centre .... *Diadesmis*

Diadesmis has small valves (4–45 μm long, 2–10 μm wide) with a hyaline area along the edge of the valve and a clear area along the axis either side of the raphe – which can be difficult to see using a light microscope. Cells can join together at their valve faces to form small ribbons or chains. Widespread in distribution on damp rocks, soils, mosses and in ponds and springs. Bacillariophyta. (Not illustrated.)

(b) Cells with two chloroplasts **190**

190 (189) (a) Cells short, round to elliptical, with blunt apices, 3–35 μm long and 2–12 μm wide .... *Staurosira*

Cells can either be attached to a substratum (e.g. sand grains) or be free-living. Cells contain plate-like chloroplasts and can form filaments which may be zig-zag in shape. Found in shallow waters and rivers. Bacillariophyta. (Not illustrated.)

(b) Cells larger, lanceolate to rhomboid, with narrowly rounded apices .... *Pseudostaurosira*

Cells elliptical, lanceolate or more rhomboid in valve view. Apices narrowly rounded to almost rostrate. Valve margin may be slightly undulate. Cells (11–30 μm long, 2–12 μm wide) contain plate-like chloroplasts and are usually linked to form filaments. Found in the epipsammon in a range of waters. Bacillariophyta. (Not illustrated.)

191 (176) (a) Cells elongate and narrow, no true raphe present on either valve. Two elongate chloroplasts. Isopolar with apices gently narrowing. May be attached to a surface by a mucilaginous pad or free-floating .... *Synedra*

Cells of *Synedra* are elongate – needled-shaped or fusiform, sometimes with capitate ends. No raphe is present and the axial clear area is narrow. The central clear area can be from margin to margin. Fine parallel striae are present either side of the axial area. Cell length from <25 μm (*S. parasitica*) to > 500 μm (*S. ulna*), with cell width ranging from 3 to 10μm. Two long chloroplasts are present although only one is usually visible in valve view. Some species are epiphytic on other algae (forming colonies with cells radiating out from a single attachment point) while other species are free-floating. Generally widespread and sometimes abundant in the plankton of lakes and slow flowing rivers. Can produce odours in drinking water and block treatment filters when in large numbers (Palmer, 1962). Bacillariophyta. Plate XXVI. Figs. 2.23 and 4.43.

(b) Cells not attached, wider in shape. True raphe present **192**

192 (191) (a) Two chloroplasts per cell, one either side of central axis. A series of dots (carinal dots) are visible along one margin .... *Nitzschia*

Nitzschia is a large genus whose cells may be elliptical, narrow linear, spindle-shaped or sigmoid in valve view. In some species, the valve centre may be slightly constricted. The raphe is displaced to one margin, with the raphe on each valve diagonally opposite the other. The raphe

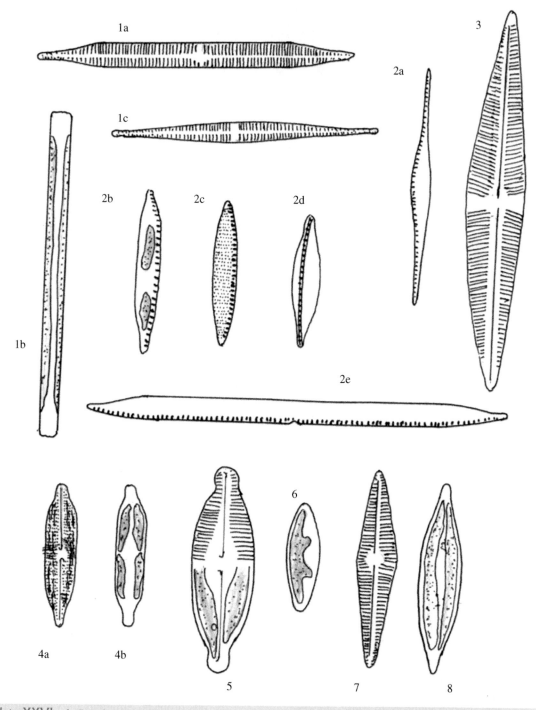

Plate XXVI **1.** *Synedra*, valve views: (a) *S. ulna* – valve view, markings, (b) *S. ulna* – girdle view, chloroplasts, (c) *S. acus.* **2.** *Nitzschia*: (a) *N. acicularis*, (b) *N. palea*, (c) *N. amphibia*, (d) *N. dissipata*, (e) *N. linearis*. **3.** *Stauroneis.* **4.** *Neidium*: (a) valve view, frustule markings, (b) chloroplasts. **5.** *Caloneis* – frustule markings/chloroplasts. **6.** *Luticola.* **7.** *Brachysira.* **8.** *Craticula.*

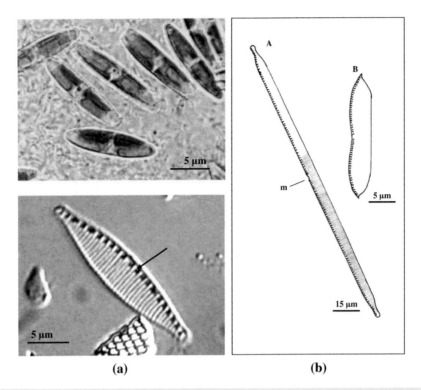

(a)   (b)

Figure 4.70 (a) *Nitzschia*. Top: Live cells, each with two olive brown chloroplasts. Bottom: Acid digest, lake sediment sample. The raphe (arrow) with punctae is displaced to the margin (keel). Reproduced with permission from M. Capstick. (b) A: *Nitzschia linearis*. Straight-sided alga with small median constriction (m). B: *Nitzschia dubia*.

structure itself is a canal supported by bars which appear as carinal dots. There is no clear central area in the striae which may be fine or coarse in appearance. Two large chloroplasts are present, one either end of the central area. Cells (20–250 μm long, 4.5–16 μm wide) are usually solitary but can form stellate colonies or a number can be present in a mucilaginous tube. May be benthic or planktonic. A very common and widespread genus found in a variety of water types. *Nitzschia* **sp.** can grow in open water treatment filters in such numbers to cause blockage. Bacillariophyta. Plate XXVI. Figs. 4.70a and 4.70b.

(b) Chloroplasts not as above. No carinal dots present **193**

193 (192) (a) One or two chloroplasts, plate-like or H-shaped. Pyrenoid may be present or obvious large droplets on the chloroplast **194**

(b) Chloroplasts not as above **195**

194 (193) (a) Two plate-like, H-shaped chloroplasts, one above and one below the middle of the cell, often with two conspicuous droplets. Striations on valve surface interrupted by one or more marginal lines. Polar terminals of the raphe forked, with central terminals hooked/bent in opposite direction. . . . . ***Neidium binodis***

Neidium binodis has two chloroplasts (other species of ***Neidium*** have four) each containing a pyrenoid. Cells 15–120 μm long, 4–30 μm wide, linear to lanceolate with widely rostrate ends. The valve surface has fine striae which may be interrupted near to the valve margin. Central ends of raphe bent in opposite directions. A widespread genus found in oligo- to mesotrophic waters. Bacillariophyta. Plate XXVI.

(b) Two chloroplasts typically present, one either side of the long axis. Central raphe terminals straight or bent the same way. Clear marginal lines present. Cells with wide rostrate apices. . . . .***Caloneis***

Caloneis cells (60–125 μm long, 25–30 μm wide) are elliptical to fusiform with rounded to rostrate or capitate ends. There is either a single plate-like chloroplast or two lying either side of the axial line. The valve face bears fine striae which are interrupted by a line just inside the valve margin. The raphe ends are either straight or both bent to the same side. Frequent on rocks and stones in streams and amongst mosses, sometimes in alkaline waters. Bacillariphyta. Plate XXVI.

195 (193) (a) Four chloroplasts per cell. . .***Neidium***

Neidium cells typically have four chloroplasts (each with pyrenoid) per cell although ***N. binodis*** (see key No 194) has two H-shaped chloroplasts. The valves are linear to lanceolate with the cell apices broadly rostrate. The valve is ornamented with rows of punctae which are interrupted by a series of small gaps near to the margin. There is a raphe whose ends are bent in

opposite directions. Cells vary in size: 15–200 μm long, 4–30 μm wide – depending on species. Widespread in nutrient-poor or mildly acidic waters. Bacillariophyta. Plate XXVI.

(b) Less than four chloroplasts per cell **196**

196 (195) (a) Two plate-like chloroplasts, one either side of long axis, sometimes with lobed margins. Stauros (see Glossary) present **197**

(b) Cells not as above. **198**

197 (196) (a) Central area of valve wider, with striae missing, and the clear area reaching nearly completely from side to side forming the stauros. Cells rhomboidal to linear-lanceolate . . . . ***Stauroneis***

The valves of ***Stauroneis*** are lanceolate to elliptical with rounded or rostrate ends. The valve surface has parallel to slightly radiate striae with a marked clear area or stauros in the central region which extends to the valve margin. Cells (8–160 μm long, 3–20 μm wide) contain two chloroplasts, one lying each side of the apical axis, with one to many pyrenoids. A widespread genus – common on damp rocks and amongst mosses, with many species occurring in more oligotrophic waters including lakes. Bacillariophyta. Plate XXVI.

(b) Valves widely lanceolate, clear central area not quite reaching the cell margins. A prominent stigma is present at one side of the central area . . . . ***Luticola***

Valves linear, elliptical or lanceolate with rounded to slightly capitate apices. Valve surface with radiate, punctate, striae. Wide central area which does not extend to the valve edge. Raphe straight. Cells

(20–50 µm long, 5–12 µm wide) contain a single chloroplast with central pyrenoid. A widespread species occurring on damp surfaces such as rocks, stones or soil. Bacillariophyta. Plate XXVI.

198 (196) (a) Valve surface covered by fine striae with irregular gaps. Cells lanceolate, some with capitate/rostrate apices. Valve face with thickened ridge along edge .... ***Brachysira***

Brachysira cells are solitary. The genus is widespread in many waters but especially in flowing systems. Raphe straight. Cells (14–115 µm long, 4–21 µm wide) have a single chloroplast. Bacillariophyta. Plate XXVI.

(b) Valves not as above **199**

199 (198) (a) Striae parallel to raphe and cross-hatched. Very small clear central area which may be completely absent. Valves widely lanceolate to slightly elliptical. Two plate-like chloroplasts .... ***Craticula***

The genus ***Navicula*** is a very large genus the species of which are sometimes hard to distinguish. With the advent of new techniques such as scanning electron microscopy several smaller genera have been divided off. These include ***Craticula***, ***Diadesmis*** and ***Luticola.***

Craticula cells (40–150 µm long, 13–30 µm wide) have two plate-like chloroplasts and two large obvious droplets. Common in a range of electrolyte-rich waters, both fresh and brackish. Bacillariophyta. Plate XXVI.

(b) Clear central area present, striae not cross hatched **200**

200 (199) (a) Striae rib-like in appearance .... ***Pinnularia***

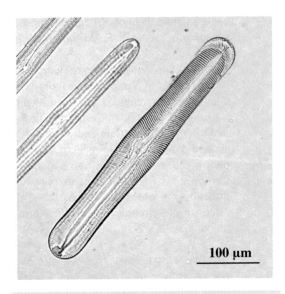

100 µm

Figure 4.7l *Pinnularia*. Valve view of this large benthic diatom. Prepared slide.

Pinnularia is a large genus. Cells linear, lanceolate or even elliptical. The poles are usually rounded, capitate or rostrate. Striae are usually coarse (but may be finer in some species). There is a central raphe whose middle ends bend in the same direction. Cells 20–200 µm long, 4–50 µm wide. Usually two plate-like chloroplasts, one either side of the midline. Some species may have other shaped chloroplasts. Widespread and common on sediments and other substrata and may be mixed in with moss clumps. Water types range from nutrient-poor to nutrient-rich and slightly alkaline to mildly acidic. Bacillariophyta. Plate XXVII. Figs. 1.13 and 4.71.

(b) Striae finer and not rib-like **201**

201 (200) (a) Raphe lying within a thickened ridge **202**

(b) Raphe not lying within a thickened ridge .... ***Navicula***

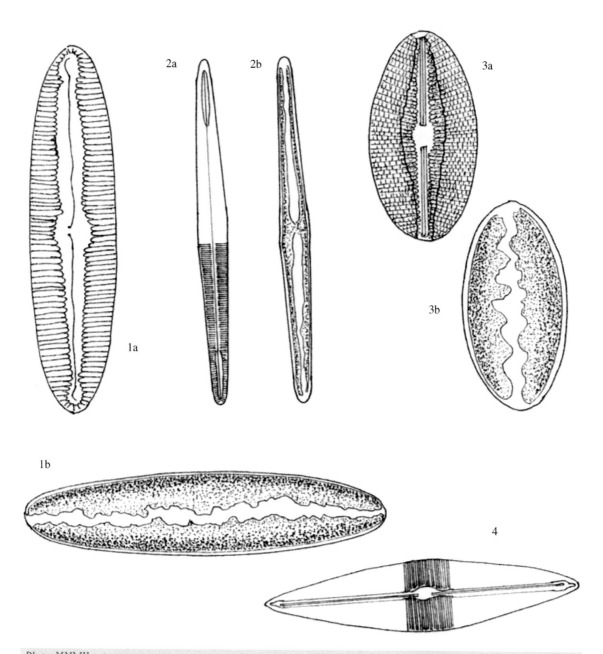

Plate XXVII **1.** *Pinnularia viridis*: (a) surface markings, (b) chloroplasts. **2.** *Amphipleura*: (a) surface markings (b) chloroplasts. **3.** *Diploneis*: (a) surface markings, (b) chloroplasts. **4.** *Frustulia*.

True **Navicula** species have lanceolate valves with a narrow axial area flanked by fine striae which are slightly radiate at the centre but parallel towards the cell apices. The raphe is hooked at the apices with the ends both pointing the same way. Cells are often very motile with a gliding motion (naviculoid movement). Cell apices may be narrowly rounded or subcapitate. Two plate-like chloroplasts are usually present lying either side of the apical axis – though some species, **N. subtilissima** and **N. bryophilia**, have a single, almost H-shaped, chloroplast. A very widespread and common genus. Found in a range of waters often occurring in benthic films in streams and rivers as well as in lakes. Bacillariophyta. Figs. 4.73a and 4.73b.

202 (201) (a) Raphe short with thickened ridges, only visible towards the poles ....**Amphipleura**

Amphipleura can be widespread but is not very common. The valves are lanceolate and the raphe is very short and located near to each pole edged by a definite rib. Cells (80–140 μm long, 7–9 μm wide) contain a single H-shaped chloroplast with pyrenoid. Found on silt and sediments, sometimes mixed with filamentous algae. More often reported from alkaline waters. Bacillariophyta. Plate XXVII.

(b) The thickened area either side of the raphe extends along the whole length of the valve **203**

203 (202) (a) Cells ovoid, two distinctly lobed chloroplasts, one either side of the long axis ....**Diploneis**

Diploneis cells are linear, rhombic or elliptical. The poles are blunt and rounded. The raphe is along the mid-line and bordered by two ridges. There are different types

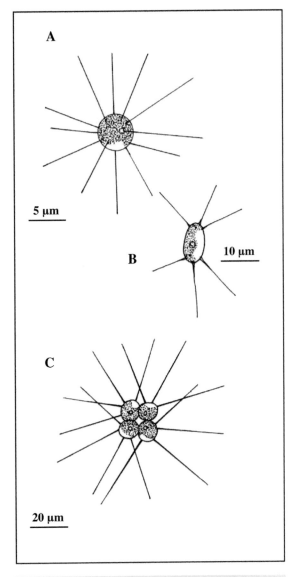

Figure 4.72 (A) *Golenkinia*, (B) *Lagerheimia*, (C) *Micractinium*.

of striae on the valve surface. Cells 10–130 μm long, 10–60 μm wide. Found on bottom deposits and silt in a wide range of waters. Bacillariophyta. Plate XXVII.

(b) Cells more elongate. Chloroplast not lobed ....**Frustulia**

Frustulia is a benthic form living on sediments. A widespread genus, found in a range of pH waters. Cells (50–160 μm long, 10–27 μm wide) rhomboid to lanceolate, contain two plate-like chloroplasts. Found singly or occasionally in mucilaginous tubes and in slightly acidic waters. Bacillariophyta. Plate XXVII.

204 (126) (a) Cells ovoid to spherical **205**

(b) Cells other shape (note cells with a median groove, although roughly oval in outline, are included here). **210**

205 (204) (a) Cells bearing spines.. . . . . **206**

(b) Cells without spines.. . . . . **209**

206 (205) (a) Cells spherical or only very slightly oval **207**

(b) Cells oval to elliptical **208**

207 (206) (a) Cells solitary, spherical, 5–21 μm in diameter. Numerous fairly long (relative to the cell diameter) spines . . . . *Golenkinia*

Golenkinia cells are solitary but may form false colonies when they fail to separate after division. Cells have several spines (12+), 24–45 μm long, and contain a single cup-shaped chloroplast with pyrenoid. Planktonic. Chlorophyta. Fig. 4.72

(b) Cells sometimes solitary though usually in colonies of 8–16 cells. Cells ovoid to spherical, 3–7 μm diameter. 1–5 spines per cell each 20–35 μm long (see Section 96, Fig. 4.72) . . . . *Micractinium*

208 (206) (a) Cells with two to many spines arising at each pole. . . . . *Lagerheimia*

Cells of *Lagerheimia* are ovate (6–18 μm in diameter, up to 23 μm long) and can join together in small groups of up to 8 cells. 1–8 chloroplasts per cell, each with a pyrenoid. Planktonic in lakes and ponds (Fig. 4.72).

(b) Cells with 3–8 spines arising from each pole. . . . . *Chodatella*

Cells of *Chodatella* are oblong to ovate (6–18 μm wide, 10–21 μm long) and contain 1–4 chloroplasts, each with a pyrenoid. The spines are fine, 10–24 μm long. Chlorophyta. Planktonic in lakes and ponds. Plate XXVIII.

NB. *Chodatella* is now usually included in with *Lagerheimia*.

209 (205) (a) Cells small and spherical (<10 μm in diameter). . . . . *Chlorella*

Chlorella cells (2–10 μm diameter) are spherical to sub-spherical with a single parietal chloroplast which nearly fills the cell. A single pyrenoid is present. Common in nutrient-rich waters but easily overlooked because of their small size. Its small size also means that it can pass through traditional water treatment sand filters, giving rise to colour problems in the treated water. Chlorophyta. Plate XXVIII. Fig. 4.74

(b) Cells ovoid and >10 μm in size *Oocystis*

Oocystis has ovoid to egg-shaped cells that are either solitary or in small groups of 2, 4, 8, or 16 cells enclosed by the mother cell wall (sometimes quite faint and difficult to see). Cells may have thickened poles (polar nodules). Cells (7–50 μm long, 6–12 μm wide) contain one to many discoid or plate-like chloroplasts. Pyrenoids may or may not

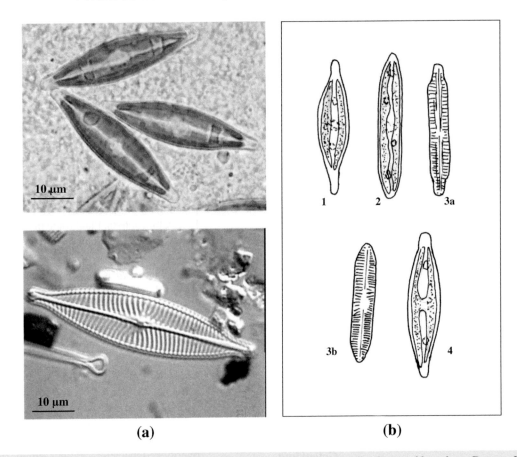

Figure 4.73 (a) *Navicula.* Top: Group of live cells, each with two elongate olive-brown chloroplasts. Bottom: Valve view of diatom frustule, showing striae radiating out from a central area. Acid digest, lake sediment sample. Reproduced with permission from M. Capstick. (b) Species of *Navicula* with two plate-like chloroplasts (1–3) or an 'H'-shaped chloroplast (4). 1. *N. rhynchocephala.* 2. *N. tripunctata.* 3a,b. *N. bryophila.* 4. *N. subtilissima.*

be present. Widespread in lakes, ditches and ponds. Typically planktonic in harder nutrient-rich waters, though some species occur in softer waters. Chlorophyta. Plate XXVIII.

210 (204) (a) Cells have a distinct median groove or isthmus around the whole-cell central region. Sometimes the two sides of the groove are quite close together, so care must be taken with the observation (see Figs 4.76–4.77). **215**

(b) Cells without a median groove around the entire cell..................... **211**

211 (210) (a) Cells angular, 4–5 sided, with angles rounded and tipped with a short spine. One side of the cell with a deep groove but not extending around the whole cell .... ***Tetraedron***

Tetraedron cells (8–22 μm diameter) are pentagonal in shape, and contain one (usually) to many chloroplasts, usually

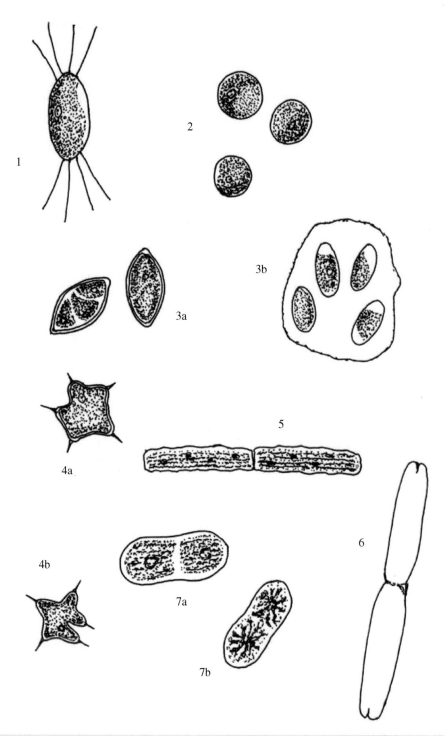

Plate XXVIII **1.** *Chodatella* **2.** *Chlorella* **3.** *Oocystis.* **4.** *Tetraedron.* **5.** *Pleurotaenium.* **6.** *Tetmemorus* **7.** *Actinotaenium.*

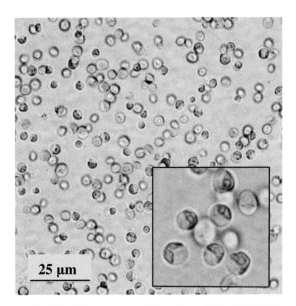

Figure 4.74 *Chlorella.* Main image: low power view of algal culture. Inset: Cell detail, showing parietal chloroplast and apical clear region.

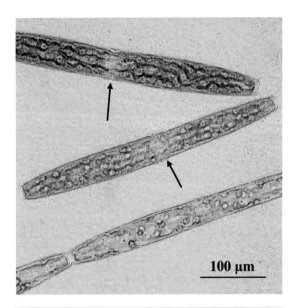

Figure 4.75 *Pleurotaenium.* These large algal cells have a median constriction (arrow). Chloroplasts are arranged in longitudinal bands (seen clearly in top cell) and have numerous pyrenoids.

with pyrenoids. Common, but rarely abundant, in lowland waters. Chlorophyta. Plate XXVIII.

(b) Cells not as above. Groove hardly or not present. If present the slight groove is equal in size on both sides of the cell or not obvious. There may be a clear area in the mid-region of the cell between two chloroplasts........................**212**

212 (211) (a) Cells with a cylindrical or near cylindrical shape **213**

(b) Cells not cylindrical in shape. **217**

Desmids are almost entirely freshwater with only a few occurring in brackish conditions. Taxonomically they are divided into two groups – (1) the Saccoderm desmids, which have a relatively simple shape lacking a median constriction and (2) the Placoderm desmids that have a median constriction, dividing the cell into two semicells.

Most desmids are single cells although a few form filaments. Some aggregate in masses, held together by a large quantity of mucilage. The study of desmids using a microscope has long been popular because of their beautiful shapes (e.g. ***Micrasterias*** and ***Euastrum***. Not in key, see Plate XXIX, Fig. 4.81).

213 (212) (a) Cells are an elongated straight cylinder, sometimes slightly tapering at ends. About ×10 as long as wide, with a shallow but noticeable constriction in the middle....................***Pleurotaenium***

In ***Pleurotaenium***, the median constriction is not deep and a swollen area may be present on either side of it. Cells (20–650 μm length, up to 75 μm wide) have numerous long band-shaped chloroplasts and pyrenoids. Often found in soft, more acid, waters and amongst mosses. Chlorophyta. Plate XXVIII. Fig. 4.75.

(b) Cells not as above but a shorter, more rounded cylinder. **214**

214 (213) (a) Cells large, 150–240 µm long, cylindrical to fusiform in shape with a distinctive notch at each apex . . . . . ***Tetmemorus***

Cells are solitary rounded cylinders, occasionally with a slight median constriction, and have apices with a characteristic deep incision. One chloroplast in each semicell with several pyrenoids. The cell wall may be covered with fine granules. Common in ponds and acid pools and amongst *Sphagnum*. Chlorophyta. Plate XXVIII.

(b) Cells are a short more rounded cylinder and only 22–100 µm long, without a notch at each end of the cell . . . .***Actinotaenium***

Actinotaenium cells are relatively short with only a shallow, wide, constriction in the centre. Cell wall very finely punctate. Chloroplast stellate with a pyrenoid. Frequent in more acidic ponds amongst other algae and mosses. Chlorophyta. Plate XXVIII.

215 (210) (a) Median groove wide, each semicell with horn-like extensions or at least a polygonal shape. Cell walls may have granulations and some planktonic species may be surrounded with mucilage . . . .***Staurastrum***

Staurastrum is a very large genus of desmids. Cells are divided into two semicells with a wide groove or isthmus between them. The outer angles of each semicell is extended into long horns or processes (up to 12 in number) which may at times be short and stumpy (giving the cell a polygonal appearance) and may end in short spines. The cell wall may be ornamented. In apical view (i.e. looking down on the cell so that the median

groove cannot be seen) rather than from the side (medial groove clearly visible), the cells are usually triangular but more arms may be present. Each semicell has a lobed chloroplast with a central pyrenoid. Cells 10–140 µm long. ***Staurastrum*** is one of the more common desmids in the plankton, also occurring on sediments. Present in nutrient-poor (most species) and moderately nutrient-rich lakes. Chlorophyta. Plate XXIX. Fig. 4.76.

(b) Median groove more narrow, or if wide then cell walls smooth and with a single spine at cell apices. **216**

216 (215) (a) Median groove marked, acute to obtuse in angle. Semicells approximately triangular, each apex with a single spine . . . .***Staurodesmus***

Staurodesmus cells are usually triradiate with a spine at the end of each angle. Cells (up to 60 µm long) are divided into two semicells with a wide groove or isthmus in the centre. A common genus in the plankton of acid, nutrient-poor waters, although sometimes reported from nutrient-rich waters, and amongst aquatic macrophytes Chlorophyta. Plate XXIX. ***Xanthidium*** is similar but with several spines at all apices and sides. Plate XXIX.

(b) Median groove very narrow, overall cell shape ovoid to rounded. Sometimes with slightly flattened sides or ends. No spines or extended processes present but cell wall may have markings. . . . .***Cosmarium***

This is the largest desmid genus and is very widespread. The cells are normally rounded (overall shape), divided into semicells with a narrow groove or isthmus in the centre between the two halves. The semicells in some species have an angular, polygonal appearance. The cell

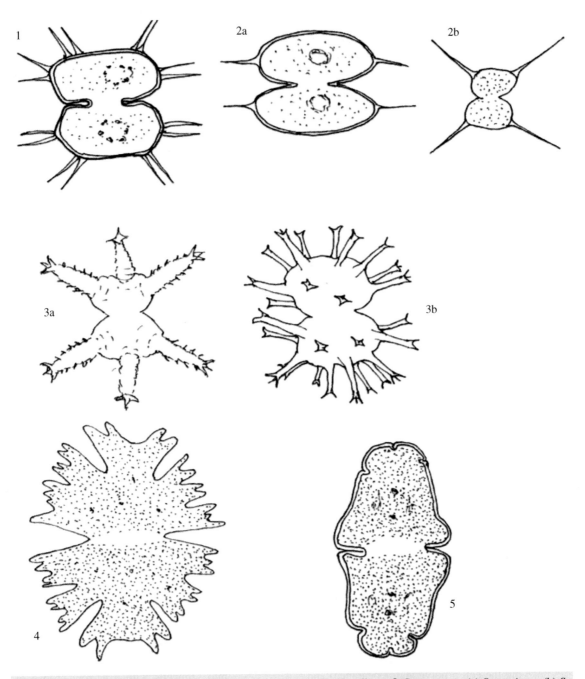

Plate XXIX   **1.** *Xanthidium.* **2.** *Staurodesmus.* (a) *S. convergens*, (b) *S. sellatus.* **3.** *Staurastrum:* (a) *S. anatinum*, (b) *S. arctison.* **4.** *Micrasterias.* **5.** *Euastrum.*

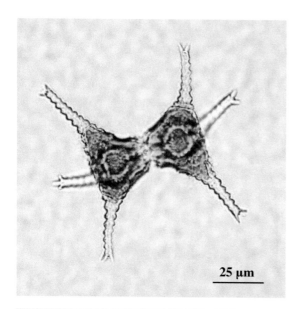

Figure 4.76 *Staurastrum*. Single cell, composed of two prominent semicells.

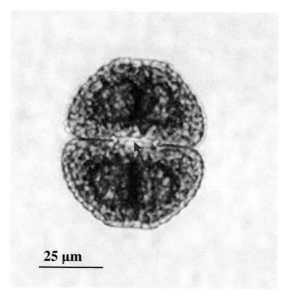

Figure 4.77 *Cosmarium*. The two semicells are joined by a narrow isthmus (arrow) at the cell equator.

walls may be ornamented with granules either scattered over the surface or in regular rows. Pyrenoids are present in the chloroplasts. Over 1000 species with sizes ranging from <10 to 200 μm. Common in upland areas (e.g. acid, low-nutrient bogs) although some species do occur in more alkaline or nutrient-rich sites. Chlorophyta. Fig. 4.77.

217 (212) (a) Cells usually in groups of 2–4–8 or more, united along some, if not all, of the lateral walls. Spines sometimes present especially on the end cells . . . . ***Scenedesmus***

(see also key No. 101)

(b) Cells not united as above **218**

218 (217) (a) Cells polygonal or tetragonal, sometimes with spines at the angles . . . . ***Tetraedron***

(see also key No. 211)

Tetraedron has angular cells which may or may not have spines at the apices. Typically a single chloroplast (sometimes more) with one to many pyrenoids. Common but not abundant in the plankton. Chlorophyta. Plate XXVIII.

(b) Cells elongate, crescent or cigar-shaped **219**

219 (218) (a) Cells with a single chloroplast almost filling the whole cell, with or without a pyrenoid. Cells needle-like or fusiform in shape **220**

(b) Cells with two axial chloroplasts, one in each half of the cell and each one having several pyrenoids in a row along its length **223**

220 (219) (a) Cells lunate to arcuate often in aggregates of 4–8–16 . . . . ***Selenastrum***

Selenastrum cells are strongly curved and frequently occur in aggregations but without a mucilaginous surround

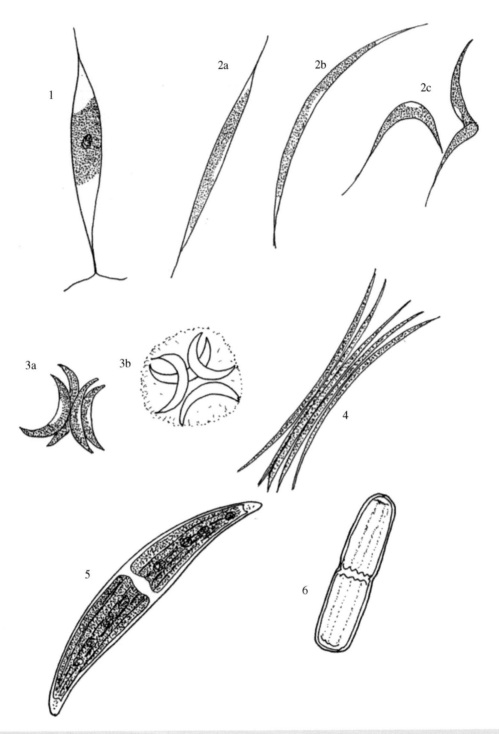

Plate XXX **1.** *Ankyra.* **2.** *Monoraphidium,* species shape variations: (a) *M. griffithii,* (b) *M. arcuatum,* (c) *M. contortum.* **3.** *Selenastrum,* variation in colony appearance: *S. bibraianum* (a) naked colony, (b) cells enclosed within a mucilaginous envelope. **4.** *Ankistrodesmus.* **5.** *Closterium.* **6.** *Penium.*

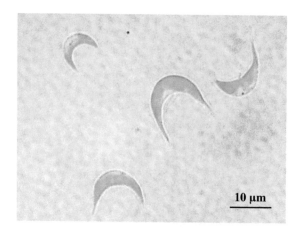

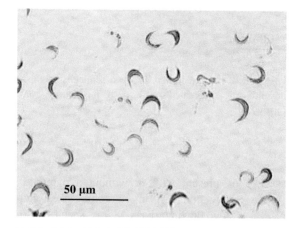

Figure 4.78 *Selenastrum.* Top: DIC image of crescent-shaped cells. Bottom: General view of culture showing range of cell size (iodine-stained preparation).

Figure 4.79 *Monoraphidium.* Top: Detailed view of elongate cells containing dispersed (iodine stained) starch grains. Bottom: General view of culture.

being present. Cells (2–8 μm wide, 13–50 μm long) with a single chloroplast which fills the whole cell. Common in plankton. Chlorophyta. Plate XXX. Fig. 4.78.

(b) Cells spindle-shaped to fusiform, straight, curved or sigmoid. Either solitary or in loose aggregations **221**

221 (220) (a) Cells fusiform with ends drawn into long narrow spines, one of which has a forked end . . . . *Ankyra*

Cells of *Ankyra* are fusiform or spindle-shaped with long drawn out apices. There is a single chloroplast with one to several pyrenoids. Common in the plankton of lakes. Chlorophyta. Plate XXX.

(b) Cells do not have either of the spines forked at the end and are not markedly fusiform **222**

222 (221) (a) Cells needle- or sickle-shaped and solitary . . . . *Monoraphidium*

Monoraphidium cells (1–5 μm wide, 7–100 μm long) have a single chloroplast,

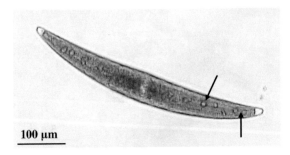

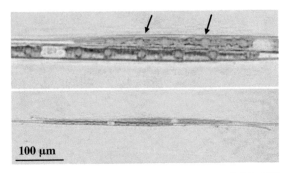

Figure 4.80 *Closterium*. Top: Crescent-shaped species (*C. moniliferum*) with two chloroplasts containing pyrenoids (arrows). Bottom pair: Very narrow (>500 μm long) celled species (*C. aciculare*) common in phytoplankton, showing a pair of cells (lower figure) and chloroplast detail with pyrenoids (arrows – upper figure).

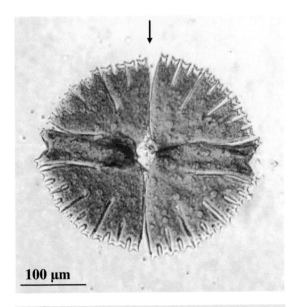

Figure 4.81 *Micrasterias*. The two semicells of this desmid are demarcated by a deep median groove (arrow).

almost filling the cell. A pyrenoid may be present. Abundant in the plankton of eutrophic lakes. Chlorophyta. Plate XXX. Fig. 4.79.

(b) Cells needle-shaped and occurring in loose, irregular bundles or tangled groups .... ***Ankistrodesmus***

The cells of this alga are very similar to *Monoraphidium* (**which used to be included in this genus**) but occur in groups or bundles. A mucilage envelope may be present. Cells (1–5 μm wide, 20–165 μm long) have a narrow spindle shape and may be curved or twisted. A single parietal chloroplast

is present with or without a pyrenoid. Very common in eutrophic lakes and slow flowing rivers. Some species also occur in more acid waters. Chlorophyta. Plate XXX.

223 (219) (a) Cells straight with rounded or truncated ends .... ***Penium***

Cells of ***Penium*** can be up to 10 times long as wide. They are cylindrical in shape, arranged in two halves each of which has a chloroplast with one to several pyrenoids. A small median constriction may be present in the truncated apices. Cell wall smooth or sometimes punctate. Cells 7–26 μm wide, 10–274 μm long. Common, often in acidic waters. Chlorophyta. Plate XXX.

(b) Cells slightly or strongly curved with tapering ends .... ***Closterium***

Cells elongate (35–1000 µm long) and tapering towards the ends. The outline may be bow-shaped, sickle-shaped or more or less straight. Cells are divided into two halves (without a median constriction) and contain two chloroplasts – each lying either side of the central area, with several prominent pyrenoids along each chloroplast. Cell walls may be smooth or have longitudinal striae. Widespread in waters ranging from acidic nutrient-poor to alkaline nutrient-rich, occurring in plankton or in amongst moss plants. Chlorophyta. Plate XXX. Fig. 4.80.

4.3 List of algae included and their occurrence in the key

The number at which particular genera key out is given below. Where a particular species of a genus would key out differently to the main genus these are listed by species in addition to the genus.

Genus or species	Key number
Acanthoceros	143
Achnanthes	175
Achnanthidium	175
Actinastrum	81
Actinocyclus	138
Actinoptychus	129
Actinotaenium	214
Anabaena	52
Amphipleura	202
Amphora	162
Ankistrodesmus	222
Ankyra	221
Apatococcus	105
Aphanizomenon	51
Aphanocapsa	67
Aphanochaete	15
Aphanothece	66
Arthrospira	53
Asterionella	78
Asterococcus	89
Asteromphalus	128
Audouinella	11
Aulacodiscus	135
Aulacoseira	27
Auliscus	135
Batrachospermum	11
Botryococcus	86
Brachysira	198
Bulbochaete	15
Caloneis	194
Calothrix	48
Campylodiscus	151
Carteria	116
Centronella	147
Ceratium	123
Chaetoceros	143
Chaetopeltis	95
Chaetophora	16
Chamaesiphon	57
Chara	2
Chlamydomonas	122
Chlorella	209
Chlorogonium	117
Chodatella	208
Chromulina	113
Chroococcus	65
Chroomonas	120
Chrysochomulina	113
Cladophora	23
Closterium	223
Cocconeis	152
Coelastrum	98
Coelosphaerium	61
Coleochaete	12
Coscinodiscus	138
Cosmarium	216
Craticula	199
Crucigenia	105
Cryptomonas	120
Cyclotella	137
Cylindrospermum	50
Cymatopleura	154
Cymbella	164
Desmidium	33
Desmococcus	105
Diadesmis	189
Diatoma	29 *and* 146
Diatoma mesodon	184
Diatoma hyemalis	185
Dictyosphaerium	85
Didymosphenia	169

Dinobryon	7	*Micrasterias*	212
Diploneis	........153 *and* 203	*Microcystis*	67
Draparnaldia	17	*Microspora*	39
Elakatothrix	84	*Microthamnion*	21
Ellerbeckia	133	*Monoraphidium*	222
Encyonema	164	*Mougeotia*	37
Enteromorpha	35	*Navicula*	201
Epipyxis	7	*Neidium*	195
Epithemia	172	*Nitella*	2
Euastrum	212	*Nitzschia*	192
Eucocconeis	175	*Nitzschia sigmoidea*	150
Eudorina	74	*Nostoc*	52
Euglena	112	*Oedogonium*	40
Eunotia	161	*Oocystis*	100 *and* 209
Fragilaria	28	*Orthosira*	132
Fragilaria crotonensis	189	*Oscillatoria*	54
Fragilaria capucina	189	*Palmella*	90
Fragilariforma	187	*Pandorina*	72
Frustulia	203	*Pediastrum*	93
Geminella	31	*Penium*	224
Glenodinium	125	*Peridinium*	124
Gloeocapsa	65	*Peronia*	168
Gloeocystis	89	*Phacus*	111
Gloeotrichia	49	*Phormidium*	55
Golenknia	207	*Pinnularia*	200
Gomphonema	169	*Pithophora*	23
Gomphosphaeria	60	*Planothidium*	175
Gongrosira	20	*Pleurotaenium*	213
Gonium	74	*Pleurococcus*	105
Gymnodinium	125	*Protoderma*	94
Gyrosigma	149	*Pseudanabaena*	52
Haematococcus	121	*Psammothidium*	175
Hannaea	163	*Pseudostaurosira*	190
Hildenbrandia	12	*Pteromonas*	122
Homoeothrix	45	*Pyramimonas*	116
Hormidium	42	*Reimeria*	163
Hyalodiscus	136	*Rhizoclonium*	23
Hydrodictyon	6	*Rhizosolenia*	142
Kirchneriella	84	*Rhodomonas*	119
Klebsormidium	42	*Rhoicosphenia*	166
Kolbesia	175	*Rhopalodia*	172
Lagerheimia	208	*Rivularia*	49
Lemanea	11	*Scenedesmus*	101
Lepocinclis	113	*Scytonema*	45
Lyngbya	55	*Selenastrum*	220*1*
Mallomonas	107	*Semiorbis*	159
Melosira	27	*Snowella*	62
Meridion	77	*Sorastrum*	99
Meridion anceps	185	*Spermatozopsis*	115
Merismopedia	58	*Sphaerocystis*	90
Micractinium	96	*Sphaerella*	121

4.4 Algal identification: bibliography

This bibliography is provided as a supplement to the main reference list at the end of the book, and is intended to help readers who wish to go into more detail with identification of both genera and species within the genera.

Anagnostidis, K., Komarek, J. (1985). Modern approach to the classification system of cyanophytes 1 – Introduction. Archiv für Hydrobiologie Suppl. 71 (Algological Studies 38–39), 291–302.

Anagnostidis, K., Komarek, J. (1988). Modern approaches to the classification system of cyanophytes 3 – Oscillatoriales. Archiv für Hydrobiologie Suppl. 80 (Algological Studies 50–53), 327–472.

Barber, H.G., Haworth, E.Y. (1981). A Guide to the Morphology of the Diatom Frustule. Ambleside, UK, Freshwater Biological Association, Scientific Publication No. 44.

Bellinger, E.G. (1992). A Key to Common Algae, Freshwater and Some Coastal Species. 4th ed. London, Institute of Water and Environmental Management.

Bourelly, P. (1966). Les Algues D'Eau Douce. Tome 1: les Algues vertes. Editions N. Paris, Boubee & Cie.

Bourelly, P. (1968). Les Algues D'Eau Douce. Tome 2: Les Algues Jaunes et Brunes. Chrysophycees, Pheophycees, Xanthophycees et Diatomees. Editions N. Paris, Boubee & Cie.

Bourelly, P. (1970). Les Algues D'Eau Douce. Tome 3: Les Algues Blues et Rouges. Les Eugleniens, Peridiniens et Cryptomonadines. Editions N. Paris, Boubee & Cie.

Brook, A.J. (1981). The Biology of Desmids. London, Blackwell.

Carr, N.G., Whitton, B.A. (eds) (1982). The Biology of Cyanophyta. London, Blackwell.

Chorus, I., Bartrum, J. (eds) (1999). Toxic Cyanophyta in Water. London and New York, Spon Press.

Cox, E.J. (1996). Identification of Freshwater Diatoms from Live Material. London, Chapman & Hall.

Croasdale, H., Flint, E.A. (1988). Flora of New Zealand, Desmids. Vol. II. Christchurch, New Zealand, Botany Division, D.S.I.R.

Desikachary, T.V. (1959). Cyanophyta. New Delhi, Indian Council of Agricultural Research.

DeYoe, H.R., Lowe, R.L., Marks, J.C. (1992). Effects of nitrogen and phosphorus on the endosymbiont load of *Rhopalodia gibba* and *Epithemia turgida* (Bacillariophyceae). Journal of Phycology 28, 773–777.

Fritsch, F.E. (1956). The Structure and Reproduction of the Algae. Vol. 1 and 2. Cambridge University Press.

Hartley, B. (1996). An Atlas of British Diatoms based on illustrations by H.G. Barber and J.R. Carter, edited by P.A. Sims. Ambleside, UK, Biopress Ltd.

Hendey, N. I. (1964). An Introductory Account of the Smaller Algae of British Coastal Waters. Part V: Bacillariophyceae (Diatoms). London, HMSO.

Hustedt, F. (1930). Bacillariophyta (Diatomeae) Heft 10. In Pascher, A. (ed.). Die Susswasser-Flora Mitteleuropas. Gustav Fischer, Jena.

John, D.M., Whitton, B.A., Brook, A.J. (2002). The Freshwater Algal Flora of the British Isles. Cambridge University Press.

Kelly, M. (2000). Identification of common benthic diatoms in rivers. Field Studies 9, 583–700.

Leedale, G.F. (1967). Euglenoid Flagellates. Biological Science series. Englewood Cliffs, NJ, Prentice-Hall.

Lind, E.M., Brook, A.J. (1980). Desmids of the English Lake District. Cumbria, UK, Freshwater Biological Association. Scientific Publication No. 42.

National Rivers Authority (1990). Toxic blue-green algae. Water Quality Series No 2. Ambleside, UK, National Rivers Authority.

Palmer, C.M. (1962). Algae in water supplies. Washington, DC, U.S. Department of Health, Education, and Welfare.

Pascher, A. (1913–1936). Die Susswasserflora Deutchlands. Osterreichs und der Schweiz. Vol. 15.

Patrick, R., Reimer, C.W. (1966, 1975). The diatoms in the United States, 2 vols. Monographs of the Academy of Natural Sciences, Philadelphia.

Pentecost, A. (1984). Introduction to freshwater algae. Richmond, Surrey, UK, Kingprint Ltd.

Prescott, G.W. (1951). Algae of the Western Great Lakes Area. Dubuque, Iowa, Wm. C. Brown Company Publishers. Reprint edition (1982) by Otto Koeltz Science Publishers, Koenigstein, Germany.

Prescott, G.W. (1964). The Freshwater Algae. Dubuque, Iowa, Wm. C. Brown Company Publishers.

Round, F.E., Crawford, R.M., Mann, D.G. (1990). The Diatoms: Biology and Morphology of the Genera. Cambridge, UK, Cambridge University Press.

Smith, G.M. (1950). Freshwater algae of the United States. McGraw-Hill.

Van Den Hoek, C., Mann, D.G., Jahns, H.M. (1995). Algae. An introduction to phycology. Cambridge, UK, Cambridge University Press.

Wehr, J.D., Sheath, R.G. (2003). Freshwater algae of North America. Ecology and classification. London, Academic Press.

West, G.S., Fritsch, F.E. (1927). A treatise on the British Freshwater Algae. Cambridge, UK, Cambridge University Press.

West, W., West, G.S. (1904). A monograph of the British Desmidiaceae, Vols I to III. London, The Ray Society.

Williams, D.M. (1985). Morphology, Taxonomy and Inter-Relationships of the Ribbed Raphid Diatoms from the Genera Diatoma and Meridion (Diatomaceae: Bacillariophyta). Cramer, Hirschberg, Germany, 255 pp.

Glossary

Acicular   Needle-shaped

Acuminate   Tapering gradually towards the apex

Akinete   A resting spore with thickened walls formed from a vegetative cell in blue-green algae, usually much larger than other vegetative cells

Alternating series   A group of cells arranged side by side but with every other cell displaced either up or down relative to the mid-line (contrast with linear series)

Annular   Shaped like a ring

Anterior   Front end, i.e. the end at the front of motile forms, or top end

Apical axis   The axis linking the two poles (ends) of the cell

Araphid   Without a raphe system on either valve

Arcuate   Bow-shape, strongly curved

Areolae   Regular perforated markings on the wall of a diatom forming a distinct pattern

Attenuate   Narrowing to a point

Auxospore   A resting spore formed by sexual activity or vegetative activity in some diatoms that allows size restitution of the cell to occur

Axial area   A clear or hyaline area running along the apical axis usually containing the raphe

Benthos   Living on the bottom, on the sediment surface or rocks

BF   Bright field light microscopy

Bifurcate   Forked into two branches

Biseriate   A double row of cells or pores on a cell wall

Bloom   A dense growth of algae that visibly discolours the water

Branching (true)   Along the length of a filament one cell gives rise to two or more cells one of which continues the main axis whilst the others grow at an angle to produce an additional filament (the branch)

(false)   Side growths occur along a filament but they are formed by a second filament growing sideways and not by a single cell giving rise to two or more others which themselves form lateral branches

Bulbous   Bulb-like, swollen at one end

Calyptra   A hood or cap-like covering usually appearing on the apical cells of some filaments

Capitate   With one or both ends of a cell swollen into a head

Carinal dots   Dots or short bar-like markings spaced evenly along the keel (at one margin) edge of some diatoms (particularly *Nitzschia*)

Central nodule   A thickening on the wall of a diatom at the centre between the raphe endings

Freshwater Algae: Identification, Enumeration and Use as Bioindicators, Second Edition. Edward G. Bellinger and David C. Sigee.
© 2015 John Wiley & Sons, Ltd. Published 2015 by John Wiley & Sons, Ltd.

Centrales   Diatoms that are radially symmetrical and usually circular in outline from the valve view – Centric Diatoms

Chaetae   Hair or bristle-like growths of the cell wall

Chloroplast   An organelle within the cell concerned with photosynthesis and containing the photosynthetic pigments. Often green in colour but can be golden to brown (or other colour) depending upon the combination of pigments present.

Cingulum   Part of the girdle region of a diatom associated with one valve

Clathrate   With obvious spaces between the cells

Clavate   Club-shaped

Coccoid   Cells that are rounded or spherical

Coenobium   A colony of cells whose number is fairly constant (often a multiple of four), the number being determined early on in its development and no further increase takes place until a new generation develops. The cells are arranged in a specific way.

Coenocytic   Having multinucleate cells or structures, e.g. in *Vaucheria* which is not divided into cells

Colony   A group of cells joined together more or less permanently, or enclosed within the same mucilage or sheath

Contractile vacuole   A small, usually spherical, body which regularly fills and then contracts, associated with osmoregulation. Often found in flagellates

Costae   An elongate thickening (rib) on a diatom valve (singular Costa)

Crenulate   Gently wavy

Crescent-shaped   Shaped like the arc of a circle and tapering towards the ends

Cruciform   Having four arms and shaped like a cross

Crustose   Crust-like. Composed of flat tiers of cells

Cuneate   Wedge-shaped

Cyst   A thick-walled resting spore whose walls may be impregnated with silica

Cytoplasm   All of the protoplasmic content of the cell excluding the nucleus

Daughter cells   Cells formed by division of the mother cell

Daughter colony   A new colony arising within a mother cell or mother colony

Dendroid   Tree-like, branching like a tree or bush. Used to describe the colony shape in *Dinobryon*, which is not truly filamentous

DIC   Differential interference contrast (light microscopy)

Dichotomous   Branching or dividing to form two equal-sized branches

Dorsal   The upper or more convex surface of a cell (as opposed to lower or ventral), e.g. in the diatom *Cymbella*

Dorsiventral   A cell with two margins showing different curvature

Dorsiventral flattening   Flattened in cross section in a cell, similar to laterally flattened or flattened in side view

Ellipsoidal   A figure with curved margins but elongate ends which are sharply rounded

Endospore   Non-motile spore produced in an indefinite number from a parent cell, e.g. in *Chamaesiphon*

Epicingulum   Girdle sections produced from the parent cell in diatoms

Epicone   The upper or anterior part of a dinoflagellate cell

Epilithic   Living on rock surfaces

Epipelic   Free living on sediments

Epiphytic   Living on a plant surface

Episammic   Living on sand grains

Epitheca   The part of the cell wall of a dinoflagellate above the transverse furrow or in diatoms the epivalve + epicingulum derived from the mother cell. The larger of the two halves of the diatom cell wall, into which the hyptheca fits.

Epivalve   Flat surface (face) of the epitheca – circular in centric diatoms, elongate in pinnate diatoms

Exospore   A spore produced by budding from another cell in blue-green algae

Eyespot   A complex of granules, usually red or brownish in colour, sensitive to light and found in some motile species

False branching   Branching formed by a different filament growing sideways and not by a single cell giving rise to two or more cells that develop into new branches (see Fig. 4.21)

Fibula   In diatoms a bridge of silica between parts of the valve either side of the raphe

Filament   A series of cells forming one or more rows in a linear arrangement (cells stacked end to end)

Flask-like   Wide at the base and abruptly narrowing to a neck

Frustule   The cell wall in diatoms which is made of silica and is composed of two halves (valves) linked together by a number of bands called collectively the girdle

Fusiform   An elongate shape widest at the centre and tapering at each end (spindle or cigar-shaped)

Gas vacuole   A protein covered gas-filled organelle found in cyanophyta

Genuflexed   Crooked in the middle, bent at an angle similar to a person bowing at the knee

Gibbous   Swollen to form a small locally occurring bump usually at the centre

Girdle   The zone of silica bands linking the two halves of a diatom frustule

Girdle view   Side view of a diatom frustule, showing the girdle or area where the two halves of a diatom cell join (as opposed to the valve view when the girdle generally cannot be seen)

Glaucous   Greyish-green or green with a white overcast

Glycogen   A starch-like storage product found in blue-green algae

Gonidia   Spore-like thick-walled reproductive cells in blue-green algae

Gullet   A canal or groove at the anterior end that opens at its base to a chamber. Found in some flagellates

Haptonema   A flagellum-like organelle linked to the basal bodies of flagella found in the Haptophyta.

Heterocyst   An enlarged cell present in some blue-green algae with thickened walls and often with a highly refractive appearance making it distinct from other cells. Associated with nitrogen fixation

Heteropolar   The two ends (poles) of a diatom cell are differently shaped

Heterovalvar   Each valve of a diatom is a different shape

H-pieces   Sections of a cell wall in a filamentous species where the wall is composed of two overlapping halves that appear shaped like an H. When the cells of the filament separate (or break) they do so at the centre rather than at the cross walls between two cells. The H-pieces are often more obvious at a filament end

Hormogonium   (pleural hormogonia) a short part of a filament in blue-green algae that can be released and form a new filament thus serving as a reproductive and dispersal mechanism

Hyaline   Transparent and colourless

Hypocingulum   Cingulum within the epitheca

Hypocone   The lower part or posterior portion of a dinoflagelate cell

Hypotheca   The smaller of the two halves of the diatom cell wall, fits into the epitheca

Hypovalve   Flat surface (face) of the hypotheca, circular in centric **diatoms**, elongate in pennate diatoms

Imbricate   Overlapping

Intercalary   Arranged within a series of cells rather than being at the end or alongside such as can occur with heterocysts in blue-green algae

Iodine test   The application of weak iodine solution to starch produces a blue-black colour. Even in algae producing starch as a food storage product the test may not always give an obvious result as it depends upon the physiological state of the cell, i.e. whether it has been producing starch and how much

Isobilateral   Each side of the cell, as divided by the longitudinal axis, is the same shape

Isopolar   Both ends (poles) of the cell are the same shape and size

Isthmus   The narrow part of a desmid cell connecting the two semicells

Keel   The channel or flange, sometimes raised on ribs or struts, in which the raphe runs in some diatoms, e.g. *Surirella*

L.A.B.   Long as wide, used to compare length and breadth dimensions of cells

Lamellate   Composed of layers (also laminate)

Lanceolate   Lance-shaped, long and narrow with a small increase in diameter in the mid-region and tapering towards the ends

Lenticular   Lens-shaped or hemispherical

Leucosin   A whitish food reserve characteristic of many chrysophytes and usually found as highly refractive rounded particles

Linear series   A row of cells arranged side by side in a straight line (for comparison see alternating series)

Lorica   A shell or open flask-like structure which the organism lies. The shape varies but there is always an opening at one end, sometimes with a collar through which a flagellum may pass

Lunate   Crescent or moon-shaped

Mantle   The marginal part of a diatom frustule, joined to the epivalve or hypovalve

Marginal lines   Lines of markings or clear areas within markings running parallel to the margin of a cell. Usually used in descriptions of markings on certain diatoms

Marginal process   Structures that occur around the margin of a cell in some diatoms

Matrix   The surrounding matter, especially mucilage, to a cell

Median groove   A groove or channel running around the mid-region of a cell (as in dinoflagellates where one of the two flagella occur in the median groove which runs around the centre of the cell or in some desmids where there is an indentation between the two semicells)

Metabolic   Able to change its shape (as in some species of *Euglena*) as the cell wall is not rigid

Monospore   A non-motile spore produced by the monosporangium in Rhodophytes

Mother cell   A cell that divides to form two or more offspring (daughter cells)

Mucilage   A material, often made of polysaccharides, which swells in water and is slimy to tough

Multiseriate   Composed of many rows

Nannoplankton   Algal cells whose size is small enough to pass through a plankton net

Obliquely flattened   Flattened at an angle, neither horizontally or vertically

Obovoid   An approximate oval but with one end (termed the anterior) wider than the other

Ocellus   'Eye-shaped' marking. A thickened rimmed plate of silica with a number of pores on its surface on a centric diatom valve

Ovoid   Egg-shaped

Paramylum   A solid starch-like food reserve found in some euglenoids

Parietal   Arranged at or around the wall of a cell

Pellicle   A thin membrane covering a cell that has no true wall

Pennate Diatoms that are bilaterally symmetrical about the apical axis

Perforation With holes or spaces between the cells

Periplast Cell membrane of euglenoids, or other bounding membrane

Phycocyanin The blue-green pigment in the cells of Cyanophyta

Polar nodule A body on the inner wall at the end (pole) of some diatoms and other algae

Polar septum A partition or cross wall near the end of the cell

Posterior The rear or back end of an organism

Prostrate Creeping or growing over a substrate and being closely adhering to it

Pseudocilium A flagellum-like structure but not an organ of locomotion, found in the **Tetrasporales**

Pseudoparenchymatous Appearing like parenchyma but in reality is made up of closely packed filaments

Pseudoraphe A false raphe in diatoms. A clear area between makings on the valve surface which forms an area in the position of a true raphe but no canal, visible as a line, is present

Punctate Small dots or pores on the surface of a diatom frustule usually forming a characteristic pattern

Pyrenoid A protein body in the cell associated with the chloroplast which may or may not have a sheath of starch (depending on genus). If starch is the storage product the pyrenoid will stain darkly with iodine

Pyriform Pear-shaped, with a wider base than top

Quadrate Square or rectangular in shape

Raphe A longitudinal canal (visible as a line on the surface) in the wall of some diatoms. Some genera have a raphe on one valve some on both and others no raphe at all. The raphe is associated with gliding movements in some genera

Raphe ribs Longitudinal ribs running alongside the raphe

Reniform Kidney-shaped

Reticulate Like a net in its arrangement

Rhomboid A parallelogram with oblique angles and adjacent sides unequal

Ribbon-like Large numbers of elongate cells joined by their sides to form a filament-like structure. In transverse section, however, a ribbon is never circular (see *Fragilaria crotonensis*)

Rostrate Narrowing or ending in a beak

Scalariform Ladder-like

Segment-shaped Shaped like the segment of an orange

SEM Scanning electron microscope

Semicell One half of a desmid cell

Septum A cross partition either completely or partly across a diatom cell dividing it into chambers (pleural Septa).

Seta A hair-like growth arising from a cell

Setiferous Bearing setae or hairs

Sheath A covering envelope, sometimes thin covering a filament or group of cells

Sickle-shaped Acutely curved or crescent-shaped

Sigmoid Shaped like an S

Siliceous wall A cell wall impregnated with silica in diatoms and often bearing distinctive markings. Its presence can be detected by digesting the organic matter with an oxidizing agent leaving the silica wall behind. Diatoms have distinctive silica cell walls but some other algae have small silica scales on the outside of the cell wall, e.g. in some members of the **Chrysophyta**

Sinus A deep furrow between the semicells of a desmid

Siphonaceous A tubular or filamentous-like thallus having no cross walls, e.g. in *Vaucheria*

Spathulate   A wide upper part tapering gradually to form a stalk

Starch test   When a solution of iodine is applied to the specimen, if starch is present a blue or blue-black colour is produced. See iodine test above.

Stauros   A thicker hyaline region on a diatom valve extending from the central nodule to the margins of the valve

Stellate   Star-shaped

Sternum   A hyaline region running along the long axis of a diatom valve in diatoms that have no raphe (araphid) with striae either side

Stigma   see eyespot

Strand   A group of cells joined side by side to form a filament-like structure (see also ribbon)

Striae   Delicate, sometimes long narrow markings or lines on the valve of a diatom

Sub-pyramidal   Shaped like a pyramid but with the top cut off

Sulcus   A narrow pair of furrows that may be found in the girdle area of some filamentous diatoms

Test   A rigid urn or pot-shaped surround to a cell with a space between it and the cell inside

Thallus   A plant body with little differentiation into tissues

Theca   see a test

Transapical   At right angles to the apical or long axis in diatoms

Trichome   A thread-like series of cells, exclusive of the sheath, found in blue-green algae

Trichotomous   Branching or dividing to form three equal-sized branches

Triradiate   With three arms radiating out from a central position

Truncate   A cell with a flat top or end

Tychoplankton   Plankton in shallow inshore waters, often in amongst plants and weeds, and detached from a surface and are therefore chance members of the plankton

Undulate   With a wavy surface or edge

Unilateral   Pertaining to one side only

Uniseriate   Formed as a single row (of pores in diatoms or cells in other algae)

Vacuole   A space in the cytoplasm lined by a membrane filled with fluid or occasionally granules

Valve   One of the two halves of a diatom frustule

Valve mantle   The margin of a diatom valve sometimes with a definite slope or structure

Valve view   The view of a diatom cell looking at the valve face and not seeing any overlapping portions (as opposed to the girdle view)

Ventral   The less convex margin of a dorsiventral cell, e.g. in the diatom *Cymbella*

Water bloom   A dense growth of planktonic algae that distinctly colours the water and may form a scum on the surface

Whorls   Several branches arising at the same point around a main axis, e.g. in *Chara*

Xanthophylls   A yellow, carotene derived, photosynthetic pigment

XRMA   X-ray microanalysis

Zoospore   A motile sexual spore

Zygospore   A thick-walled resting spore formed from the fusion of two gametes

References

Adams, S.M. (2005). Using multiple response bioindicators to assess the health of estuarine ecosystems: an operational framework. In Bortone, S.A. (ed.). Estuarine Indicators. Boca Raton, FL, CRC Press, pp. 5–18.

Agatz, M., Asmus, R.M., Deventer, B. (1999). Structural changes in the benthic diatom community along a eutrophication gradient on a tidal flat. Helgoland Marine Research 53, 92–101.

Anderson, N.J., Rippey, B., Stevenson, A.C. (1990). Change to a diatom assemblage in a eutrophic lake following point-source nutrient re-direction: a palaeolimnological approach. Freshwater Biology 23, 205–217.

Asaeda, T., Sultana, M., Manatunga, J., et al. (2004). The effect of epiphytic algae on the growth and production of Potamogeton perfoliatus L. in two light conditions. Environmental and Experimental Botany 52, 225–238.

Augspurger, C., Gleixner, G., Kramer, C., et al. (2008). Tracking carbon flow in a 2-week-old and 6-week-old stream biofilm food web. Limnology and Oceanography 53, 642–650.

Baker, P.D. (1999). Role of akinetes in the development of cyanobacterial populations in the lower Murray River, Australia. Marine and Freshwater Research 50, 265–279.

Barber, H.G., Haworth, E.Y. (1981). The Diatom Frustule. Ambleside, UK, Freshwater Biological Association, p. 112.

Barbour, M.T., Swietlik, W.F., Jackson, S.K., et al. (2000). Measuring the attainment of biological integrity in the USA: a critical element of ecological integrity. Hydrobiologia 422/423, 453–464.

Battarbee, R.W., Charles, D.F., Dixit, S.S., et al. (1999). Diatoms as indicators of surface water acidity. In Stoermer, E.F., Smol, J.P. (eds.). The Diatoms: Applications for the Environmental and Earth Sciences. Cambridge, UK, Cambridge University Press, pp. 85–127.

Bellinger, E.G. (1974). A note on the use of algal sizes in estimates of population standing crops. British Phycological Journal 9, 157–161.

Bellinger, E.G. (1977). Seasonal size changes in certain diatoms and their possible significance. European Journal of Phycology 12, 233–239.

Bellinger, E.G. (1992). A Key to Common Algae, 4th edn. London, The Institution of Water and Environmental Management, p. 138.

Bennion, H., Battarbee, R. (2007). The European Union Framework Directive: opportunities for palaeolimnology. Journal of Palaeolimnology 38, 285–295.

Bennion, H., Duigan, C.A., Haworth, E.Y., et al. (1996). The Anglesey lakes, Wales, UK – changes in trophic status of three standing waters as inferred from diatom transfer functions and their implications for conservation. Aquatic Conservation: Marine and Freshwater Ecosystems 6, 81–92.

Bennion, H., Fluin, J., Simpson, G.L. (2004). Assessing eutrophication and reference conditions for Scottish freshwater lochs using subfossil diatoms. Journal of Applied Ecology 41, 124–138.

Bold, H.C., Wyne, M.J. (1985). Introduction to the Algae, 2nd edn. Englewood Cliffs, NJ, Prentice Hall.

Boon, P.J., Pringle, C.M. (2009). Background, philosophy and context. In Boon, P.J., Pringle, C.M. (eds.). Assessing the Conservation Value of Freshwaters. Cambridge, UK, Cambridge University Press, pp. 6–38.

Borstad, G.A. (1982). The influence of the meandering Guiana Current and the Amazon River discharge on surface salinity near Barbados. Journal of Marine Research 40, 421–434.

Bortone, S.A. (2005). Estuarine Indicators. Boca Raton, FL, CRC Press.

Brettum, P. (1989). Alger som Indikator pa Vannkvalitet i Norske Innsjoer: Planteplankton. Report O-86116, Oslo, Norway, Norsk Institutt for Vannforskning (NIVA), p. 2344.

Buesseler, K.O., Antia, A.N., Chen, M., et al. (2007). An assessment of the use of sediment traps for estimating upper ocean particle fluxes. Journal of Marine Research 65, 345–416.

Bura, R., Cheung, M., Liao, B., et al. (1998). Composition of extracellular polymeric substances in the activated sludge floc matrix. Water Science Technology 37, 325–333.

Burkholder, J.M., Sheath, R.G. (1984). The seasonal distribution, abundance and diversity of desmids (Chlorophyta) in a softwater, north temperate stream. Journal of Phycology 20(34), 159–172.

Byron, E., Eloranta, P. (1984). Recent historical changes in the diatom community of Lake Tahoe, California–Nevada, USA. Verhandlungen der Internationale Vereinigung fur Theoretische und Angewandte Limnologie 22, 1372–1376.

Carney, H., Hunter, D.A., Goldman, C.R. (1994). Seasonal, interannual, and long-term dynamics of plankton diatoms in oligotrophic Lake Tahoe. 11th International Diatom Symposium, San Francisco, CA, California Academy of Sciences, pp. 621–629.

Caron, D.A., Lim, E.L., Dennett, M.R., et al. (1999). Molecular phylogenetic analysis of the heterotrophic chrysophyte genus Paraphysomonas (Chrysophyceae), and the design of rRNA-targeted oligonucleotide probes for two species. Journal of Phycology 35, 824–837.

Carvalho, L., Moss, B. (1995). The current status of a sample of English Sites of Special Scientific Interest subject to eutrophication. Aquatic Conservation: Marine and Freshwater Ecosystems 5, 191–204.

Castenholz, R.W. (1992). Species usage, concept, and evolution in the cyanobacteria (blue-green algae). Journal of Phycology 28, 737–745.

Cattaneo, A., Methot, G., Pinel-Alloul, B., et al. (1995). Epiphyte size and taxonomy as biological indicators of ecological and toxicological factors in lake Saint-Francois (Quebec). Environmental Pollution 87, 357–372.

Cattaneo, A., Asioli, A., Comoli, P., et al. (1998). Organisms' response in a chronically polluted lake supports hypothesized link between stress and size. Limnology and Oceanography 43, 1938–1943.

Cavendar-Bares, K.K., Frankael, S.L., Chisholm, S.W. (1998). A dual sheath flow cytometer for shipboard analyses of phytoplankton communities from the oligotrophic oceans. Limnology and Oceanography 43, 1383–1388.

Cemagref. (1982). Etude des methodes biologiques quantitatives d'appreciation de la qualite des eaux. Rapport Division Qualite des Eaux Lyon, Pierre-Benite, France 8, Agence financiere de Bassin Rhone-Mediterránee-Corse. Lyon, Cemagref, 118, 218 pp.

CEN. (2003). EN 13946, Water Quality – Guidance Standard for the Routine Sampling and Pre-treatment of Benthic Diatoms from Rivers. Geneva, Switzerland, Comité Européen de Normalisation.

CEN. (2004). prEN 14007, 2004 Water Quality – Guidance Standard for the Identification, Enumeration and Interpretation of Benthic Diatom Samples from Rivers. Geneva, Switzerland, Comité Européen de Normalisation.

Charles, D.F., Smol, J.P. (1994). Long term chemical changes in lakes: quantitative inferences using biotic remains in the sediment record. Advances in Chemistry 237, 3–31.

Condie, S.A., Bormans, M. (1997). The influence of density stratification on particle settling, dispersion and population growth. Journal of Theoretical Biology 187, 65–75.

Coste, M., Ayphassorho, H. (1991). Etude de la qualite des eaux du Bassin Artois-Picardie a l'aide des communautes de diatomees benthiques (Application des indices diatomiques). Raport Cemagref. Bordeaux, France, Agence de l'Eau Artois-Picardie, Douai, pp. 277.

Cowles, T.J. (2004). Planktonic layers: physical and biological interactions on the small scale. In Seuront, L., Strutton, P.G. (eds.). Handbook of Scaling Methods in Aquatic Biology. Boca Raton, FL, CRC Press, pp. 31–49.

Cumming, B.F., Smol, J.P., Birks, H.J. (1992). Scaled chrysophytes (Chrysophyceae and Synurophyceae) from Adirondack drainage lakes and their relationship to environmental variables. Journal of Phycology 28, 162–178.

Darley, W.M. (1982). Algal Biology: A Physiological Approach. Oxford, UK, Blackwell Scientific.

Davey, M.C., Crawford, R.M. (1986). Filament formation in the diatom Melosira granulata. Journal of Phycology 22, 144–150.

Davidson, T., Sayer, C., Bennion, H., et al. (2005). A 250 year comparison of historical, macrofossil and pollen records of aquatic plants in a shallow lake. Freshwater Biology 50, 1671–1686.

Dean, A.P. (2004). Interactions of phytoplankton, zooplankton and planktonic bacteria in two contrasting lakes. Life Sciences PhD thesis, University of Manchester, p. 386.

Dean, A.P., Sigee, D.C. (2006). Molecular heterogeneity in *Aphanizomenon flos-aquae* and *Anabaena flos-aquae* (Cyanophyta): a synchrotron-based Fourier-transform infrared study of lake populations. European Journal of Phycology 41, 201–212.

Dean, A.P., Martin, M.C., Sigee, D.C. (2007). Resolution of codominant phytoplankton species in a eutrophic lake using synchrotron-based Fourier transform infrared spectroscopy. Phycologia 46, 151–159.

De Angelis, D.L., Wolf, M.M., Basset, A. (2004). The importance of spatial scale in the modelling of aquatic organisms. In Seuront, L., Strutton, P.G. (eds.). Handbook of Scaling Methods in Aquatic Biology. Boca Baton, FL, CRC Press, pp. 383–400.

Defarge, C., Trichet, J., Jaunet, A.-M., et al. (1996). Texture of microbial sediments revealed by cryo-scanning electron microscopy. Journal of Sediment Research 66, 935–947.

De Jonge, M., Van de Vijver, B., Blust, R., et al. (2008). Responses of aquatic organisms to metal pollution in a lowland river in Flanders: a comparison of diatoms and macroinvertebrates. Science of the Total Environment 407, 615–629.

Dela-Cruz, J., Pritchard, T., Gordon, G., et al. (2006). The use of periphytic diatoms as a means of assessing impacts of point source inorganic nutrient pollution in south-eastern Australia. Freshwater Biology 139, 951–972.

De Pauw, N., Hawkes, H.A. (1993). Biological monitoring of river water quality. In Walley, W.J., Judd, S. (eds.). River Water Quality Monitoring and Control. Current Practices and Future Directions. Birmingham, UK, Aston University Press, pp. 87–111.

Descy, J.P. (1979). A new approach to water quality estimation using diatoms. Nova Hedwigia 64, 305–323.

Descy, J.P., Coste, M. (1991). A test of methods for assessing water quality based on diatoms. Verhandlungen der Internationale Vereinigung fur Theoretische und Angewandte Limnologie 24, 2112–2116.

Droppo, I.G., Ross, N., Skafel, M., et al. (2007). Biostabilisation of cohesive sediment beds in a freshwater wave-dominated environment. Limnology and Oceanography 52, 577–589.

Duff, K.E., Smol, J.P. (1995). Chrysophycean cyst assemblages and their relationship to water chemistry in 71 Adirondack Park (New York, USA) lakes. Archives of Hydrobiology 134, 307–337.

Duker, L., Palmer, M. (2009). Methods for assessing the conservation value of lakes. In Boon, P.J., Pringle, C.M. (eds.). Assessing the Conservation Value of Freshwaters. Cambridge, UK, Cambridge University Press, pp. 166–199.

Eaton, A.D., Franson, M.A., Clesceri, L.S. (2005). Standard Methods for the Examination of Water and Waste Water, 21st edn. Franson, MA, & Baltimore, MD, American Public Health Association.

Edlund, M.B., Stoermer, E.F. (1997). Ecological, evolutionary and systematic significance of diatom life histories. Journal of Phycology 33(52), 897–918.

Eloranta, P. (2000). Use of littoral algae in lake monitoring. In Heinonen, P., Ziglio, G., Van der Beken, A. (eds.). Hydrological and Limnological Aspects of Lake Monitoring. Chichester, UK, John Wiley & Sons, pp. 97–104.

Eloranta, P., Soininen, J. (2002). Ecological status of some Finnish rivers evaluated using benthic diatom communities. Journal of Applied Phycology 14, 1–7.

European Community. (1991). Council directive of 21 May 1991 concerning urban waste water treatment (91/271/EEC). Official Journal of the European Community Series L 135, 40–524.

European Union. (2000). Directive 2000/60/EEC of the European Parliament and of the Council of 23 October 2000 on establishing a framework for community action in the field of water policy. Journal of the European Community 327, 1–72.

Evans, J.H. (1972). A modified sedimentation system for counting algae with an inverted microscope. Hydrobiologia 40(2), 247–250.

Evans, V. (2013). The microbial ecology and biogeochemistry of nuclear waste storage facilities in the UK. PhD thesis, University of Manchester.

Facher, E., Schmidt, R. (1996). A siliceous chrysophycean cyst-based transfer function for Central European Lakes. Journal of Paleolimnology 16(27), 275–321.

Falkowski, P.G., Raven, J.A. (1997). Aquatic Photosynthesis. Malden, MA, Blackwell Science.

Foster, R.A., Subramaniam, A., Mahaffey, C., et al. (2007). Influence of the Amazon River plume on distributions of free-living and symbiotic bacteria in the western tropical north Atlantic Ocean. Limnology and Oceanography 52, 517–532.

Fracker, S.B., Brischle, H.A. (1944). Measuring the local distribution of *Ribes*. Ecology 25, 283–303.

Fritsch, F.E. (1956). Structure and Reproduction of the Algae. Cambridge, UK, Cambridge University Press.

Gainswin, B.E., House, W.A., Leadbeater, B.S., et al. (2006). The effects of sediment size fraction and associated biofilms on the kinetics of phosphorus release. Science of the Total Environment 360, 142–157.

Gaiser, E., Wachnika, A., Ruiz, P., *et al.* (2005). Diatom indicators of ecosystem change in subtropical coastal wetlands. In Bortone, S.A. (ed.). Estuarine Indicators. Boca Raton, FL, CRC Press, pp. 127–144.

Galand, P.E., Lovejoy, C., Pouliot, J., *et al.* (2008). Microbial community diversity and heterotrophic production in a coastal arctic ecosystem: a stamukhi lake and its source waters. Limnology and Oceanography 53, 813–823.

Gelin, F., Boogers, I., Noordelos, A.A., *et al.* (1997). Resistant biomolecules in marine microalgae of the classes Eustigmatophyceae and Chlorophyceae: geochemical implications. Organic Geochemistry 26, 659–675.

Gervais, F. (1997). Light-dependent growth, dark survival, and glucose uptake by cryptophytes isolated from a freshwater chemocline. Journal of Phycology 33, 18–25.

Gillott, M. (1990). Cryptophyta (cryptomonads). In Margulis, L., Corliss, J.O., Melkonian, M., Chapman, D.J. (eds.). Handbook of Protoctista. Boston, MA, Jones and Bartlett Publishers, pp. 139–151.

Goericke, R. (1998). Response of phytoplankton community structure and taxon-specific growth rates to seasonally varying physical forcing in the Sargasso Sea off Bermuda. Limnology and Oceanography 43, 921–935.

Gosselain, V., Coste, M., Campeau, S., *et al.* (2005). A large scale stream benthic diatom database. Hydrobiologia 542, 151–163.

Graham, L.E., Wilcox, L.W. (2000). Algae. Upper Saddle River, NJ, Prentice Hall.

Graham, L.E., Graham, J.M., Russin, W.A., *et al.* (1994). Occurrence and phylogenetic significance of glucose utilisation by charophycean algae: glucose enhancement of growth in *Coleochaete orbicularis*. American Journal of Botany 81, 423–432.

Graham, J.M., Arancibia-Avila, P., Graham, L.E. (1996a) Physiological ecology of a species of the filamentous green alga *Mougeotia* under acidic conditions: light and temperature effects on photosynthesis and respiration. Limnology and Oceanography 41, 253–262.

Graham, J.M., Arancibia-Avila, P., Graham, L.E. (1996b) Effects of pH and selected metals on growth of the filamentous alga *Mougeotia* under acidic conditions. Limnology and Oceanography 41, 263–270.

Gunnison, D., Alexander, M. (1975). Resistance and susceptibility of algae to decomposiiton by natural microbial communities. Limnology and Oceanography 20, 64–70.

Hall, R.I., Smol, J.P. (1999). Diatoms as indicators of lake eutrophication. In Stoermer, E.F., Smol, J.P. (eds.). The Diatoms: Applications for the Environmental and Earth Sciences. Cambridge, UK, Cambridge University Press, pp. 128–168.

Hamilton, P.B., Proulx, M., Earle, C. (2001). Enumerating phytoplankton with an upright compound microscope using a modified settling chamber. Hydrobiologia 444, 171–175.

Hasle, G.R., Syvertsen, E.E. (1997). Marine diatoms. In Tomas, C.R. (ed.). Identifying Marine Phytoplankton. New York, Academic Press, pp. 5–385.

Heinonen, P. (1980). Quantity and composition of phytoplankton in Finnish inland waters. Publications of the Water Research Institute (Helsinki) 37, 1–917.

Hillebrand, H., Durselen, C.-D., Kirschtel, D., *et al.* (1999). Biovolume calculation for pelagic and benthic microalgae. Journal of Phycology 35, 403–424.

Hochman, A. (1997). Programmed cell death in prokaryotes. Critical Reviews in Microbiology 23, 207–214.

Hofman, G. (1996). Recent developments in the use of benthic diatoms for monitoring eutrophication and organic pollution in Germany and Austria. In Whitton, B.A., Rott, E. (eds.). Use of Algae for Monitoring Rivers II. Innsbruck, Austria, Dr. Eugen Rott, pp. 73–77.

Hoham, R., Duval, B. (2001). Microbial Ecology of Snow and Freshwater Ice with Emphasis on Snow Algae. Cambridge, UK, Cambridge University Press.

Horne, A.J., Goldman, C.R. (1994). Limnology, 2nd edn. New York, McGraw-Hill.

Hornstrom, E. (1981). Trophic characterisation of lakes by means of qualitative phytoplankton analysis. Limnologica (Berlin) 13, 246–261.

Hornstrom, E. (1999). Long-term phytoplankton changes in acid and limed lakes in SW Sweden. Hydrobiologia 394(23), 93–102.

Iliopoulou-Georgudaki, J., Kantzaris, V., Katharios, P., *et al.* (2003). An application of different bioindicators for assessing water quality: a case study in the rivers Alfeios and Pineios. Ecological Indicators 2(106), 345–360.

Jarvi, H.P., Neal, C., Warwick, A., *et al.* (2002). Phosphorus uptake into algal biofilms in a lowland chalk river. Science of the Total Environment 282/283, 353–373.

Jeffrey, S.W., Humphrey, G.F. (1975). New spectrophotometric equations for determining chlorophylls a, b, c1 and c2 in higher plants, algae and natural phytoplankton. Biochemie und Physiologie der Pflanzen 167, 191–194.

Jespersen, A., Christoffersen, K. (1987). Measurements of chlorophyll-a from phytoplankton using ethanol as extraction solvent. Archiv fur Hydrobiologie 109, 445–454.

John, D.M., Whitton, B.A., Brook, A.J. (2002). The Freshwater Algal Flora of the British Isles. Cambridge, UK, Cambridge University Press, p. 702.

Jones, J.G. (1979). A Guide to Methods for Estimating Microbial Numbers and Biomass in Fresh Water. Scientific Publication No.39. Freshwater Biological Association.

Jones, R. (1984). Application of a primary production model to epiphytic algae in a shallow eutrophic lake. Ecology 65, 1895–1903.

Jones, N.L., Thompson, J.K., Arrigo, K.R., et al. (2009). Hydrodynamic control of phytoplankton loss to the benthos in an estuarine environment. Limnology and Oceanography 54, 952–969.

Kalff, J.K. (2002). Limnology. Upper Saddle River, NJ, Prentice Hall.

Karlsson, I. (2003). Benthic growth of Gloeotrichia echninulata Cyanobacteria. Hydrobiologia 506–509, 189–193.

Karlsson-Elfgren, I., Brunberg, A.-K. (2004). The importance of shallow sediments in the recruitment of Anabaena and Aphanizomenon (Cyanophyceae). Journal of Phycology 40, 831–836.

Kelly, M.G. (2001). Use of similarity measures for quality control of benthic diatom samples. Water Research 35, 2784–2788.

Kelly, M.G. (2002). Role of benthic diatoms in the implementation of the Urban Wastewater Treatment Directive in the River Wear, north-east England. Journal of Applied Phycology 14, 9–18.

Kelly, M.G., Whitton, B.A. (1995). The trophic diatom index: a new index for monitoring eutrophication in rivers. Journal of Applied Phycology 121, 433–444.

Kelly, M.G., Penny, C.J., Whitton, B.A. (1995). Comparative performance of benthic diatom indices used to assess river water quality. Hydrobiologia 302, 179–188.

Kelly, M.G., Cazaubon, A., Coring, E., et al. (1998). Recommendations for the routine sampling of diatoms for water quality assessments in Europe. Journal of Applied Phycology 10, 215–224.

King, L., Clarke, G., Bennion, H., Kelly, M., Yallop, M. (2006). Recommendations for sampling littoral diatoms in lakes for ecological status assessments. Journal of Applied Phycology 18, 15–25.

Kitting, C.L., Fry, B., Morgan, M.D. (1984). Detection of inconspicuous epiphytic algae supporting food webs in seagrass meadows. Oecologia 62, 145–149.

Komarek, J., Kling, H., Komárková, J. (2003a). Filamentous cyanobacteria. In Wehr, J.D., Sheath, R.G. (eds.). Freshwater Algae of North America. Amsterdam, The Netherlands, Academic Press, pp. 117–196.

Komarek, J., Anagnostidis, K. (2003b). Coccoid and colonial cyanobacteria. In Wehr, J.D., Sheath, R.G. (eds.). Freshwater Algae of North America. Amsterdam, The Netherlands, Academic Press.

Kopacek, J., Vesely, J., Stuchlik, E. (2001). Sulphur and nitrogen fluxes and budgets in the Bohemian Forest and Tatra mountains during the industrial revolution (1850–2000). Hydrology and Earth System Sciences 20, 391–405.

Krembs, C., Juhl, A.R., Strickler, J.R. (1998). The Spatial information preservation method: sampling the nanoscale spatial distribution of microorganisms. Limnology and Oceanography 43, 298–306.

Kressel, M., Groscurth, P. (1994). Distinction of apoptic and necrotic cell death by in situ labelling of fragmented DNA. Cell and Tissue Research 278, 549–556.

Kristiansen, J. (2005). Golden Algae: A Biology of Chrysophytes. A.R.G. Gantner Verlag Kommanditgesellschaft, p. 167.

Kwandrans, J., Eloranta, P., Kawecka, B., et al. (1998). Use of benthic diatom communities to evaluate water quality in rivers of southern Poland. Journal of Applied Physiology 10(107), 193–201.

Lackey, J.B. (1968). Ecology of Euglena. In Buetow, D.E. (ed.). The Biology of Euglena. New York, Academic Press, pp. 28–44.

Laird, K.R., Das, B., Kingsbury, M., et al. (2013). Paleolimnological assessment of limnological change in 10 lakes from northwest Saskatchewan downwind of the Athabasca oils sands based on analysis of siliceous algae and trace metals in sediment cores. Hydrobiologia 720, 55–73.

Leclercq, L., Maquet, B. (1987). Deux nouveaux indices chimique et diatomique de qualite d'eau courante. Application au Samson et a ses affluents (bassin de la Meusebelge). Comparaison avec d'autres indices chimiques, biocenotiques et diatomiques. Document de travail. Institut Royal des Sciences Naturelles de Belqique 38, 113.

Lecointe, C., Coste, M., Prygiel, J. (1993). 'OMNIDIA': a software for taxonomy, calculation of diatom indices and inventories management. Hydrobiologia 269–270, 509–523.

Lee, R. (1997). Phycology. Cambridge, UK, Cambridge University Press.

Lehman, P.W., Marr, K., Boyer, G.L. (2013). Long-term trends and causal factors associated with Microcystis abundance and toxicity in San Francisco Estuary and implications for climate change impacts. Hydrobiologia 718, 141–158.

Lembi, C.A., O'neal, S.W., Spencer, D.F. (1988). Algae as weeds; economic impact, ecology, and management alternatives. In Lembi, C.A., Waaland, J.R. (eds.). Algae and Human Affairs. Cambridge, UK, Cambridge University Press, pp. 455–481.

Lewin, R.A. (1976). Naming the blue-greens. Nature 259, 360.

Lewitus, A.J., Kana, T.M. (1994). Responses of estuarine phytoplankton to exogenous glucose: stimulation versus inhibition of photosynthesis and respiration. Limnology and Oceanography 39, 182–189.

Liebmann, H. (1962). Handbuch der Frischwasser- und Abwasser-biologie. Munchen, Germany, R. Oldenbourg, p. 113.

Lin, C., Blum, J. (1977). Recent invasion of a red alga (*Bangia atropurpurea*) in Lake Michigan. Journal of the Fisheries Research Board of Canada 24, 13–161.

Litaker, R., Vandersea, M.W., Kibler, S.R., *et al.* (2003). Identification of *Pfeisteria piscida* and *Pfeisteria*-like organisms using internal transcribed spacer-specific PCR assays. Journal of Phycology 39, 754–761.

Livingstone, D. (1984). The preservation of algal remains in lake sediments. In Haworth, E.Y., Lund, J.W. (eds.). Lake Sediments and Environmental History. Leicester, UK, Leicester University Press, p. 185.

Livingstone, D., Reynolds, C.S. (1981). Algal sedimentation in relation to phytoplankton periodicity in Rostherne Mere. British Phycological Journal 16, 195–206.

Lowe, R.L. (2003). Keeled and canalled raphid diatoms. In Wehr, J.D., Sheath, R.G. (eds.). Freshwater Algae of North America. Amsterdam, The Netherlands, Academic Press, pp. 669–684.

Lund, J.W. (1951). A sedimentation technique for counting algae and other organisms. Hydrobiologia 3, 390–394.

Lund, J.W., Kipling, G., Le Cren, E.D. (1958). The inverted microscope method of estimating algae numbers and the statistical basis of estimation by counting. Hydrobiologia 11, 143–170.

Lyon, D.R., Ziegler, S.E. (2009). Carbon cycling within epilithic biofilm communities across a nutrient gradient of headwater streams. Limnology and Oceanography 54, 439–449.

Margalef, R. (1958). Temporal succession and spatial heterogeneity in phytoplankton. In Buzzati-Traverso, A. (ed.). Perspectives in Marine Biology. Berkeley, CA, University of California Press.

Martens, K. (1997). Speciation in ancient lakes. TREE 12, 177–182.

Martineau, E., Wood, S.A., Miller, M.R., *et al.* (2013). Characterisation of Antarctic cyanobacteria and comparison with New Zealand strains. Hydrobiologia 711, 139–154.

Mason, C.F. (2002). Biology of Freshwater Pollution. Harlow, UK, Longman.

McAlice, B.J. (1971). Phytoplankton sampling with the Sedgwick Rafter cell. Limnology and Oceanography 16, 19–28.

McCourt, R.M., Hoshaw, R.W., Wang, J-C. (1986). Distribution, morphological diversity and evidence for polyploidy in North American Zygnemataceae (Chlorophyta). Journal of Phycology 22(31), 307–313.

McNabb, C.D. (1960). Enumeration of freshwater phytoplankton concentrated on the membrane filter. Limnology and Oceanography 5, 57–61.

McQuatters-Gollop, A., Raitsos, D., Edwards, M., *et al.* (2007). A long-term chlorophyll data set reveals regime shift in North Sea phytoplankton biomass unconnected to nutrient trends. Limnology and Oceanography 52, 635–648.

Miller, D.N., Bryant, J.E., Madsen, E.L., *et al.* (1999). Evaluation and optimization of DNA extraction and purification procedures for soil and sediment samples. Applied and Environmental Microbiology 65, 4715–4724.

Muller, U. (1999). The vertical zonation of adpressed diatoms and other epiphytic algae on *Phragmites australis*. European Journal of Phycology 34, 487–496.

Nature Conservancy Council. (1991). Nature Conservation and Pollution from Farm Waste. Peterborough, UK.

Nayar, S., Goh, B.P., Chou, L.M. (2004). Environmental impact of heavy metals from dredged and resuspended sediments on phytoplankton and bacteria assessed in in situ mesocosms. Ecotoxicology and Environmental Safety 59, 349–369.

Nedbalova, L., Vrba, J., Fott, J., *et al.* (2006). Biological recovery of the Bohemian forest lakes from acidification. Biologia Bratislava 61(Suppl. 20), S453–S465.

Neustupa, J., Veselá, J., St'astny, J. (2013). Differential cell size structure of desmids and diatoms in the phytobenthos of peatlands. Hydrobiologia 709, 159–171.

Norton, T.A., Melkonian, M., Andersen, R.A. (1996). Algal biodiversity. Phycologia 35 (50) 308–326.

Novarino, G., Lucas, I.A. (1995). Some proposals for a new classification system of the Cryptophyceae. Botanical Journal of the Linnean Society 111, 3–21.

Nygaard, G. (1949). Hydrobiological studies on some Danish ponds and lakes, II: the quotient hypothesis and some little known or new phytoplankton organisms. Kunglige Danske Vidensk, Selskab 7, 1–242.

OECD. (1982). Eutrophication of waters, assessment and control. OECD Cooperative Programme on monitoring inland waters (eutrophication control). Paris, France, Organisation for Economic Cooperation and Development.

Paddock, S.W. (2007). Microscopy. In Walker, K.W. (ed.). Principles and Techniques of Biochemistry and Molecular Biology, 6th edn. Cambridge, UK, Cambridge University Press.

Paerl, H.W., Dyble, J., Pinckney, J.L., *et al.* (2005). Using microalgal indicators to assess human- and climate-induced ecological change in estuaries. In Bortone, S.A. (ed.). Estuarine Indicators. Boca Raton, FL, CRC Press, pp. 145–174.

Palmer, G. (1969). A composite rating of algae tolerating organic pollution. Journal of Phycology 5, 78–82.

Patrick, R., Hohn, M.H., Wallace, J.H. (1954). A new method for determining the pattern of the diatom flora. Notulae Naturae (Philadelphia) 259, 12.

Pechar, L., Prikryl, L., Faina, R. (2002). Hydrobiological evaluation of Třeboň fishponds since the end of the nineteenth century. Hydrobiologia 40, 247–250.

Penell, R.I., Lamb, C. (1997). Programmed cell death in plants. The Plant Cell 9, 1157–1168.

Pfandl, K., Chatzinotas, A., Dyal, P., *et al.* (2009). SSU rRNA gene variation resolves population heterogeneity and ecophysiological differentiation within a morphospecies (*Stramenopiles*, Chrysophyceae). Limnology and Oceanography 54, 171–181.

Pokorny, J., Kvet, J., Eiseltová, M., *et al.* (2002a). Role of macrophytes and filamentous algae in fishponds. In Kvet, J., Jenik, J., Soukupova, L. (eds.). Freshwater Wetlands and Their Sustainable Future. New York, Parthenon Publishing, pp. 97–124.

Pokorny, J., Kvet, J., Cerovská, K. (2002b) The role of wetlands in energy and material flows in the landscape. In Kvet, J., Jenik, J., Soukupova, L. (eds.). Freshwater Wetlands and Their Sustainable Future. New York, Parthenon Publishing, pp. 445–462.

Potapova, M., Charles, D.F. (2007). Diatom metrics for monitoring eutrophication in rivers of the United States. Ecological Indicators 7, 48–70.

Prartano, T., Wolff, G.A. (1998). Organic geochemistry of lacustrine sediments: a record of changing trophic status of Rostherne Mere. Organic Geochemistry 28, 729–747.

Prygiel, J., Coste, M. (1996). Recent trends in monitoring French rivers using algae, especially diatoms. In Whitton, B.A., Rott, E. (eds.). Use of Algae for Monitoring Rivers II. Innsbruck, Austria, Dr. Eugen Rott, pp. 87–94.

Prygiel, J., Leveque, L., Iserentant, R. (1996). Un nouvel indice diatomique pratique pour l'evaluation de la qualite des eaux en reseau de surveillance. Revue des Sciences de L'eau 1, 97–113.

Qari, H. (2006). Studies on the lipid composition of cultured algae, phytoplankton and zooplankton from Rostherne Mere. Life Sciences PhD thesis, University of Manchester.

Reynolds, C.S. (1980). Phytoplankton assemblages and their periodicity in stratifying lake systems. Holarctic Ecology 3, 141–159.

Reynolds, C.S. (1990). The Ecology of Freshwater Phytoplankton. Cambridge, UK, Cambridge University Press.

Reynolds, C.S., Bellinger, E.G. (1992). Patterns of abundance and dominance of the phytoplankton of Rostherne Mere, England: evidence from an 18-year data set. Aquatic Sciences 54, 10–36.

Reynolds, C.S., Jaworski, G.H. (1978). Enumeration of natural *Microcystis* populations. British Phycological Journal 13(1), 1269–1277.

Richardson, D.C., Kaplan, L.A., Newbold, J.D., *et al.* (2009). Temporal dynamics of seston: a recurring nighttime peak and seasonal shifts in composition in a stream ecosystem. Limnology and Oceanography 54, 344–354.

Rimet, F., Cauchie, H-M., Hoffmann, L., *et al.* (2005). Response of diatom indices to simulated water quality improvements in a river. Journal of Applied Phycology 17, 119–128.

Romani, A.M., Amalfitano, S., Artigas, J., *et al.* (2013). Microbial biofilm structure and organic matter use in Mediterranean streams. Hydrobiologia 719, 43–58.

Ronaghi, M. (2001). Pyrosequencing sheds light on DNA sequencing. Genome Research 11, 3–11.

Rosen, G. (1981). Phytoplankton indicators and their relations to certain chemical and physical factors. Limnologica (Berlin) 13, 263–290.

Rott, E. (1981). Some results from phytoplankton counting intercalibrations. Schweizer Zeitung Hydrologia 214, 34–62.

Round, F.E. (1993). A Review and Methods for the Use of Epilithic Diatoms for Detecting and Monitoring Changes in River Water Quality 1993. Methods for the Examination of Waters and Associated Materials. London, UK, Her Majesty's Stationery Office.

Round, F.E., Crawford, R.M., Mann, D.G. (1990). The Diatoms – Biology and Morphology of the Genera. Cambridge, UK, Cambridge University Press.

Rudi, K., Skulberg, O.M., Skulberg, R., *et al.* (2000). Application of sequence-specific labeled 16 S rRNA gene oligonucleotide probes for genetic profiling of cyanobacterial abundance and diversity by array hybridisation. Applied Environmental Microbiology 66, 4004–4011.

Sahin, B., Ozdemir, T. (2008). Epiphytic algae on mosses in the Altindere valley National Park (MackaTrabzon/Turkey). Pakistan Journal of Biological Sciences 11, 2278–2281.

Schiefele, S., Schreiner, C. (1991). Use of diatoms for monitoring nutrient enrichment acidification and impact salts in rivers in Germany and Austria. In Whitton, B.A., Rott, E., Friedrich, G. (eds.). Use of Algae for Monitoring Rivers. Innsbruck, Austria, Institut for Botanik, Universitat Innsbruck, pp. 103–110.

Scholin, C., Miller, P., Buck, K., et al. (1997). Detection and quantitation of Pseudonitzschia australis in cultured and natural populations using LSU rRNA-targeted probes. Limnology and Oceanography 42, 1265–1272.

Schopf, J.W. (1993). Microfossils of the Early Archean Apex Chert: new evidence of the antiquity of life. Science 260, 124.

Seidl, M., Huang, V., Mouchel, J.M. (1998). Toxicity of combined sewer overflows on river phytoplankton. Environmental Pollution 101, 107–116.

Semenova, E., Kuznedolov, K. (1998). A study of the biodiversity of Baikal picoplankton by comparative analysis of the 16 s rRNA gene 5-terminal regions. Molecular Biology 32, 754–760.

Serra, T., Vidal, J., Casamitjana, X., et al. (2007). The role of surface vertical mixing in phytoplankton distribution in a stratified reservoir. Limnology and Oceanography 52, 620–634.

Shapiro, J. (1990). Current beliefs regarding dominance by blue-greens: the case for the importance of CO_2 and pH. Verhandlungen der Internationale Vereinigung fur Theoretische und Angewandte Limnologie 24, 38–54.

Sheath, R.G., Cole, K.M. (1992). Biogeography of stream macroalgae in North America. Journal of Phycology 28, 449–460.

Sheath, R.G., Hambrook, J.A. (1990). Freshwater ecology. In Cole, K.M., Sheath, R.G. (eds.). Biology of the Red Algae. Cambridge, UK, Cambridge University Press, pp. 423–453.

Sheldon, R.B., Boylen, C.W. (1975). Factors affecting the contribution by epiphytic algae to the primary productivity of an oligotrophic freshwater lake. Applied Microbiology 30, 657–667.

Sherman, B.S., Webster, I.T., Jones, G.J., et al. (1998). Transitions between Aulacoseira and Anabaena dominance in a turbid river wide pool. Limnology and Oceanography 43, 1902–1915.

Sigee, D.C. (1984). Some observations on the structure, cation content and possible evolutionary status of dinoflagellate chromosomes. Botanical Journal of the Linnean Society 88, 127–147.

Sigee, D.C. (2004). Freshwater Microbiology: Diversity and Dynamic Interactions of Microorganisms in the Aquatic Environment. Chichester, UK, John Wiley & Sons, p. 524.

Sigee, D.C., Levado, E. (2000). Cell surface elemental composition of Microcystis aeruginosa: high-Si and low-Si subpopulations within the water column of a eutrophic lake. Journal of Plankton Research 22, 2137–2153.

Sigee, D.C., Morgan, A.J., Sumner, A.T., et al. (1993). X-ray Microanalysis in Biology: Experimental Techniques and Applications. Cambridge, UK, Cambridge University Press, p. 105.

Sigee, D.C., Teper, J., Levado, E. (1999). Elemental composition of the cyanobacterium Anabaena flos-aquae collected from different depths within a stratified lake. European Journal of Phycology 34, 477–486.

Sigee, D.C., Dean, A., Levado, E., et al. (2002). Fourier-transform infrared spectroscopy of Pediastrum duplex: characterisation of a micro-population isolated from a eutrophic lake. European Journal of Phycology 37, 19–26.

Sigee, D.C., Selwyn, A., Gallois, P., et al. (2007). Patterns of cell death in freshwater colonial cyanobacteria during the late summer bloom. Phycologia 46, 284–292.

Simpson, A.G. (1997). The identity and composition of the Euglenozoa. Archiv fur Protistenkunde 148, 318–328.

Skácelová, O., Lepš, J. (2014). The relationship of diversity and biomass in phytoplankton communities weakens when accounting for species proportions. Hydrobiologia 724, 67–77.

Sladecek, V. (1986). Diatoms as indicators of organic pollution. Acta Hydrochimica et Hydrobiologia 14, 555–566.

Smol, J.P. (1985). The ratio of diatom frustules to chrysophycean statospores: a useful paleolimnological index. Hydrobiologia 123(24) 199–208.

Soininen, J., Paavola, R., Muotka, T. (2004). Benthic diatom communities in boreal streams: community structure in relation to environmental and spatial gradients. Ecography 27, 330–342.

Sosnovskaya, O.A., Kharchenko, G.V., Klochenko, P.D. (2008). Primary production of phytoepiphyton in water bodies of Kiev. Hydrobiological Journal/Gidrobiologicheskiy Zhurnal 44, 35–41.

Spaulding, S.A., McKnight, D.M., Smith, R.L., et al. (1994). Plankton population dynamics in perpetually ice-covered Lake Fryxell, Antarctica. Journal of Plankton Research 16, 527–541.

Stal, L.J. (1995). Physiological ecology of cyanobacteria in microbial mats and other communities. New Phytologist 131, 1–32.

Stancheva, R., Sheath, R.G., Read, B.A., *et al.* (2013). Nitrogen-fixing cyanobacteria (free-living and diatom endosymbionts): their use in southern California stream bioassessment. Hydrobiologia 720, 111–127.

Stanier, R.Y., Sistrom, W.R., Hansen, T.A., *et al.* (1978). Proposal to place the nomenclature of the cyanobacteria (blue-green algae) under the rules of the International Code of Nomenclature of Bacteria. International Journal of Systematic Bacteriology 28, 335–336.

Stenger-Kovács, C., Toth, L., Toth, F., *et al.* (2014). Stream order dependent diversity metrics of epilithic diatom assemblages. Hydrobiologia 721, 67–75.

Stephen, D. (1997). The role of macrophytes in shallow lake systems: whole lake, mesocosm and laboratory studies. PhD thesis, University of Liverpool.

Stević, F., Mihaljević, M., Špoljavić, D. (2013). Changes of phytoplankton functional groups in a floodplain lake associated with hydrological perturbance. Hydrobiologia 709, 143–158.

Stockner, J. (1972). Paleolimnology as a means of assessing eutrophication. Verhandlungen der Internationale Vereinigung fur Theoretische und Angewandte Limnologie 18, 1018–1030.

Strickland, J., Parsons, T. (1972). A Practical Handbook of Sea Water Analysis. Ottawa, Canada, Fisheries Research Board of Canada, p. 311.

Sutherland, T.F., Amos, C.L., Grant, J. (1998). The effect of buoyant biofilms on the erodibility of sublittoral sediments of a temperate microtidal estuary. Limnology and Oceanography 43, 225–235.

Takamura, N., Hatakeyama, S., Sugaya, Y. (1990). Seasonal changes in species composition and production of periphyton in an urban river running through an abondoned copper mining region. Japanese Journal of Limnology 51, 225–235.

Taylor, D., Dalton, C., Leira, M., *et al.* (2006). Recent histories of six productive lakes in the Irish Ecoregion based on multiproxy palaeolimnological evidence. Hydrobiologia 571, 237–259.

Taylor, J.C., Prygiel, J., Vosloo, A., *et al.* (2007). Can diatom-based pollution indices be used for biomonitoring in South Africa? A case study of the Crocodile West and Marico water management area. Hydrobiologia 592, 455–464.

Technical Standard Publication. (1982). Utilisation and Protection of Waterbodies. Standing Inland Waters. Classification. Technical Standard 27885/01, Berlin, Germany.

Thunmark, S. (1945). Zur Soziologie des susswasserplanktons: eine methodologisch-okologische Studie. Folia Limnologica Scandinavia 3, 1–66.

Tison, J., Park, Y-S., Coste, M., *et al.* (2005). Typology of diatom communities and the influence of hydroecoregions: a study on the French hydrosystem scale. Water Research 39, 3177–3188.

Tittel, J., Zippel, B., Geller, W. (1998). Relationships between plankton community structure and plankton size distribution in lakes of northern Germany. Limnology and Oceanography 43, 1119–1132.

Tuchman, N. (1996). The role of heterotrophy in algae. In Stevenson, R.J. (ed.). Algal Ecology. New York, Academic Press, pp. 299–318.

Turner, M.A., Howell, E.T., Summerby, M., *et al.* (1991). Changes in epilithon and epiphyton associated with experimental acidification of a lake to pH5. Limnology and Oceanography 36(35), 1390–1405.

Underwood, G.J., Kromkamp, J. (1998). Distribution of estuarine benthic diatom species along salinity and nutrient gradients. European Journal of Phycology 33, 173–183.

Underwood, G.J., Kromkamp, J. (1999). Primary production by phytoplankton and microphytobenthos in estuaries. In Nedwell, D.B., Raffaelli, D.G. (eds.). Advances in Ecological Research. New York, Academic Press, pp. 93–153.

Urbach, E., Chisholm, S.W. (1998). Genetic diversity in *Prochlorococcus* populations flow cytometrically sorted from the Sargasso Sea and Gulf Stream. Limnology and Oceanography 43, 1615–1630.

Vairappan, C. (2006). Seasonal occurrences of epiphytic algae on the commercially cultivated red alga *Kappaphycus alvarezii* (Solieraceae, Gigartinales, Rhodophyta). Journal of Applied Phycology 18, 611–617.

Van Dam, H., Mertens, A., Sinkeldam, J. (1994). A coded checklist and ecological indicator values of freshwater diatoms from the Netherlands. Netherlands Journal of Aquatic Ecology 26, 117–133.

Van Den Hoek, C., Mann, D.G., Jahns, H.M. (1995). Algae: an Introduction to Phycology. Cambridge, UK, Cambridge University Press.

Vanormelingen, P., Vyverman, W., De Bock, D., *et al.* (2009). Local genetic adaptation to grazing pressure of the green alga *Desmodesmus armatus* in a strongly connected pond system. Limnology and Oceanography 54, 503–511.

Verspagen, J.M., Snelder, E.O., Visser, P.M., *et al.* (2005). Benthic–pelagic coupling in the population dynamics of the harmful cyanobacterium *Microcystis*. Freshwater Biology 50, 854–867.

Vollenweider, R.A. (1969). A Manual on Methods for Measuring Primary Production in Aquatic Environments. Philadelphia, PA, FA Davis.

Vrba, J., Kopacek, J., Fott, J., *et al.* (2003). Long-term studies (1871–2000) on acidification and recovery of lakes in the Bohemian Forest (central Europe). The Science of the Total Environment 310, 73–85.

Walne, P.L., Kivic, P.A. (1990). Euglenida. In Margulis, L., Corliss, O., Melkonian, M., Chapman, D.J. (eds.). Handbook of Protoctista. Boston, MA, Jones & Bartlett Publishers, pp. 270–287.

Webster, K.E., Frost, T.M., Watras, C.J., *et al.* (1992). Complex biological response to the experimental acidification of Little Rock Lake, Wisconsin, USA. Environmental Pollution 78, 673–678.

Wehr, J.D. (2003). Brown algae. In Wehr, J.D., Sheath, R.G. (eds.). Freshwater Algae of North America. Amsterdam, The Netherlands, Academic Press, pp. 757–773.

Wehr, J.D., Sheath, R.G. (2003). Freshwater Algae of North America. Amsterdam, The Netherlands, Academic Press, p. 918.

Welker, M., Walz, N. (1998). Can mussels control the plankton in rivers? – a planktological approach applying a Lagrangian sampling strategy. Limnology and Oceanography 43, 753–762.

Welker, M., Sejnohova, L., Nemethova, D., *et al.* (2007). Seasonal shifts in chemotype composition of *Microcystis* sp. communities in the palagial and the sediment of a shallow reservoir. Limnology and Oceanography 52, 609–619.

Wetzel, R.G. (1983). Limnology, 2nd edn. Philadelphia, PA, Saunders.

Wetzel, R.G., Likens, G.E.Â. (1991). Limnological Analyses. New York, Springer-Verlag, p. 391.

Willen, E. (2000). Phytoplankton in water quality assessment – an indicator concept. In Heinonen, P., Ziglio, G., Van der Beken, A. (eds.). Hydrological and Limnological Aspects of Lake Monitoring. Chichester, UK, John Wiley & Sons, pp. 57–808.

Woelkerling, W.J. (1976). Wisconsin desmids. 1. Aufwuchs and plankton communities of selected acid bogs, alkaline bogs, and closed bogs. Hydrobiologia 48, 209–232.

Wolk, F., Seuront, L., Yamazaki, H., *et al.* (2004). Comparison of biological scale resolution from CTD and microstructure measurements. In Seuront, L., Strutton, P.G. (eds.). Handbook of Scaling Methods in Aquatic Ecology. Boca Raton, FL, CRC Press, pp. 3–15.

Yopp, J.H., Tindall, D.R., Miller, D.M., *et al.* (1978). Isolation, purification and evidence for a halophilic nature of the blue-green alga *Aphanothece halophytica* Fremy (Chroococcales). Phycologia 17, 172–178.

Zelinka, M., Marvan, P. (1961). Zur Prazisierung der biologischen Klassifikation des Reinheit fliessender Gewasser. Archives of Hydrobiology 57, 389–407.

Zubkov, M.V., Fuchs, B.M., Tarran, G.A., *et al.* (2003). High uptake of organic nitrogen compounds by *Prochlorococcus* cyanobacteria as a key to their dominance in oligotrophic ocean waters. Applied Environmental Microbiology 69, 1299–1304.

Index

Note: References to Figures, Plates and Tables are all indicated by ***bold italic*** page numbers. See also the list of algae included in the key on pp. 249–251 and a glossary of terms on pp. 253–258.

Freshwater Algae: Identification, Enumeration and Use as Bioindicators, Second Edition. Edward G. Bellinger and David C. Sigee.
© 2015 John Wiley & Sons, Ltd. Published 2015 by John Wiley & Sons, Ltd.